TREATISE ON ASTRONOMY.

BY

ELIAS LOOMIS, LL.D.

A

TREATISE

ON

ASTRONOMY.

BY

ELIAS LOOMIS, LL.D.,

'ROFESSOR OF NATURAL PHILOSOPHY AND ASTRONOMY IN YALE COLLEGE; AUTHOR OF "AN INTRODUCTION TO PRACTICAL ASTRONOMY," AND OF A SERIES OF MATHEMATICS FOR SCHOOLS AND COLLEGES.

NEW YORK:
HARPER & BROTHERS, PUBLISHERS,
327 AND 335 PEARL STREET.
1868.

PREFACE.

THE design of the following treatise is to furnish a text-book for the instruction of college classes in the first principles of Astronomy. My aim has accordingly been to limit the book to such dimensions that it might be read entire without omissions, and to make such a selection of topics as should embrace every thing most important to the student. I have aimed to express every truth in concise and simple language; and when it was necessary to introduce mathematical discussions, I have limited myself to the elementary principles of the science. The entire book is divided into short articles, and each article is preceded by a caption, which is designed to suggest the subject of the article. Whenever it could be done to advantage, I have introduced simple mathematical problems, designed to test the student's familiarity with the preceding principles. At the close of the book will be found a collection of miscellaneous problems, many of them extremely simple, which are to be used according to the discretion of the teacher.

I have dwelt more fully than is customary in astronomical text-books upon various physical phenomena, such as the constitution of the sun, the condition of the moon's surface, the phenomena of total eclipses of the sun, the laws of the tides, and the constitution of comets. I have also given a few of the results of recent researches respecting binary stars. It is hoped that the discussion of these topics will enhance the interest of the subject with a class of students who might be repelled by a treatise exclusively mathematical.

My special acknowledgments are due to Professor H. A. Newton, who has read all the proofs of the work, and to whom I am indebted for numerous important suggestions.

CONTENTS.

CHAPTER I.

THE EARTH—ITS FIGURE AND DIMENSIONS.—DENSITY.—ROTATION.

CHAPTER II.

THE PRINCIPAL ASTRONOMICAL INSTRUMENTS.

CHAPTER III.

ATMOSPHERIC REFRACTION.—TWILIGHT.

CHAPTER IV.

EARTH'S ANNUAL MOTION.—EQUATION OF TIME.—CALENDAR.

CHAPTER V.

PARALLAX.—ASTRONOMICAL PROBLEMS.

CHAPTER VI.

THE SUN—ITS PHYSICAL CONSTITUTION.

CHAPTER VII.

PRECESSION OF THE EQUINOXES.—NUTATION.—ABERRATION.

CHAPTER VIII.

THE MOON—ITS MOTION—PHASES—TELESCOPIC APPEARANCE.

CHAPTER IX.

CENTRAL FORCES.—GRAVITATION.—LUNAR IRREGULARITIES.

CHAPTER X.

ECLIPSES OF THE MOON.

CHAPTER XI.

ECLIPSES OF THE SUN.

CHAPTER XII.

METHODS OF FINDING THE LONGITUDE OF A PLACE.

CHAPTER XIII.

THE TIDES.

CHAPTER XIV.

THE PLANETS—ELEMENTS OF THEIR ORBITS.

CHAPTER XV.

THE INFERIOR PLANETS.

CHAPTER XVI.

THE SUPERIOR PLANETS.

CHAPTER XVII.

QUANTITY OF MATTER IN THE SUN AND PLANETS.—PLANETARY PERTURBATIONS.

CHAPTER XVIII.

COMETS.—COMETARY ORBITS.—SHOOTING STARS.

CHAPTER XIX.

THE FIXED STARS—THEIR LIGHT, DISTANCE, AND MOTIONS.

CHAPTER XX.

DOUBLE STARS.—CLUSTERS.—NEBULÆ.

ASTRONOMY.

CHAPTER I.

GENERAL PHENOMENA OF THE HEAVENS.—FIGURE AND DIMENSIONS OF THE EARTH.—DENSITY OF THE EARTH.—PROOF OF THE EARTH'S ROTATION.—ARTIFICIAL GLOBES.

1. *Astronomy is the science which treats of the heavenly bodies.* The heavenly bodies consist of *the sun*, *the planets with their satellites*, *the comets*, and *the fixed stars.*

Astronomy is divided into Spherical and Physical. Spherical Astronomy treats of the appearances, magnitudes, motions, and distances of the heavenly bodies. Physical Astronomy applies the principles of Mechanics to explain the motions of the heavenly bodies, and the laws by which they are governed.

2. *Diurnal motion.*—If we examine the heavens on a clear night, we shall soon perceive that the stars constantly maintain the same position relative to each other. A map showing the relative position of these bodies on any night, will represent them with equal exactness on any other night. They all seem to be at the same distance from us, and to be attached to the surface of a vast hemisphere, of which the place of the observer is the centre. But, although the stars are relatively fixed, the hemisphere, as a whole, is in constant motion. Stars rise obliquely from the horizon in the east, cross the meridian, and descend obliquely to the west. The whole celestial vault appears to be in motion round a certain axis, carrying with it all the objects visible upon it, without disturbing their relative positions. The point of the heavens which lies at the extremity of this axis of rotation is fixed, and is called the *pole.* There is a star called the *pole star*, distant about $1\frac{1}{2}°$ from the pole, which moves in a small circle round the pole as a centre. All other stars appear

also to be carried around the pole in circles, preserving always the same distance from it.

3. *Axis of the celestial sphere.* — This motion of rotation is perfectly uniform, as may be proved by observations with a telescope. Suppose the telescope of a theodolite to be directed to the pole star; the star will appear to move in a small circle whose diameter is about three degrees; and the telescope may be so pointed that the star will move in a circle around the intersection of the spider-lines as a centre. This point of intersection is then the pole. The surface of the visible heavens to which all the heavenly bodies appear to be attached, is called *the celestial sphere.*

4. *Use of a telescope mounted equatorially.*—Having determined the axis of the celestial sphere, a telescope may be mounted capable of revolving upon a fixed axis which points toward the celestial pole, in such a manner that the telescope may be placed at any desired angle with the axis, and there may be attached to it a graduated circle by which the magnitude of this angle may be measured. A telescope thus mounted is called an *equatorial telescope*, and it is frequently connected with clock-work, which gives it a motion round the axis corresponding with the rotation of the celestial sphere.

5. *Diurnal paths of the heavenly bodies.*—Let now the telescope be directed to any star so that it shall be seen in the centre of the field of view, and let the clock-work be connected with it so as to give it a perfectly uniform motion of rotation from east to west. The star will follow the telescope, and the velocity of motion may be so regulated, that the star shall remain in the centre of the field of view from rising to setting, the telescope all the time maintaining the same angle with the axis of the heavens. The same will be true of every star to which the telescope is directed; from which we conclude that all objects upon the firmament describe circles at right angles to its axis, each object always remaining at the same distance from the pole.

6. *Time of one revolution of the celestial sphere.*—If the telescope be detached from the clock-work, and, having been pointed upon

a star, be left fixed in its position, and the exact time of the star's passing the central wire be noted, on the next night at about the same hour the star will again arrive upon the central wire. The time elapsed between these two observations will be found to be 23h. 56m. 4s., expressed in solar time.

This, then, is the time in which the celestial sphere makes one revolution; and this time is always the same, whatever be the star to which the telescope is directed.

7. *A sidereal day.*—The time of one complete revolution of the firmament is called a *sidereal day.* This interval is divided into 24 sidereal hours, each hour into 60 minutes, and each minute into 60 seconds.

Since the celestial sphere turns through 360° in 24 sidereal hours, it turns through 15° in one sidereal hour, and through 1° in four sidereal minutes.

8. *The diurnal motion is never suspended.*—With a telescope of considerable power, all the brighter stars can be seen throughout the day, unless very near the sun; and by the method of observation already described, we find that the same rotation is preserved during the day as during the night.

All the heavenly bodies, without exception, partake of this diurnal motion; but the sun, the moon, the planets, and the comets appear to have a motion of their own, by which they change their position among the stars from day to day.

9. The *celestial equator* is the great circle in which a plane passing through the earth's centre, and perpendicular to the axis of the heavens, intersects the celestial sphere.

10. If a plummet be freely suspended by a flexible line and allowed to come to a state of rest, this line is called a *vertical line.* The point where this line produced meets the visible half of the celestial sphere, is called the *zenith;* and the point where it meets the invisible hemisphere, which is under the plane of the horizon, is called the *nadir.*

Every plane passing through a vertical line is called *a vertical plane*, or *a vertical circle.*

That vertical circle which passes through the celestial pole is called *the meridian.* The vertical circle at right angles to the meridian is called *the prime vertical.*

11. A *horizontal plane* is a plane perpendicular to a vertical line.

The *sensible horizon* of a place is the circle in which a plane passing through the place, and perpendicular to the vertical line at the place, cuts the celestial sphere.

The *rational horizon* is the circle in which a plane passing through the earth's centre, and parallel to the sensible horizon, cuts the celestial sphere. On account of the distance of the stars, these two planes intersect the celestial sphere sensibly in the same great circle.

The meridian and prime vertical meet the horizon in four points, called the cardinal points; or the *north*, *south*, *east*, and *west* points.

12. The *altitude* of a heavenly body is its elevation above the horizon measured on a vertical circle. The zenith distance of a body is its distance from the zenith measured on a vertical circle. The zenith distance is the complement of the altitude.

The *azimuth* of a body is the arc of the horizon intercepted between the north or south point of the horizon, and a vertical circle passing through the body. Altitudes and azimuths are measured in degrees, minutes, and seconds. The *amplitude* of a star is its distance from the east or west point at the time of its rising or setting.

13. *Consequences of the diurnal motion.*—If an observer could watch the whole apparent path of any star in the sky, he would see it describe a circle around the line PP′; but as only half the celestial sphere is visible, it is evident that a part of the path of a star may lie below the horizon and be invisible. Thus, in Fig. 1, let PP′ be the axis of rotation of the celestial sphere; NILSMK be the horizon produced to intersect

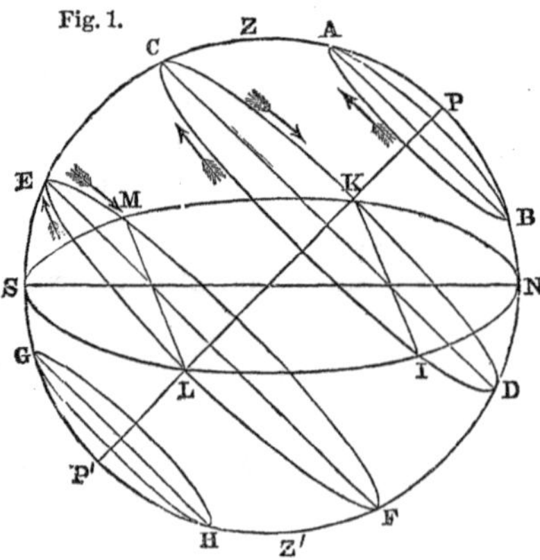

the sphere, and dividing it into two hemispheres, NS being the north and south line. If the parallel circles passing through A, C, E, and G be the apparent diurnal paths of four stars, then it is evident that

1st. The star which describes the circle AB will never descend below the horizon.

2d. The star which describes the circle GH will never come above the horizon.

3d. The star which describes the circle ICKD will be above the horizon while it moves through ICK, and below the horizon through the portion KDI.

4th. The star which describes the circle LEMF will be above the horizon through the portion of the circle LEM, and below the horizon through the portion MFL.

These stars are said to rise at I and L, and to set at K and M. They rise in the eastern part of the horizon, and set in the western.

With the star C, the visible portion of its path ICK is greater than the invisible portion KDI; while with the star E, the visible portion of its path LEM is less than the invisible portion MFL.

14. *Culminations of the heavenly bodies.*—When stars cross the meridian above the pole they are said to *culminate*, or attain their greatest altitude. All stars cross the meridian twice every day; once above the pole, and once below the pole. The former is called their *upper culmination*, the latter is called their *lower culmination*. Thus the star which describes the circle AB has its upper culmination at A, and its lower culmination at B.

It is evident from the figure that all stars which lie to the north of the equator, will remain above the horizon for a longer period than below it; all stars south of the equator will remain above the horizon for a shorter time than below it; and stars situated in the plane of the equator will remain above the horizon and below it for equal periods of time.

15. *How the pole star may be found.*—Among the most remarkable of the stars which never set in the latitude of New York, is the group of stars known as Ursa Major, shown in Fig. 2, which also represents the constellations Ursa Minor and Cassiopea. The constellation Ursa Major (represented on the left), is easily

recognized by its resemblance to the figure of a dipper, and may be used to find the pole star by drawing a line through β and α

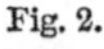
Fig. 2.

(called the Pointers), which will pass through the pole star α Ursæ Minoris. A line drawn through δ Ursæ Majoris and the pole star, will pass nearly through β Cassiopeæ (represented on the right).

16. *What stars never set.*—If a circle were drawn through N, the north point of the horizon, parallel to the equator, it would cut off a portion of the celestial sphere having P for its centre, all of which would be above the horizon; and a circle drawn through S, the south point of the horizon, parallel to the equator, would cut off a portion having P′ for its centre, which would be wholly below the horizon. Stars which are nearer to the visible pole than the point N never set, and those which are nearer to the invisible pole than the point S never rise.

17. *Why a knowledge of the dimensions of the earth is important.*—The bodies of which astronomy treats are all (with the exception of the earth) *inaccessible.* Hence, for determining their distances, we are obliged to employ *indirect* methods. The eye can only judge of the *direction* of objects, and is unable to determine directly their distances; but by measuring the bearings of an inaccessible object from two points whose distance from each other is known, we may compute the distance of that object by the methods of trigonometry. In all our observations for determining the distance of the celestial bodies, the base line must be drawn upon the earth. It is therefore necessary to determine with the utmost precision the form and dimensions of the earth.

18. *Proof that the earth is globular.*—The figure of the earth is nearly *globular*. This is proved,

1st. By its having been many times *sailed round* in different directions. This fact can only be explained by supposing that the earth is rounded; but it does not alone furnish sufficiently precise information of its exact figure.

2d. By the phenomena of *eclipses of the moon*. These eclipses are caused by the earth coming between the sun and moon, so as to cast its shadow upon the latter. The form of this shadow is always such as one globe would project upon another. This argument is conclusive, but its force would not be admitted by those who deny the Copernican theory of the universe.

3d. By our *seeing the top-mast of a ship*, as it recedes from the observer, *after the hull has disappeared*. If the earth was a plane surface, the top-mast, having the smallest dimensions, should disappear first, while the hull and sails, having the greatest dimensions, should disappear last; but, in fact, the reverse takes place.

Fig. 3.

Land is visible from the top-mast when it can not be seen from the deck. The tops of mountains can be seen from a distance when their base is invisible. The sun illumines the summits of mountains long after it has set in the valleys. An æronaut, ascending in his balloon after sunset, has seen the sun reappear with all the effects of sunrise; and on descending, he has witnessed a second sunset.

4th. If we travel northward, following a meridian, we shall find the altitude of the pole to increase continually at the rate of one degree for a distance of about 69 miles. This proves that a section of the earth made by a meridian plane is very nearly a circle, and also affords us the means of determining its dimensions, as shown in Art. 20.

19. *First method of determining the earth's diameter.*—The facts just stated not only demonstrate that the earth is globular, but afford us a rude method of computing its diameter. For this purpose we measure the height of some mountain, and also the distance at which it can be seen at sea. Let BD represent a mountain (Chimborazo, for example), 4 miles in height; and suppose the distance, AB, at which it can be seen at sea, is 179 miles. Then, in the triangle ABC, representing the radius of the earth by R, we shall have

Fig. 4.

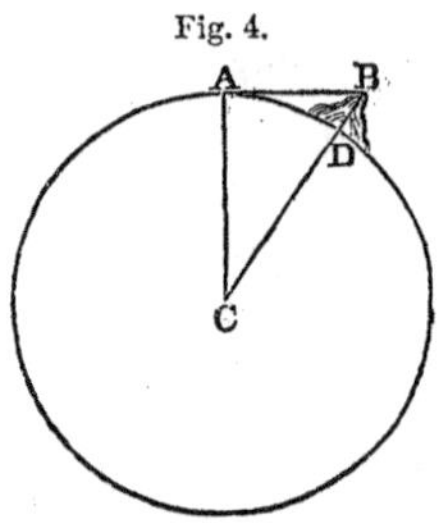

$$(R+4)^2=R^2+179^2,$$

from which we find that $R=4000$ miles nearly. Thus we learn that the radius of the earth is about 4000 miles. Similar observations made in all parts of the earth, give nearly the same value for the radius, which can only be explained by supposing that the earth is nearly a sphere.

The earth is known to be globular by the most accurate measurements, as will be more fully explained hereafter.

20. *Second method of determining the earth's diameter.*—Having ascertained the general form of the earth, we wish to determine, as accurately as we can, its diameter. For this purpose we first ascertain the length of one degree upon its surface; that is, the distance between two points on the earth's surface so situated that the lines drawn from them to the centre of the earth may make with each other an angle of one degree.

Fig. 5.

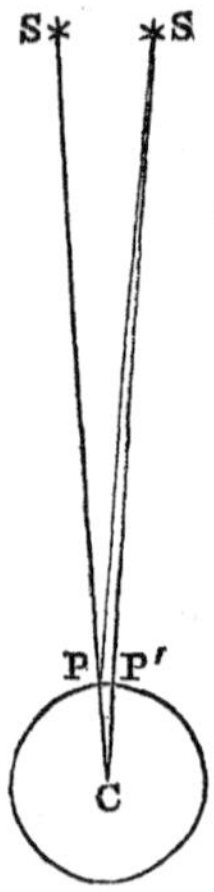

Let P and P′ be two places on the earth's surface, distant from each other about 70 miles, and let C be the centre of the earth. Suppose two persons at the places P and P′ observe two stars S and S′, which are at the same instant vertically over the two places—that is, in the direction of plumb-lines suspended at those places. Let the directions of these plumb-lines be continued downward so as to intersect at C the centre of the earth. The angle which the directions of these stars make at P is SPS′, and the angle as seen from C is SCS′; but, on account of the distance of the stars, these angles are sensibly equal to each other. If, then, the angle SPS′ be measured, and the distance between the places P and

P′ be also measured by the ordinary methods of surveying, the length of one degree can be computed. In this way it has been ascertained that the length of a degree of the earth's surface is about 69 statute miles, or 365,000 feet.

Since a second is the 3600th part of a degree, it follows that the length of one second is one hundred feet very nearly.

Since the plumb-line is perpendicular to the earth's surface, its change of direction in passing from one place to another may be found by allowing one second for every hundred feet, or more exactly by allowing 365,000 feet for each degree.

21. The *circumference* of the earth may be found approximately by the proportion

1 degree : 360 degrees :: 69 miles : 24,840 miles;

and hence the diameter is found to be about 7900 miles; which results are a little too small, but may be employed as convenient numbers for illustration.

The earth being globular, it is evident that the terms *up* and *down* can not every where denote the same absolute direction. The term *up* simply denotes *from* the earth's centre, while *down* denotes *towards* the earth's centre; but the absolute direction denoted by these terms at New York is very different from that denoted by the same terms at London or Canton.

22. *Irregularities of the earth's surface.*—The highest mountain peaks do not exceed five miles in height, which is about $\frac{1}{1600}$ of the earth's diameter. Accordingly, on a globe 16 inches in diameter, the highest mountain peak would be represented by a protuberance having an elevation of $\frac{1}{100}$ inch, which is about twice the thickness of an ordinary sheet of writing-paper. The general elevation of the continents above the sea would be correctly represented by the thinnest film of varnish. In other words, the irregularities of the earth's surface are quite insignificant in comparison with its absolute dimensions.

23. *Cause of the diurnal motion.*—The apparent diurnal rotation of the heavens may be caused either by a real motion of the celestial sphere, or by a real motion of the earth in a contrary direction. The former supposition is felt to be absurd as soon as we learn the distances and magnitudes of the celestial bodies.

B

The latter supposition is in itself not improbable, and perfectly explains all the phenomena. Moreover, we find direct proof of the rotation of the earth, in the descent of a body falling from a great height, which falls a little to the eastward of a vertical line.

The figure of the earth, which is not that of a perfect sphere, affords independent proof of its rotation.

Analogy also favors the same conclusion. All the planets which we have been able satisfactorily to observe, rotate on their axes, and their figures are such as correspond to the time of their rotation.

The rotation of the earth gives to the celestial sphere the appearance of revolving in the contrary direction, as the forward motion of a boat on a river gives to the banks an appearance of backward motion; and since *the apparent motion of the heavens* is from east to west, *the real rotation of the earth* which produces that appearance must be from west to east.

24. The *earth's axis* is the diameter around which it revolves once a day. The extremities of this axis are the terrestrial *poles;* one is called the *north pole*, and the other the *south pole.*

The terrestrial *equator* is a great circle of the earth perpendicular to the earth's axis.

Meridians are great circles passing through the poles of the earth.

25. The *latitude* of a place is the arc of the meridian which is comprehended between that place and the equator. Latitude is reckoned north and south of the equator, from 0 to 90°.

A *parallel of latitude* is any small circle on the earth's surface parallel to the terrestrial equator. These parallels continually diminish in size as we proceed from the equator to the pole.

The polar distance of a place is its distance from the nearest pole, and is the complement of the latitude.

The *longitude* of a place is the arc of the equator intercepted between the meridian of that place and some assumed meridian to which all others are referred. The English reckon longitude from the observatory of Greenwich, the French from the observatory of Paris, and the Germans from the observatory of Berlin, or from the island of Ferro, which is assumed to be 20° west of the observatory in Paris. In the United States we sometimes reckon longitude from Washington, and sometimes from Greenwich. Longitude is usually reckoned east and west of the first meridian, from 0 to 180°. The longitude and latitude of a place determine its position on the earth's surface.

26. *The latitude of a place.*—Let SÆNQ represent the earth surrounded by the distant starry sphere HZOK. The diameter of the earth being insignificant in comparison with the distance of the stars, the appearance of the heavens will be the same

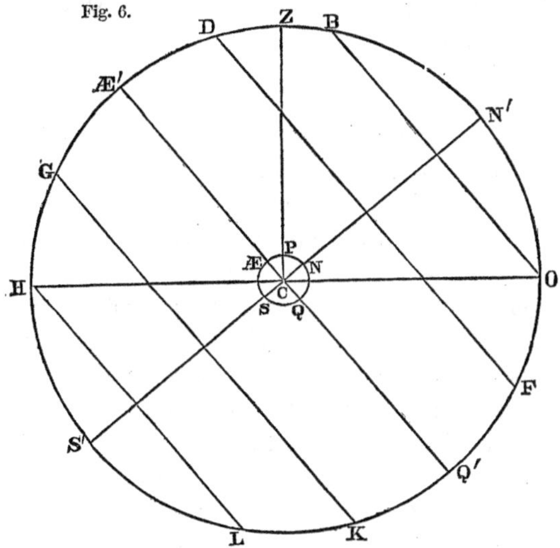

whether they are viewed from the centre of the earth, or from any point on its surface. Suppose the observer to be at P, a point on the surface between the equator Æ and the north pole N. The latitude of this place is ÆP, or the angle ÆCP. If the line PC be continued to the firmament, it will pass through the point Z, which is the zenith of the observer. If the terrestrial axis NS be continued to the firmament, it will pass through the celestial poles N′ and S′. If the terrestrial equator ÆQ be continued to the heavens, it will constitute the celestial equator Æ′Q′. The observer at P will see the entire hemisphere HZO, of which his zenith Z is the pole. The other hemisphere will be concealed by the earth.

The arc N′O contains the same number of degrees as Æ′Z, and

the arc ZN′ is the complement of ON′; that is, *the altitude of the visible pole is equal to the latitude of the place*, and the zenith distance of the visible pole is the complement of the latitude.

27. *How the latitude of a place may be determined.*—If there were a star situated precisely at the pole, its altitude would be the latitude of the place. The pole star describes a small circle around the pole, and crosses the meridian twice in each revolution, once above and once below the pole. The half sum of the altitudes in these two positions is equal to the altitude of the pole; that is, to the latitude of the place. The same result would be obtained by observing any circumpolar star on the meridian both above and below the pole.

28. Circles which pass through the two poles of the celestial sphere are called *hour circles.* If two such circles include an arc of 15° of the celestial equator, the interval between the instants of their coincidence with the meridian will be one hour.

29. The *right ascension* of a star is the arc of the celestial equator comprehended between a certain point on the equator called the first point of Aries, and an hour circle passing through that star. Right ascension is sometimes expressed in degrees, minutes, and seconds of arc, but generally in hours, minutes, and seconds of time. It is reckoned eastward from zero up to 24 hours, or 360 degrees. If the hands of the sidereal clock be set to 0h. 0m. 0s. when the first point of Aries is on the meridian, the clock (if it neither gains nor loses time) will afterward indicate at each instant the right ascension of any object which is then on the meridian, for the motion of the hands of the clock corresponds exactly with the apparent diurnal motion of the heavens. While 15° of the equator pass the meridian, the hands of the clock move through one hour.

The sidereal day therefore begins when the first point of Aries crosses the meridian, and the sidereal clock should always indicate 0h. 0m. 0s. when the first of Aries is on the meridian.

30. The distance of an object from the celestial equator, measured upon the hour circle which passes through it, is called its *declination*, and is north or south according as the object is on

the north or south side of the equator. North declination is indicated by the sign +, and south declination by the sign −.

The position of an object on the firmament is indicated by its declination and right ascension. Its declination expresses its distance north or south of the celestial equator, and its right ascension expresses the distance of the hour circle upon which it is situated, from a fixed point upon the celestial equator.

The *north polar distance* of a star is its distance from the north pole.

31. *A right sphere.*—The celestial sphere presents different appearances to observers in different latitudes. If the observer were situated at the terrestrial equator, the poles would lie in the horizon, the celestial equator would be perpendicular to the plane of the horizon, and hence the horizon would bisect the equator and all circles parallel to it. Therefore all celestial objects would be for equal periods above and below the horizon, and they would appear to rise perpendicularly on the eastern side of the horizon, and set perpendicularly on the western side. Such a sphere is called a *right sphere*, the diurnal motion being at right angles to the horizon.

32. *A parallel sphere.*—At one of the poles of the earth, the celestial pole being in the zenith, the celestial equator would coincide with the horizon, and by the diurnal motion all celestial objects would move in circles parallel to the horizon. This is called a *parallel sphere*. In a parallel sphere, an object upon the equator will be carried by the diurnal motion round the horizon, without either rising or setting.

33. *An oblique sphere.*—At all latitudes between the equator and the pole, the celestial equator is inclined to the horizon at an angle equal to the distance of the pole from the zenith; that is, equal to the complement of the latitude. The parallels DF, GK, Fig. 6, are unequally divided by the horizon; that is, all objects between the celestial equator and the visible pole are longer above than below the horizon, and all objects on the other side of the equator are longer below than above the horizon.

A parallel, BO, whose distance from the visible pole is equal to the latitude, is entirely above the horizon; and the same is true of all parallels still nearer to that pole. Also the parallel HL, whose distance from the invisible pole is equal to the latitude, is entirely below the horizon; and the same is true of all parallels still nearer to that pole. Hence, in the United States, stars within a certain distance of the north pole never set, and stars at an equal distance from the south pole never rise.

The circle BO is called the circle of *perpetual apparition*, because the stars which are included within it never set. The radius of this circle is equal to the latitude of the place.

The circle HL is called the circle of *perpetual occultation*, because the stars which are included within it never rise. The radius of this circle is also equal to the latitude of the place.

The celestial sphere here described is called an *oblique sphere*, the diurnal motion being oblique to the horizon.

Whether the sphere be right or oblique, one half of the celestial equator will be below the horizon, and the other half above it. Every object on the equator will therefore be above the horizon during as long a time as it is below, and will rise and set at the east and west points.

34. *Effects of centrifugal force.*—We have discovered that the earth has a globular figure, and that it rotates upon its axis once in 24 sidereal hours. But, since the earth rotates upon an axis, its form *can not be that of a perfect sphere;* for every body revolving in a circle acquires a centrifugal force which tends to make it recede from the centre of the circle. Every particle, P, upon the earth's surface acquires, therefore, a force which acts in a direction, EP, perpendicular to the axis of rotation. This centrifugal force, which we will represent by PA, may be resolved into two other forces PB and PD, one acting in the direction of a radius of the earth, and the other at right angles to the radius. The former, being opposed to the earth's attraction, has the effect of diminishing the weight of the body; the latter, being directed toward the equator, tends to produce motion in the direction of the equator.

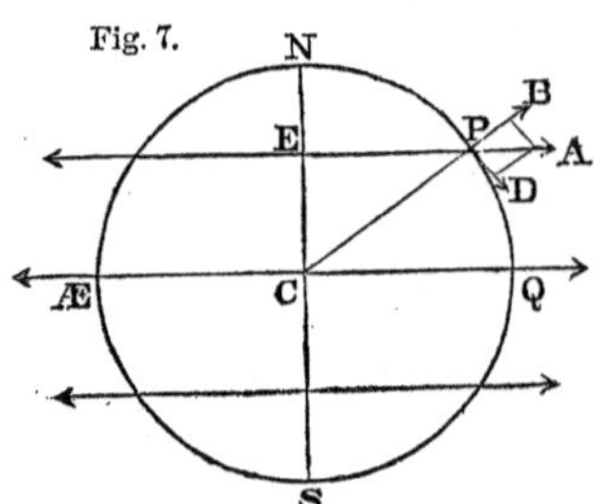

The intensity of the centrifugal force increases with the radius of the circle described, and is therefore greatest at the equator.

Moreover, the nearer the point is to the equator, the more directly is the centrifugal force opposed to the weight of the body.

The effects, therefore, produced by the rotation of the earth are,

1st. All bodies decrease in weight in going from the pole to the equator; and,

2d. All bodies which are free to move, tend from the higher latitudes toward the equator.

35. *The effect of centrifugal force computed.*—Let A be a ball attached to a string AS; let S be a fixed point, and ACE the circle in which the ball revolves, and AC the arc which the ball describes in a given time. When the ball was at A, it was moving in the direction of the tangent AB, and it would continue in this direction if it were acted upon by no other force than the first impulse; but we find it deflected into the diagonal AC, and this diagonal is the resultant of two forces represented by AB, AD. Now AB represents the path which the ball would describe under the first impulse, and therefore AD represents the motion impressed upon it by the tension of the string, and which deflects the ball from the tangent to the circle, and this is equal to the centrifugal force generated by the revolution of the ball.

Fig. 8.

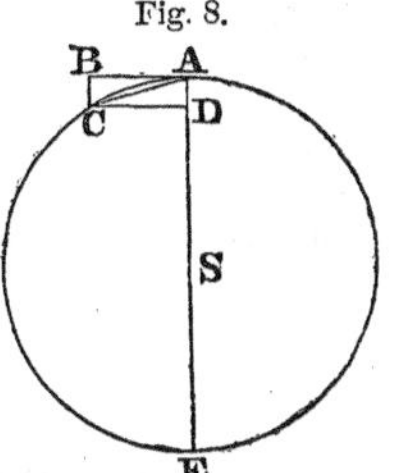

Now $\quad AD : AC :: AC : AE$; whence $AD = \frac{AC^2}{2AS}$.

But AC represents the velocity of the revolving body. If, then, C represents the centrifugal force of a revolving body, V its velocity in feet per second, and R the length of the string in feet,

we shall have $$C = \frac{V^2}{2R}.$$

36. *Centrifugal force compared with the force of gravity.*—We may compare the centrifugal force of a body with the force of gravity, by comparing the spaces through which the body would move in a given time, under the operation of these two forces.

Let W = the weight of the revolving body, and $g = 16$ feet, the space through which W would fall freely in one second. Then

we shall have $$W : C :: g : \frac{V^2}{2R};$$

whence $$C = \frac{W.V^2}{2Rg}.$$

We may also express the centrifugal force of a revolving body by reference to the number of revolutions made in a given time. Let N represent the number of revolutions, or the fraction of a revolution performed by the body in one second. The circumference of the circle which the body describes will be $2\pi R$. The space through which the body moves in one second, that is, its velocity, is $2\pi R.N$. Hence we have

$$C=\frac{W.(2\pi RN)^2}{2R.g}=\frac{2\pi^2}{g}\times R.N^2.W=1.2275\times R.N^2.W.$$

The amount of the loss of weight produced at the equator by centrifugal force, may be computed as follows:

The radius of the equator is 20,923,600 feet; and since the time of one rotation is 23h. 56m. 4s., or 86164 seconds, $N=\frac{1}{86164}$.

Hence $$C=1.2275\times 20{,}923{,}600\times\frac{1}{(86164)^2}\times W,$$

or $$C=\frac{W}{289}.$$

Thus we find that at the equator the centrifugal force of a body arising from the earth's rotation, is $\frac{1}{289}$ part of the weight; and since this force is directly opposed to gravity, the weight must sustain a loss of $\frac{1}{289}$ part.

37. *Centrifugal force at any latitude.*—The centrifugal force at the equator is to the centrifugal force in any other latitude as radius to the cosine of the latitude. But the entire centrifugal force at any latitude is to that part of the centrifugal force which is opposed to the weight of the body, as radius to the cosine of the latitude; that is, the loss of weight of a body caused by the centrifugal force at any latitude, is $\frac{1}{289}$ of the weight multiplied by the square of the cosine of the latitude.

38. *Effect of centrifugal force upon the form of a body.*—A portion, PD, of the centrifugal force causes a tendency to move toward the equator. If the surface of the globe were entirely solid, this tendency would be counteracted by the cohesion of the particles. But since a portion of the earth's surface is fluid, this portion must yield to the centrifugal force, and flow toward the equator. Thus the water must recede from the higher latitudes in either hemisphere, and accumulate around the equator. The earth, therefore, instead of being an exact sphere, must become

an oblate spheroid. A globe consisting of any plastic material would be reduced to such a figure by causing it to rotate rapidly upon an axis. The amount of the ellipticity of the earth must depend upon the centrifugal force, and the attraction exerted by the earth upon bodies placed on its surface.

39. *Weight of a body at the pole and the equator.*—We have found that at the equator the loss of weight due to centrifugal force is $\frac{1}{289}$. From a comparison of observations of the length of the seconds' pendulum made in different parts of the globe, it is found that the weight of a body at the pole actually exceeds its weight at the equator, by $\frac{1}{194}$.

The difference between these fractions is $\frac{1}{194}-\frac{1}{289}=\frac{1}{590}$; that is, the actual attraction exerted by the earth upon a body at the equator is less than at the pole, by the 590th part of the whole weight. This difference is due to the elliptic form of the meridians, by which the distance of the body at the equator from the centre of the earth is increased.

40. *How an arc of a meridian is measured.*—Numerous arcs of the meridian have been measured, for the purpose of accurately determining the figure and dimensions of the earth. These arcs are measured in the following manner:

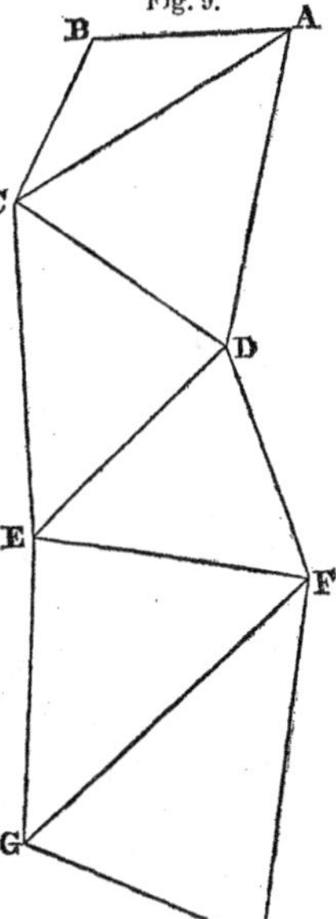

A level spot of ground is selected, where a *base* line, AB, from five to ten miles in length, is measured with the utmost precision. A third station, C, is selected, forming with the base line a triangle as nearly equilateral as is convenient. The angles of this triangle are measured with a theodolite, and the two remaining sides may then be computed. A fourth station, D, is now selected, forming with two of the former stations a second triangle, in which all the angles are measured; and since one side is already known, the others may be computed. A fifth station, E, is then selected, forming a third triangle; and thus we proceed forming a series of triangles, following nearly the direction of a meridian, by means of which we can compute the distance of the extreme stations from each other. The latitude of the most northerly and also that of the most southerly station must be determined, whence we obtain the difference of latitude corresponding to the arc measured.

This method is the most accurate known for determining the distance between two remote points on the earth's surface, because we may choose the most favorable site for measuring accurately the base line; and after this, nothing is required but the measurement of angles, which can be done with much less labor, and with much greater accuracy, than the measurement of distances.

41. *Verification of the work.*—In order to verify the entire work, a second base line is measured near the end of the series of triangles, and we compare its measured length with the length as computed from the first base, through the intervention of the series of triangles.

In the great arc measured in France between the years 1792 and 1799, the base of verification was distant 400 or 500 miles from the first base, and was seven miles in length, yet the difference between its observed and computed length did not amount to twelve inches.

In the Ordnance Survey of Great Britain and Ireland, six base lines have been measured, the longest being 7.88 miles in length, and the shortest 4.64 miles. In one instance the observed length of a base differs from its computed length by 19 inches. In each of the other cases, the discrepancy is less than three inches.

42. *Results of measurements.*—In this manner, arcs of the meridian have been measured in nearly every country of Europe. One has also been measured in India, one in South America, and one in South Africa. The operations for the survey of the coast of the United States will ultimately furnish several other arcs of a meridian.

The following is a list of the principal arcs already completed:

The	Peruvian - - -	arc is	214	miles	in length	= 3°	7′;
"	East Indian - -	"	1468	"	"	=21	21;
"	French - - - -	"	854	"	"	=12	22;
"	English - - - -	"	756	"	"	=10	56;
"	Russian - - - -	"	1753	"	"	=25	20;
"	Cape of Good Hope	"	275	"	"	= 4	0.

The sum of these arcs exceeds 60 degrees, not counting double measurements of the same part of the meridian; that is, we have measured nearly two thirds of the distance from the equator to the north pole. We can therefore compute the remaining distance with but small liability to error. The result is that

a degree at the equator = 68.702 miles.
a degree at the pole = 69.396 "
the difference = .694 "

43. *Conclusion from these results.*—If the earth were perfectly spherical, a terrestrial meridian would be an exact circle, and every part of it would have the same curvature; that is, a degree of latitude would be every where the same. But we have found that the length of a degree increases as we proceed from the equator toward the poles, and the amount of this difference affords a measure of the departure of a meridian from the figure of a circle.

The plumb-line must every where be perpendicular to the surface of tranquil water, and can not, therefore, every where point exactly toward the earth's centre. Let A, B be two plumb-lines suspended on the same meridian near the equator, and at such a distance from each other as to be inclined at an angle of 1°. Let C and D be two other plumb-lines on a meridian near one of the poles, also making with each other an angle of 1°. The distance from A to B is found to be less than from C to D, from which we conclude that the meridian curves more rapidly near A than near C.

Fig. 10.

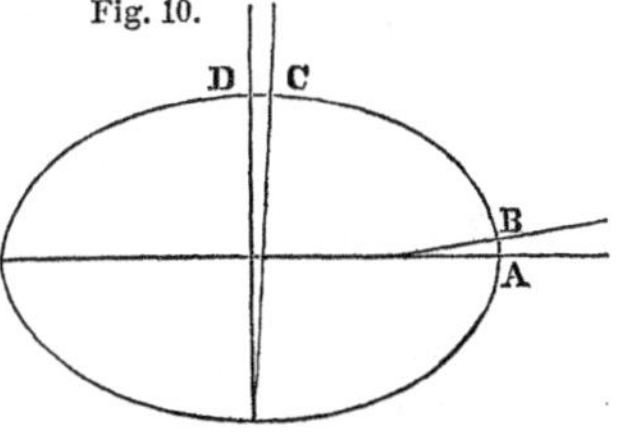

It is found that all the observations in every part of the world are very accurately represented by supposing the meridian to be an ellipse, of which the polar diameter is the minor axis.

The equatorial diameter of this ellipse is 7926.708 miles.
the polar diameter " " " 7899.755 "
the difference is 26.953 "

That is, the equatorial diameter exceeds the polar diameter by $\frac{1}{294}$th of its length. This difference is called the *ellipticity* of the earth.

The meridional circumference of the earth is 24,857.5 miles.

From measurements which have been made at right angles to the meridian, it appears that the equator and parallels of latitude are very nearly, if not exactly, circles. Hence it appears that the form of the earth is that of an *oblate spheroid;* which is a solid generated by the revolution of a semi-ellipse about its minor axis.

44. *Loss of weight at the equator explained.*—It has been mathematically proved that a spheroid whose ellipticity is $\frac{1}{294}$, and whose mean density is double the density at the surface, exerts an attraction upon a particle placed at its pole, greater by $\frac{1}{590}$th part than the attraction upon a particle at its equator; and this we have seen is the fraction which must be added to the loss of weight by centrifugal force, to make up the total loss of weight at the equator, as shown by experiments with the seconds' pendulum.

This coincidence may be regarded as demonstrating that the earth *does* rotate upon its axis once in 24 hours.

45. *Equatorial protuberance.*—If a sphere be conceived to be inscribed within the terrestrial spheroid, having the polar axis NS for its diameter, a spheroidal shell will be included between its surface and that of the spheroid, having a thickness, AB, of 13 miles at the equator, and becoming gradually thinner toward the poles. This shell of protuberant matter, by means of its attraction, gives rise to many important phenomena, as will be explained hereafter.

Fig. 11.

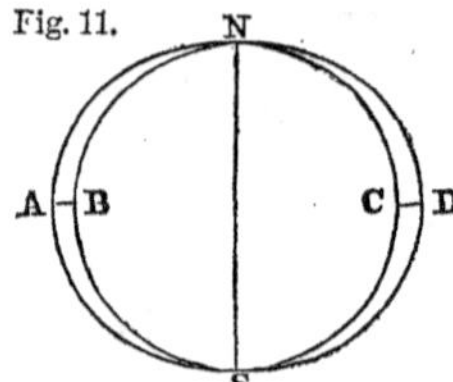

The Density of the Earth.

46. Three methods have been practiced for determining the average density of the earth. These methods are all founded upon the principle of comparing the attraction which the earth exerts upon any object, with the attraction which some other body, whose mass is known, exerts upon the same object.

First method.—By comparing the attraction of the earth with that of a small mountain.

In 1774, Dr. Maskelyne determined the ratio of the mean density of the earth to that of a mountain in Scotland, called Schehallien, by ascertaining how much the local attraction of the mountain deflected a plumb-line from a vertical position. This mountain stands alone on an extensive plain, so that there are no neighboring eminences to affect the plumb-line. Two stations were selected, one on its northern and the other on its southern side, and both nearly in the same meridian. A plumb-line, attached to an instrument called a zenith sector, designed for measuring small zenith distances, was set up at each of these stations, and the distance from the direction of the plumb-line to a certain star was measured at each station, the instant that the star was on the meridian. The difference between these distances gave the angle formed by the two directions of the plumb-lines AE, CG. Were it not for the mountain, the plumb-lines would take the positions AB, CD; and the angle which they would, in that case, form with each other, is found by measuring the distance between the two stations, and allowing about one second for every hundred feet.

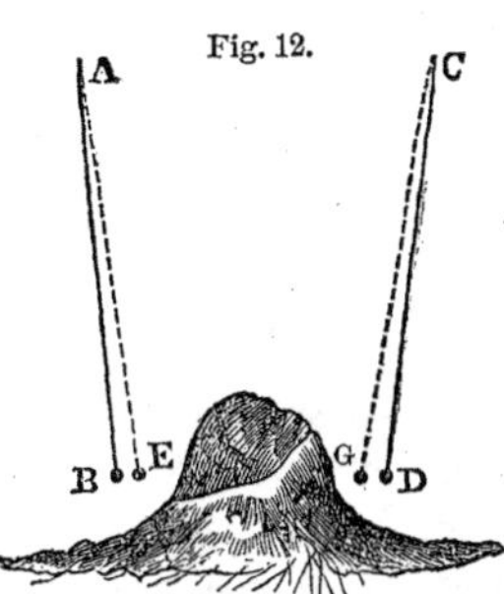

Fig. 12.

In Dr. Maskelyne's experiment, the distance between the two stations was 4000 feet; so that if the direction of gravity had not been influenced by the mountain, the inclination of the plumb-lines at the two places would have been 41 seconds. The inclination was actually found to be 53″. The difference, or 12″, is to be ascribed to the attraction of the mountain. It was computed that if the mountain had been as dense as the interior of the earth, the disturbance would have been about 21″. Therefore, the ratio of the density of the mountain to that of the entire earth, was that of 12 to 21.

The mean density of the mountain was ascertained by numerous borings to be 2.75 times that of water. Hence the mean density of the earth was concluded to be 4.95 times that of water.

In the year 1855, observations were made for ascertaining the deviation of the plumb-line produced by the attraction of Arthur's Seat, a hill 822 feet high, near Edinburg, from which the mean density of the earth was computed to be 5.32.

47. *Second method.*—The mean density of the earth has been determined by experiments with the torsion balance.

In the year 1798, Cavendish compared the attraction of the

earth with the attraction of two lead balls, each of which was one foot in diameter. The bodies upon which their attraction was exerted were two leaden balls, each about two inches in diameter. They were attached to the ends of a slender wooden rod six feet in length, which was supported at the centre by a fine wire 40 inches long. The balls, if left to themselves, will come to rest when the supporting wire is entirely free from torsion, but a very slight force is sufficient to turn it out of this plane. The position of the supporting rod was accurately observed with a fixed telescope. The large balls were then brought near the small ones, but on opposite sides, so that the attraction of both balls might conspire to twist the wire in the same direction, when it was found that the small balls were sensibly attracted by the larger ones, and the amount of this deflection was carefully measured. The large balls were then moved to the other side of the small ones, when the rod was found to be deflected in the contrary direction, and the amount of this deflection was recorded. This experiment was repeated seventeen times.

These experiments furnish a measure of the attraction of the large balls for the small ones, and hence we can compute what would be their attraction if they were as large as the earth. But we know the attraction actually exerted by the earth upon the small balls, it being measured by the weight of the balls. Thus we know the attractive force of the earth compared with that of the lead balls; and since we know the density of the lead, we can compute the average density of the earth. From these experiments, Cavendish concluded that the mean density of the earth was 5.45.

These experiments were repeated by Dr. Reich, at Freyberg, in Saxony, in the year 1836, and the mean of 57 trials gave a result of 5.44.

In the years 1841–'2 a similar series of experiments was conducted with the greatest care by Sir Francis Baily in England, and from over 2000 trials he concluded the mean density of the earth to be 5.67.

48. *Third method.*—The mean density of the earth may be determined by means of pendulum experiments at the top and bottom of a deep mine. The rate of vibration of a pendulum depends upon the intensity of the earth's attraction, and thus be-

comes a measure of this intensity. If we vibrate the same pendulum at the top and bottom of a mine whose depth is 1000 feet, we shall have a measure of the force of gravity at the bottom of the mine compared with the force at the top. Now at the top of the mine the pendulum is attracted by every particle of matter in the globe; but, since a spherical shell may be shown to exert no influence upon a point situated within it, the pendulum at the bottom of the mine will only be influenced by a sphere whose radius is 1000 feet less than that of the earth. We thus obtain the attraction of this external shell, whose thickness is 1000 feet, compared with the attraction of the entire globe; and since the volumes of both these bodies may be computed, we are able to deduce the average density of the globe, compared with that of the external shell. Now, by actual examination, we can determine the density of the strata penetrated by the mine, and hence we are able to compute the mean density of the globe.

This method was applied in one of the mines of England, near Newcastle, in the year 1854. The depth of the mine was 1256 feet; and it was found that a pendulum which vibrated seconds at the top of the mine, when transferred to the bottom of the mine gained $2\frac{1}{4}$ seconds per day. From this it was computed that the force of gravity at the bottom of the mine was $\frac{1}{19286}$ greater than at the top of the mine; and hence it was computed that the average density of the globe was 2.62 times that of the external shell. By actual examination, it was found that the average density of the rocks penetrated by the mine was 2.5, whence it follows that the mean density of the earth is 6.56.

49. *Fourth method.*—In a somewhat similar manner, we may determine the density of the earth by comparing the length of the pendulum vibrating seconds on the summit of a mountain, with that at the base of the mountain. In 1824, the vibrations of a pendulum on Mount Cenis, in Italy, at an elevation of 6734 English feet, were compared with the vibrations near the level of the sea, and the density of the earth was hence deduced to be 4.84.

The average of these seven determinations is 5.46, which must be a tolerable approximation to the truth.

These results verify, in a remarkable manner, the conjecture of Newton, who, in 1680, estimated that the average density of the earth was 5 or 6 times greater than that of water.

50. *Volume and weight of the earth.*—Having determined the dimensions of the earth, we can easily compute its volume, and, knowing its density, we can also compute its weight. Its volume is found to contain

259,400 millions of cubic miles.

Also the total weight of the earth is 6 sextillions of tons—a number expressed by the figure 6 with 21 ciphers annexed.

51. *Direct proof of the earth's rotation.*—A direct proof of the earth's rotation is derived from observations of a pendulum. If a heavy ball be suspended by a flexible wire from a fixed point, and the pendulum thus formed be made to vibrate, its vibrations will all be performed in the *same plane.* If, instead of being suspended from a fixed point, we give to the point of support a slow movement of rotation around a vertical axis, the plane of vibration will still remain unchanged. This may be proved by holding in the fingers a pendulum composed of a simple ball and string, and causing it to vibrate. Upon twirling the string between the fingers, the ball will be seen to rotate on its axis, without, however, changing its plane of vibration.

Suppose, then, a heavy ball to be suspended by a wire from a fixed point directly over the pole of the earth, and made to vibrate; these vibrations will continue to be made in the same invariable plane. But the earth meanwhile turns round at the rate of 15° per hour; and since the observer is unconscious of his own motion of rotation, it results that the plane of vibration of the pendulum *appears* to revolve at the same rate in the opposite direction.

If the pendulum be removed to the equator, and set vibrating in the direction of a meridian, the plane of vibration will still remain unchanged; and since, notwithstanding the earth's rotation, this plane always coincides with a meridian, the plane of vibration *appears* to remain unchanged.

52. *Phenomena in the middle latitudes.*—At places intermediate between the pole and the equator, the apparent motion of the plane of vibration is less than 15° per hour, and diminishes as we recede from the pole. This may be proved in the following manner:

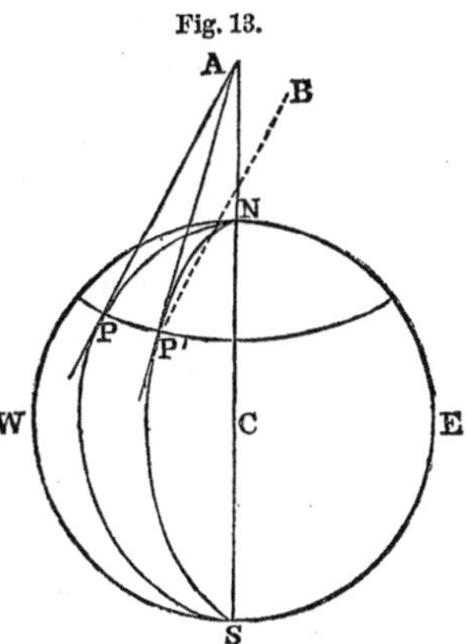

Let NPSE represent a meridian of the earth, AP a tangent to this circle at P, meeting the earth's axis produced in A. Suppose a pendulum to be set up at the point P, and vibrated in the plane of the meridian. When by the rotation of the earth the point P is brought to P′, the plane of vibration will tend to preserve its parallelism with the plane ACP; but the meridian of the place will have the position AP′C; that is, the plane of vibration will now make an angle AP′B, or PAP′, with the plane of the meridian. The angle PAP′, being taken very small, may be considered equal to $\frac{PP'}{AP}$, or $\frac{\text{cos. Lat.}}{\text{cot. Lat.}}$, which equals sin. Lat.; that is, the apparent motion of the plane of vibration is every where proportional to the *sine of the latitude.*

The hourly motion of the plane of vibration of a pendulum set up at New Haven, is therefore equal to $15° \times \sin. 41° 18'$, which is a little less than 10° per hour.

When this experiment is performed with the greatest care, the observed rate of motion coincides very accurately with the computed rate; and this coincidence may be regarded as a direct proof that the earth makes one rotation upon its axis in 24 sidereal hours.

53. *Second proof of the earth's rotation.*—A second proof of the earth's rotation is derived from the motion of falling bodies. If the earth had no rotation upon an axis, a heavy body let fall from any elevation would descend in the direction of a vertical line. But if the earth rotates on an axis, then, since the top of a tower describes a larger circle than the base, its easterly motion must be more rapid than that of the base. And if a ball be dropped from the top of the tower, since it has already the easterly motion which belongs to the top of the tower, it will retain this easterly motion during its descent, and its deviation to the east of the vertical line will be nearly equal to the excess of the motion of the top of the tower above that of the base, during the time of fall.

Let AB represent a vertical tower, and AA′ the space through which the point A would be carried by the earth's rotation in the time that a heavy body would descend through AB. A body let fall from the top of the tower will retain the horizontal velocity which it had at starting, and, when it reaches the earth's surface, will have moved over a horizontal space, BD, nearly equal to AA′. But the foot of the tower will have moved only through BB′, so that the body will be found to the east of the tower by a space equal to B′D nearly. This space B′D, for an elevation of 500 feet, in the latitude of New Haven, is a little over one inch, so that it must be impossible to detect this deviation except from experiments conducted with the greatest care and from an elevation of several hundred feet.

Fig. 14.

54. *Results of experiments.*—In the year 1791 this method was first tried at Bologna, in Italy, Lat. 44° 30′, from a tower whose height was 256 English feet. According to a mean of 12 trials, the deviation amounted to 0.74 inch to the east of a vertical line, and 0.47 inch to the south of a vertical. According to theory, the easterly deviation should have been 0.43 inch, and the southerly deviation should have been zero. The result of these experiments was not therefore satisfactory.

This discrepancy has been ascribed to a possible change in the position of the tower, due to the change of temperature, inasmuch as the experiments were made in summer, but the exact position of the plumb-line was not determined until the subsequent winter.

In the year 1802 the experiment was repeated by Benzenberg, at Hamburg, Lat. 53° 33′, from a tower whose height was 250 English feet. The mean of 31 experiments gave a deviation of 0.35 inch to the east of a vertical, and 0.11 inch to the south. According to theory, the easterly deviation should have been 0.34 inch, which accords remarkably well with the observations. The southerly deviation of 0.11 inch is not in accordance with theory.

This experiment was again repeated in 1804 by Benzenberg in Germany, Lat. 51° 25′, in a coal mine near Düsseldorf, whose depth was 280 English feet. According to 28 experiments, the average deviation to the east was 0.45 inch, while the computed deviation was 0.41 inch; showing a discrepancy of only 0.04 inch, which is quite satisfactory. The experiments also showed a deviation of 0.06 inch to the north, which is not explained by theory.

In the year 1832 these experiments were repeated with great care by Prof. Reich, at Freyberg, Saxony, Lat. 50° 53′, in a mine whose depth was 520 English feet. The balls employed were of metal, 1.59 inch in diameter, and had a specific gravity of 7.88. According to the mean of 106 trials, the easterly deviation was 1.12 inch, while the deviation by theory should have been 1.08 inch. The experiments also showed a southerly deviation of 0.17 inch, which is not accounted for by theory.

These experiments must be regarded as proving that the earth does rotate upon an axis, although the results exhibit discrepancies greater than might have been anticipated, and which, perhaps, are not fully explained.

ARTIFICIAL GLOBES.

55. Artificial globes are either terrestrial or celestial. The former exhibits a miniature representation of the earth, the latter exhibits the relative position of the fixed stars. The mode of mounting is usually the same for both, and many of the circles are the same for both globes. An artificial globe is mounted on an axis which is supported by a brass ring, which represents a meridian, and is called the *brass meridian*. This ring is supported in a vertical position by a frame in such a manner that the axis of the globe can be inclined at any angle to the horizon. The brass meridian is graduated into degrees, which are numbered from the equator toward either pole. The horizon is represented by a broad ring, whose plane passes through the centre of the globe. It is also graduated into degrees, which are numbered in both directions from the north and south points, to denote azimuths; and there is usually another set of numbers which begin from the east and west points, to denote amplitudes. It also usually contains the signs of the ecliptic, showing the sun's place for every day in the year.

On the terrestrial globe, hour circles are represented by great circles drawn through the poles of the equator; and on the celestial globe corresponding circles are drawn through the poles of the ecliptic, and a series of small circles parallel to the ecliptic are drawn at intervals of ten degrees. These are for determining celestial latitude and longitude. The ecliptic, tropics, and polar circles are drawn upon the terrestrial globe, as well as upon the celestial.

About the north pole is a small circle, graduated so as to indicate hours and minutes, while a small index, attached to the brass meridian, points to one of the divisions upon this hour circle. This index can be moved so as to be set in any required position.

There is usually a flexible strip of brass, equal in length to one quarter of the circumference of the globe, which is graduated into degrees, and may be applied to the surface of the globe so as to measure the distance between two places, or the altitude of any point above the wooden horizon. Hence it is usually called *the quadrant of altitude.*

PROBLEMS ON THE TERRESTRIAL GLOBE.

56. *To find the latitude and longitude of a given place.*

Turn the globe so as to bring the place to the graduated side of the brass meridian; then the degree of the meridian directly over the place will indicate the latitude, and the degree on the equator under the brass meridian will indicate the longitude.

Example. What are the latitude and longitude of Cape Horn?

57. *Given the latitude and longitude, to find the place.*

Bring the degree of longitude on the equator under the brass meridian, then under the given latitude on the brass meridian will be found the place required.

Example. Find the place which is situated in Lat. 30° N. and Long. 90° W.

58. *To find the bearing and distance from one place to another on the earth's surface.*

Elevate the north pole to the latitude of the first-mentioned place, and bring this place to the brass meridian. Screw the quadrant of altitude to this point of the brass meridian, and make it pass through the other place. Then the bearing of the second place from the first will be indicated on the wooden horizon, and the number of degrees on the quadrant of altitude will show the distance between the two places in degrees, which may be reduced to miles by multiplying them by 69½, because 69½ miles make nearly one degree.

Example. What is the bearing and distance of Liverpool from New York?

59. *To find the antipodes of a given place.*

Bring the given place to the wooden horizon, and the opposite point of the horizon will indicate the antipodes. The one place will be as far from the north point of the wooden horizon, as the other is from the south point.

Example. Find the antipodes of London.

60. *Given the hour of the day at any place, to find the hour at any other place.*

Bring the first-mentioned place to the brass meridian, and set the hour index to the given time. Turn the globe till the other place comes to the meridian; the hour circle will show the required time.

Example. What time is it at San Francisco when it is 10 A.M. in New York?

61. *To find the time of the sun's rising and setting at a given place, on a given day.*

Elevate the pole to the latitude of the place. On the wooden horizon find the day of the month, and against it is given the sun's place in the ecliptic, expressed in signs and degrees. Bring the sun's place to the meridian, and set the hour index to 12. Turn the globe till the sun's place is brought down to the eastern horizon; the hour index will show the time of rising. Turn the globe till the sun's place comes to the western horizon; the hour index will tell the time of setting.

Example. Required the time of rising and setting of the sun at Washington, August 18th.

CHAPTER II.

INSTRUMENTS FOR OBSERVATION. — THE CLOCK. — TRANSIT INSTRUMENT. — MURAL CIRCLE. — ALTITUDE AND AZIMUTH INSTRUMENT, AND THE SEXTANT.

62. *Why observations are chiefly made in the meridian.*—Whenever circumstances allow an astronomer to select his own time of observation, almost all his observations of the heavenly bodies are made when they are upon the meridian, because a large instrument can be more accurately and permanently adjusted to describe a *vertical* plane than any plane oblique to the horizon; and there is no other vertical plane which combines so many advantages as the meridian. The places of the heavenly bodies are most conveniently expressed by right ascension and declination, and the right ascension is simply the time of passing the meridian, as shown by a sidereal clock. Moreover, when a heavenly body is at its upper culmination, its refraction and parallax are the least possible; and in this position refraction and parallax do not affect the right ascension of the body, but simply its declination; while for every position out of the meridian, they affect both right ascension and declination.

63. *The Clock.*—The standard instruments of an astronomical observatory are the clock, the transit instrument, and the mural circle.

In a stationary observatory, a pendulum clock is used for measuring time. The clock should be so regulated that if a star be observed upon the meridian at the instant when the hands point to 0h. 0m. 0s., they will point to 0h. 0m. 0s. when the same star is next seen on the meridian. This interval is called a *sidereal day*, and is divided into 24 sidereal hours. If the clock were perfect, the pendulum would make 86,400 vibrations in the interval between two successive returns of the same star to the meridian. But no clock is perfect, and it is therefore necessary to determine the *error* and *rate* of the clock daily, and in all our observations to make an allowance for the error of the clock.

The *error* of a clock at any time is its difference from true sidereal time. The *rate* of the clock is the *change* of its error in 24 hours. Thus, if, on the 8th of January, when Aldebaran passed the meridian, the clock was found to be 30.84s. slow, and on the 9th of January, when the same star passed the meridian, the clock was 31.66s. slow, the clock lost 0.82s. per day. In other words, the error of the clock January 9th was −31.66s, and its daily rate −0.82s.

The Transit Instrument.

64. Most of the observations of the heavenly bodies are made when they are upon the celestial meridian; and, in many cases, the sole business of the observer is to determine the exact instant when the object is brought to the meridian, by the apparent diurnal motion of the firmament. This phenomenon of passing the meridian is called a *transit*, and an instrument, mounted in such a manner as to enable an observer, supplied with a clock, to ascertain the exact time of transit, is called a *transit instrument.*

65. *Description of the Transit Instrument.*—Such an instrument consists of a telescope, TT, mounted upon an axis, AB, at right angles to the tube, which axis occupies a horizontal position, and points east and west. The tube of the telescope, when horizontal, will therefore be directed north and south; and if the telescope be revolved on its axis through 180°, the central line of the tube will move in the plane of the meridian, and may be directed to any point on the celestial meridian.

Fig. 15.

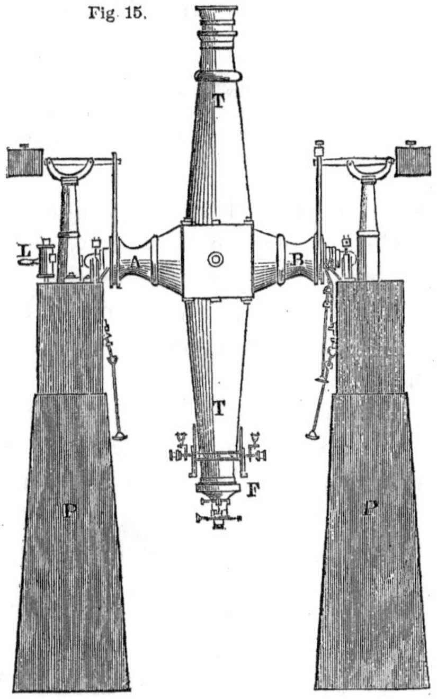

For a large transit instrument, two stone piers, PP, are erected on a solid foundation, standing on an east and west line. On the top of each of the piers is secured a metallic support, in the form of the letter Y, to receive the extremities of the axis of the telescope. At the left end of the axis there is a screw, by which the Y of that extremity may be raised or lowered a little, in order that the axis may be made perfectly horizontal. At the right end of the axis is a screw, by which the Y of that extremity may be moved backward or forward, in order to enable us to bring the telescope into the plane of the meridian. In order that the pivots of the axis may be relieved from a portion of the weight of the instrument, there is raised upon the top of each pier a brass pillar supporting a lever, from one end of which hangs a hook passing under one extremity of the axis, while a counterpoise sliding on the other end of the lever may be made to support as much of the weight of the instrument as is desired.

66. *The Spirit Level.*—When the instrument is properly adjusted, its axis will be horizontal, and directed due east and west. If the axis be not exactly horizontal, its deviation may be ascertained by placing upon it a *spirit level.* This consists of a glass tube, AB, nearly filled with alcohol or ether. The tube forms a portion of a ring of a very large radius, and when it is placed horizontally, with its convexity upward, the bubble, CD, will occupy the highest position in the middle of its length. A graduated scale is attached to the tube, by which we may measure any deviation of the bubble from the middle of the tube.

Fig. 16.

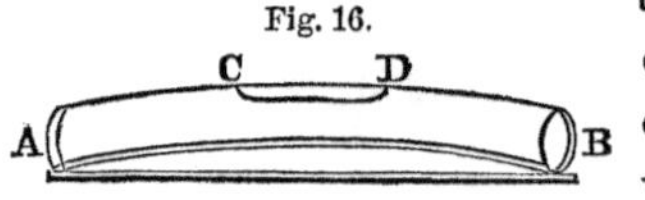

To ascertain whether the axis of the telescope is horizontal, apply the level to it, and see if the bubble occupies the middle of the tube. If it does not, one end must be elevated or depressed. In order to accomplish this, one of the supports of the axis is constructed so as to be moved vertically through a small space by means of a fine screw. The level must now be taken up and reversed end for end, and this operation must be repeated until the bubble rests in the middle of the tube in both positions of the level.

67. *Method of observing transits.*—In the focus of the eye-piece of the transit instrument, at F, is placed a system of 5 or 7 equi-

distant and vertical wires, intersected by 1 or 2 horizontal wires. When the instrument has been properly adjusted, the middle wire, MN, will be in the plane of the meridian, and when an object is seen upon it, this object will be on the celestial meridian. The fixed stars appear in the telescope as bright points of light without sensible magnitude, and by the diurnal motion of the heavens a star is carried successively over each of the wires of the transit instrument. The observer, just before the star enters the field of view, writes down the hour and minute indicated by the clock, and proceeds to count the seconds by listening to the beats of the clock, while his eye is looking through the telescope. He observes the instant at which the star crosses each of the wires, estimating the time to the nearest tenth of a second; and by taking a mean of all these observations, he obtains with great precision the instant at which the star passed the middle wire, and this is regarded as the true time of the transit. The mean of the observations over several wires, is considered more reliable than an observation over a single wire.

Fig. 17.

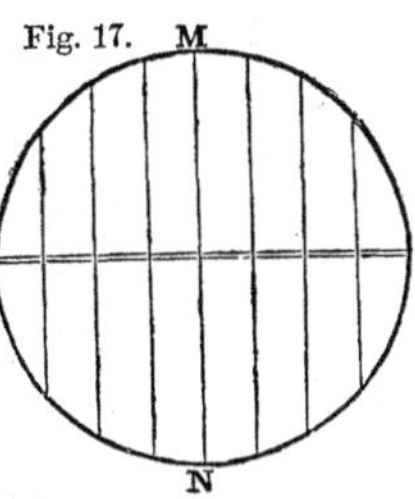

In many observatories it is now customary to employ the electric circuit to record transit observations. By pressing the finger upon a key at the instant a star is seen to pass one of the wires of the transit, a mark is made upon a sheet of paper which is graduated into seconds by the pendulum of the observatory clock, according to the mode more fully explained in Art. 337.

During the day, the wires are visible as fine black lines stretched across the field of view. At night they are rendered visible by a lamp, L, by which the field of view is faintly illumined.

When we observe the sun or any object which has a sensible disc, the time of transit is the instant at which the centre of the disc crosses the middle wire. This time is obtained by observing the instants at which the eastern and western edges of the disc touch each of the wires in succession, and taking the mean of all the observations. When the visible disc is not circular, special methods of reduction are employed.

68. *Rate of the diurnal motion.*—Since the celestial sphere revolves at the rate of 15° per hour, or 15 seconds of arc in one second of time, the space passed over between two successive

beats of the pendulum will be 15″ of arc. When the sun is on the equator, and its apparent diameter is 32′ of arc, the interval between the contacts of the east and west limbs with the middle wire will be 2m. 8s.

69. *To adjust a transit instrument to the meridian.*—A transit instrument may be adjusted to describe the plane of the meridian, by observations of the pole star. Direct the telescope to the pole star at the instant of its crossing the meridian, as near as the time can be ascertained. The transit will then be *nearly* in the plane of the meridian. Having leveled the axis, turn the telescope to a star about to cross the meridian, near the zenith. Since every vertical circle intersects the meridian at the zenith, a zenith star will cross the field of the telescope at the same time, whether the plane of the transit coincide with the meridian or not. At the moment the star crosses the central wire, set the clock to the star's right ascension which is given by the star catalogues, and the clock will henceforth indicate nearly sidereal time. The approximate times of the upper and lower culminations of the pole star are then known. Observe the pole star at one of its culminations, following its motion until the clock indicates its right ascension, or its right ascension plus 12 hours. Move the whole frame of the transit so that the central wire shall coincide nearly with the star, and complete the adjustment by means of the azimuth screw. The central wire will now coincide almost precisely with the meridian of the place.

70. *Final verification.*—The axis being supposed perfectly horizontal, if the middle wire of the telescope is exactly in the meridian, it will bisect the circle which the pole star describes in 24 sidereal hours round the polar point. If, then, the interval between the upper and lower culminations is exactly equal to the interval between the lower and upper, the adjustment is complete. But if the time elapsed while the star is traversing the eastern semicircle, is greater than that of traversing the western, the plane in which the telescope moves is westward of the true meridian on the north horizon; and *vice versa* if the western interval is greatest. This error of position must be corrected by turning the azimuth screw. The adjustment must then be verified by further observations, until, by continued approximations, the instrument is fixed correctly in the meridian.

Other methods of adjusting a transit instrument to the plane of the meridian, will be found in works specially devoted to Practical Astronomy.

The Mural Circle.

71. The mural circle is a graduated circle, *aaaa*, usually made of brass, and having an axis passing through its centre. This axis should be exactly horizontal; and it is supported by a stone pier or wall, so as to be directed due east and west. To the circle is attached a telescope, MM, so that the entire instrument, including the telescope, turns in the plane of the meridian.

Fig. 18.

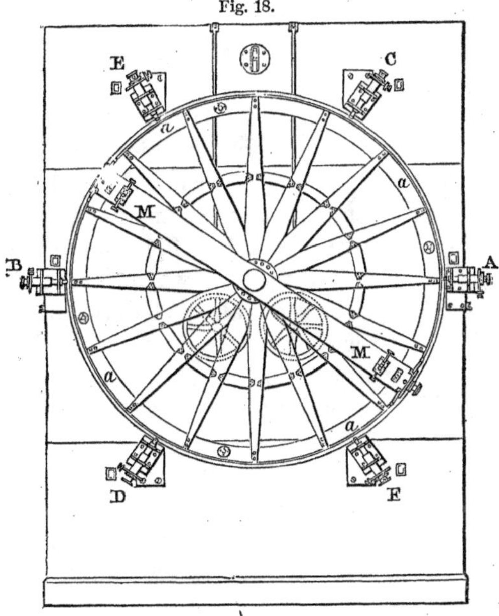

Mural circles have been made eight feet in diameter, but generally they have been made six feet; and at present astronomers are pretty well agreed that a circle of five feet is better than any larger size, being less liable to change of form from its great weight. At the great Russian observatory at Pulkova, the largest circle employed is only four feet in diameter. The circle is divided into degrees, and subdivided into spaces of five minutes, and sometimes of two minutes, the divisions being numbered from 0° to 360° round the entire circle. The smallest spaces on the limb are further subdivided to single seconds, sometimes by a *vernier*, but generally by a *reading microscope.*

72. *Use of the Vernier.*—A *vernier* is a scale of small extent, graduated in such a manner that, being moved by the side of a fixed scale, we are enabled to measure minute portions of this scale. The length of this movable scale is equal to a certain number of parts of that to be subdivided; but it is divided into parts either one more, or one less, than those of the primary scale taken for the length of the vernier. Thus, if we wish to measure hundredths of an inch, as in the case of a barometer, we first divide an inch into ten equal parts. We then construct a vernier equal in length to 11 of these divisions, but divide it into 10 equal parts, by which means each division on the vernier is $\frac{1}{10}$th longer than a division of the primary scale.

Fig. 19.

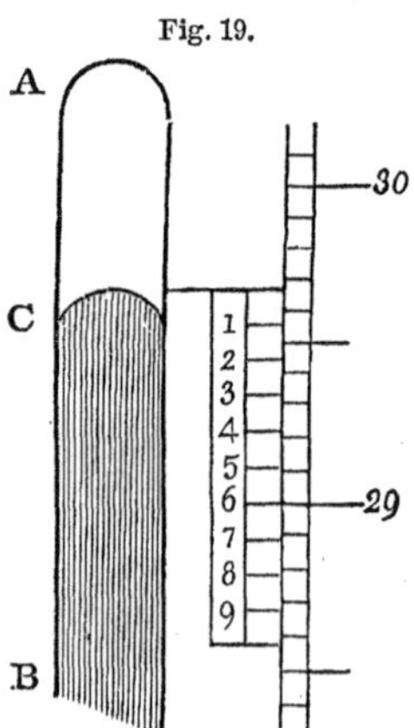

Thus, let AB be the upper end of a barometer tube, the mercury standing at the point C; the scale is divided into inches and tenths of an inch, and the middle piece, numbered from 1 to 9, is the vernier, that may be slid up or down, and having 10 of its divisions equal to 11 divisions of the scale; that is, to $\frac{11}{10}$ths of an inch. Therefore, each division of the vernier is $\frac{11}{100}$ths of an inch; or one division of the vernier exceeds one division of the scale, by $\frac{1}{100}$th of an inch. Now, as the sixth division of the vernier (in the figure) coincides with a division of the scale, the fifth division of the vernier will stand $\frac{1}{100}$th of an inch above the nearest division of the scale; the fourth division $\frac{2}{100}$ths of an inch; and the top of the vernier will be $\frac{6}{100}$ths of an inch above the next lower division of the scale; *i. e.*, the top of the vernier coincides with 29.66 inches upon the scale. In practice, therefore, we observe what division of the vernier coincides with a division of the scale; this will show the hundredths of an inch to be added to the tenths next below the vernier at the top.

73. *Vernier applied to graduated circles.*—A similar contrivance is often applied to graduated circles to obtain the value of an arc

with greater accuracy. If a circle is graduated to half degrees, or 30′, and we wish to measure single minutes by the vernier, we take an arc equal to 31 divisions upon the limb, and divide it into 30 equal parts. Then each division of the vernier will be equal to $\frac{31}{60}$ths of a degree, while each division of the scale is $\frac{30}{60}$ths of a degree; that is, each space on the vernier exceeds one on the limb by 1′.

In order, therefore, to read an angle for any position of the vernier, we pass along the vernier until a line is found coinciding with a line of the limb. The number of this line from the zero point, indicates the minutes which are to be added to the degrees and half degrees taken from the graduated scale.

74. *The reading Microscope.*—The large circles employed in astronomical observations are divided into spaces as small as 5′, and sometimes as small as 2′. By a vernier these spaces are sometimes subdivided so as to give single seconds. The vernier is generally employed in instruments made by German artists, but upon large circles made by English artists the subdivisions are usually effected by the *reading microscope.* Fig. 20 represents the appearance of one of these microscopes. It is a compound microscope, consisting of three lenses, one of which is the object lens at L, and the other two are formed into a positive eye-piece, GH.

Fig. 20.

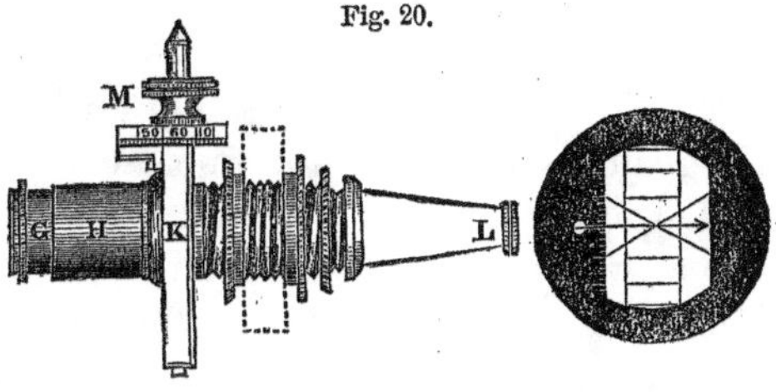

In the common focus of the object lens and the eye-piece at K, is placed the spider-line micrometer. It consists of a small rectangular frame, across which are stretched two spider-lines forming an acute cross, and is moved laterally by means of a screw, M. The figure on the right shows the field of view, with the magnified divisions on the instrument, as seen through the microscope. When the microscope is properly adjusted, the image of the divided limb and the spider-lines are distinctly visible together; and also five revolutions of the screw must exactly measure one of the 5′ spaces on the limb. One revolution of the head of the screw will therefore carry the spider-lines over a space of 1′. The circumference of the circle attached to the head, M, is divided into

60 equal parts, so that the motion of the head through one of these divisions, advances the spider-lines through a space of 1″. There are six of these microscopes, A, B, C, D, E, F, placed at equal distances round the circle, and firmly attached to the pier.

75. *To determine the horizontal point.*—In order to ascertain the horizontal point upon the limb of the circle, we direct the telescope upon any star which is about crossing the meridian, and bring its image to coincide with the horizontal wire which passes through the centre of the field of the telescope. The graduation is then read off by the fixed microscopes. On the next night, we place a vessel containing mercury in a convenient position near the floor, so that, by directing the telescope of the mural circle toward it, the same star may be seen reflected from the surface of the mercury, and we bring the reflected image to coincide with the horizontal wire of the telescope. The graduation is then read off as before. Now, by a law of optics, the reflected image will appear as much below the horizon as the star is really above the horizon; therefore half the sum of the two readings at either of the microscopes, will be the reading at the same microscope when the telescope is horizontal.

76. *To determine the altitude of any object.*—Having determined the reading of each of the microscopes when the telescope is directed to the horizon, if we wish to determine the altitude of any object, we direct the telescope to it, so that it may be seen on the horizontal wire as the star passes the meridian, and then read off the microscopes. The difference between the last reading, and the reading when the telescope is horizontal, is the altitude required.

The zenith distance of an object is found by subtracting its altitude from 90°.

The pole star crosses the meridian, above and below the pole, at intervals of 12 hours sidereal time; and the true position of the pole is exactly midway between the two points where the star crosses the meridian; therefore half the sum of the readings of either microscope when the pole star makes its transit above and below the pole, will be the reading for the pole itself.

The readings for the pole being determined, those which correspond to the point where the celestial equator crosses the meridian, are easily found, since the equator is 90° from the pole.

Having determined the position of the celestial equator, the declination of any star is easily determined, since its declination is simply its distance from the equator.

77. *The Transit Circle.*—Since the mural circle has a short axis, its position in the meridian is unstable, and therefore it can not be relied upon to give the right ascension of stars with great accuracy. It was formerly thought necessary at Greenwich to have two instruments for determining a star's place; viz., a transit instrument to determine its right ascension, and a mural circle to determine its declination. The German astronomers have, however, combined both instruments in one, under the name of meridian circle, which is essentially the transit instrument already described, with a large graduated circle attached to its axis; and a large transit circle is now in use at the Greenwich Observatory.

Fig. 21.

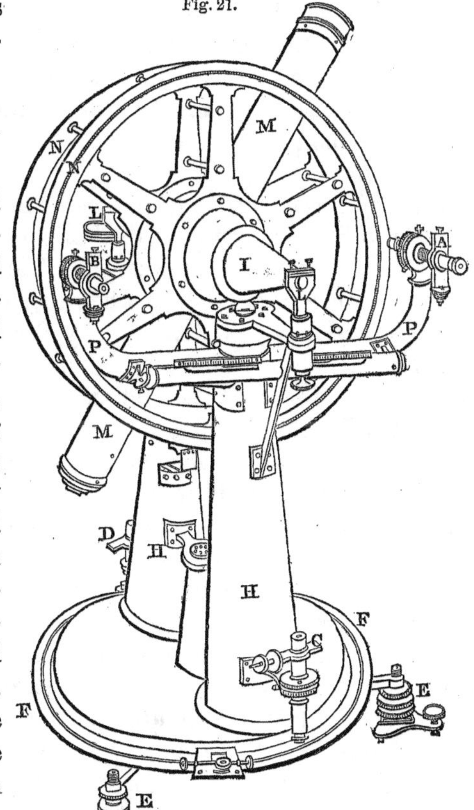

Altitude and Azimuth Instrument.

78. The altitude and azimuth instrument consists of one graduated circle confined to a horizontal plane; a second graduated circle perpendicular to the former, and capable of being turned into any azimuth; and a telescope firmly fastened to the second circle, and turning with it in altitude. The appearance of this instrument will be learned from the annexed figure.

EE are two legs of the tripod upon which the instrument rests; and in close contact with the tripod is placed the azimuth circle, FF. Above the azimuth circle, and concentric with it, is placed a strong circular plate, which sustains the whole of the upper part of the instrument, and also a pointer, to show the degree and nearest five minutes to be read off on the azimuth circle; the remaining minutes and seconds being obtained by means of the two reading microscopes C and D. The pillars, HH, support the transit axis I by means of the projecting pieces LL. The telescope, MM, is connected with the horizontal axis in a manner similar to that of the transit instrument. Upon the axis, as a centre, is fixed the double circle NN, each circle being placed close against the telescope. The circles are fastened together by small brass pillars, and the graduation is made on a narrow ring of silver, inlaid on one of the sides, which is usually termed the *face* of the instrument. The reading microscopes, AB, for the vertical circle, are carried by two arms, PP, attached near the top of one of the pillars.

In the principal focus of the telescope, are stretched spider lines, as in the transit instrument, and the illumination is effected in a similar manner.

79. *Adjustments of the instrument.*—Before commencing observations with this instrument, the horizontal circle must be leveled, and also the axis of the telescope. The meridional point on the azimuth circle is its reading when the telescope is pointed north or south, and may be determined by observing a star at equal altitudes east and west of the meridian, and finding the point midway between the two observed azimuths; or the instrument may be adjusted to the meridian in the same manner as a transit. The horizontal point of the altitude circle is its reading when the axis of the telescope is horizontal, and may be found, as with the mural circle, by alternate observations of a star directly and reflected from the surface of mercury.

This instrument has the advantage over the transit instrument and mural circle, in its being able to determine the place of a star in any part of the visible heavens; but we ordinarily require the place of a star to be given in right ascension and declination instead of altitude and azimuth, and to deduce the one from the other requires a laborious computation. Hence the altitude and azimuth instrument is but little used in astronomical observations, except for special purposes, as, for example, to investigate the laws of refraction.

The Sextant.

80. The arc of a sextant, as its name implies, contains sixty degrees, but, on account of the double reflection, is divided into 120 degrees. The annexed figure represents a sextant, the frame being generally made of brass; the handle, H, at its back, is made of wood. When observing, the instrument is to be held with one hand by the handle, while the other hand moves the index G. The arc, AB, is divided into 120 or more degrees, numbered from A toward B, and each degree is divided into six equal parts of 10′ each, while the vernier shows 10″. The divisions are also continued a short distance on the other side of zero toward A, forming what is called the arc of excess. The microscope, M, is movable about a centre, and may be adjusted to read off the divisions on the graduated limb. A tangent screw, D, is fixed to the index, for the purpose of making the contacts more accurately than can be done by hand. When the index is to be moved a considerable distance, the screw I must be loosened; and when the index is brought nearly to the required division, the screw I must be tightened, and the index be moved gradually by the tangent screw. The upper end of the index G terminates in a circle, across which is fixed the silvered index glass C, over the centre of motion, and perpendicular to the plane of the instrument. To the frame at N is attached a second glass, called the horizon glass, the lower half of which only is silvered. This must also be perpendicular to the plane of the instrument, and in such a position that its plane shall be parallel to the plane of the index glass C, when the vernier is set to zero on the limb AB.

Fig. 22.

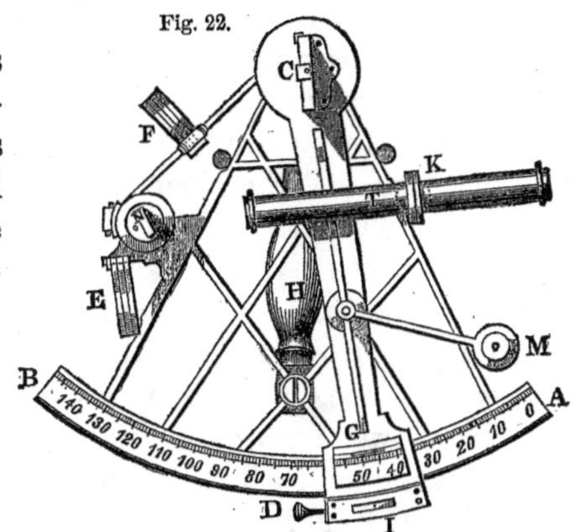

The telescope, T, is carried by a ring, K; and in the focus of the object glass are placed two wires parallel to each other, and equidistant from the axis of the telescope. Four dark glasses, of different depths of shade and color, are placed at F, between the index and horizon glasses; also three more at E, any one or more of which can be turned down, to moderate the intensity of the light before reaching the eye, when a bright object, as the sun, is observed.

81. *To measure the altitude of the sun by reflection from mercury.*—Set the index near zero. Hold the instrument with the right hand in the vertical plane of the sun, with the telescope pointed toward the sun. Two images will be seen in the field of view, one of which, viz., that formed by reflection, will apparently move downward when the index is pushed forward. Follow the reflected image as it travels downward, until it appears to be as far below the horizon as it was at first above, and the image of the sun reflected from the mercury also appears in the field of view. Fasten the index, and, by means of the tangent screw, bring the upper or lower limb of the sun's image reflected from the index glass, into contact with the opposite limb of the image reflected from the artificial horizon. The angle shown on the instrument, when corrected for the index error, will be double the altitude of the sun's limb above the horizontal plane; to the half of which, if the semi-diameter, refraction and parallax be applied, the result will be the true altitude of the centre.

If the observer is at sea, the natural horizon must be employed. Direct the sight to that part of the horizon beneath the sun, and move the index till you bring the image of its lower limb to touch the horizon directly underneath it.

82. *To measure the distance between two objects.*—To find the distance between the moon and sun, hold the sextant so that its plane may pass through both objects. Look directly at the moon through the telescope, and move the index forward till the sun's image is brought nearly into contact with the moon's nearest limb. Fix the index by the screw under the sextant, and make the contact perfect by means of the tangent screw. The index will then show the distance of the nearest limbs of the sun and moon. In a similar manner may we measure the distance between the moon and a star.

83. *Dip of the horizon.*—In observing an altitude at sea with the sextant, the image of an object is made to coincide with the

visible horizon; but since the eye is elevated above the surface of the sea, the visible horizon will be below the true horizontal plane.

Let AC be the radius of the earth, AD the height of the eye above the level of the sea, EDH a horizontal plane passing through the place of the observer; then HDB will be the *dip* or depression of the horizon, which may be found as follows:

Fig. 23.

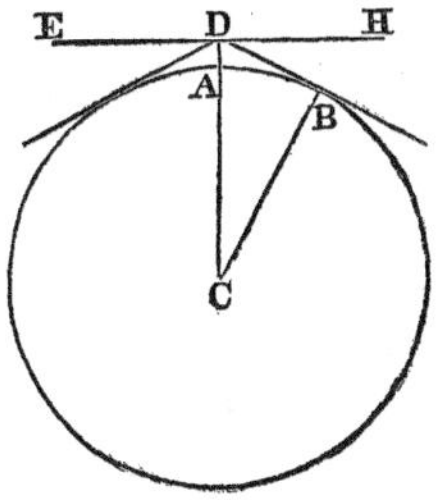

The angle HDB is equal to the angle BCD; and in the right-angled triangle BCD, $BD^2 = CD^2 - BC^2 = (AC + AD)^2 - AC^2$. Whence BD becomes known. Then, in the same triangle,

$$CD : \text{rad.} :: BD : \sin. BCD (= HDB),$$

the depression of the horizon.

The depression thus obtained is the true depression; but this must be lessened by the amount of terrestrial refraction, which is very uncertain. About $\frac{1}{8}$th or $\frac{1}{10}$th of the whole quantity is usually allowed.

The following table shows the dip or apparent depression of the horizon for different elevations of the eye, allowing $\frac{1}{10}$th for terrestrial refraction:

Height.	Depression.	Height.	Depression.
5 feet.	2′ 9″	30 feet.	5′ 15″
10 "	3 2	50 "	6 46
15 "	3 42	70 "	8 1
20 "	4 17	100 "	9 35

CHAPTER III.

ATMOSPHERIC REFRACTION.—TWILIGHT.

84. The air which surrounds the earth decreases gradually in density as we ascend from the surface. At the height of 4 miles, the density is only about half as great as at the earth's surface; at the height of 8 miles about one fourth as great; at the height of 12 miles about one eighth as great, and so on. From this law it follows that at the height of 50 miles, its density must be extremely small, so as to be nearly or quite insensible.

85. *Law of atmospheric refraction.*—According to a law of optics, when a ray of light passes obliquely from a rarer to a denser medium, it is bent toward the perpendicular to the refracting surface. Let SA be a ray of light coming from any distant object, S, and falling on the surface of a series of layers of air, increasing in density downward. The ray SA, passing into the first layer, will be deflected in the direction AB, toward a perpendicular to the surface, MN. Passing into the next layer, it will be again deflected in the direction BC, more toward the perpendicular; and passing through the lowest layer, it will be still more deflected, and will enter the eye at D, in the direction of CD; and, since every object appears in the direction from which the visual ray enters the eye, the object S will be seen in the direction DS′, instead of its true direction AS.

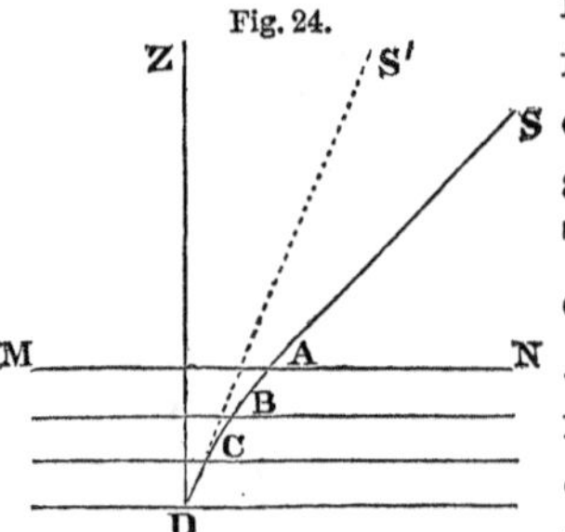

Fig. 24.

Since the density of the earth's atmosphere increases *gradually* from its upper surface to the earth, when a ray of light from any of the heavenly bodies enters the atmosphere obliquely, its path is not a broken line, as we have here supposed, but a curve, concave toward the earth. The density of the upper parts of the atmosphere being very small, the curve at first deviates very little from a straight line, but the deviation increases as it approaches the earth. Both the straight and curved parts of the ray lie in the same vertical plane; that is, the refraction of the atmosphere makes an object appear to be nearer the zenith than it really is, but does not affect its azimuth.

86. *How the refraction may be computed.*—It is a difficult problem to compute the exact amount of the refraction of the atmosphere; but for altitudes exceeding 10 degrees, the entire refraction may be assumed to take place at a single surface, as MN, and may be computed approximately in the following manner:

Let z denote the apparent zenith distance of a star, and r the effect of refraction; then, if there were no refraction, the zenith distance would be $z+r$. But we have found in Optics, Art. 691, that the sine of incidence$=m\times$sine of refraction, where m represents the index of refraction. Hence

$$\sin.(z+r)=m\ \sin.z.$$

But by Trigonometry, Art. 72,

$$\sin.(z+r)=\sin.z\ \cos.r+\cos.z\ \sin.r.$$

For zenith distances less than 80°, r is less than 6′, and therefore its sine may be considered equal to the arc, and its cosine equal to unity. Hence we find

$$\sin.z+r\ \cos.z=m\ \sin.z.$$

Dividing by cos. z, and putting $\frac{\sin.}{\cos.}$=tangent (Trigonometry, Art. 28), we obtain $\quad \text{tang.}z+r=m\ \text{tang.}z,$

or $\quad r=(m-1)\ \text{tang.}z.$

r is here expressed in parts of radius. If we wish to have its value expressed in seconds, we must multiply it by 206265, which is the number of seconds in an arc equal to radius. If r'' represents the refraction expressed in seconds, then

$$r''=206265\ r.$$

At the temperature of 50°, and pressure 29.96 inches, the refractive index of air is 1.0002836. Hence we have

$$r''=0.0002836\times206265\times\text{tang.}z.$$

or $\quad r''=58''.49,\ \text{tang.}z;$

that is, the refraction is equal to 58″.49 × tangent of the zenith distance. For altitudes exceeding ten degrees, this formula will furnish the refraction pretty nearly; but near the horizon the law of refraction is exceedingly complicated.

The following table shows the average amount of refraction for different altitudes:

Altitude.	Refraction.	Altitude.	Refraction.	Altitude.	Refraction.
0°	34′ 54″	6°	8′ 23″	20°	2′ 37″
1	24 25	7	7 20	30	1 40
2	18 9	8	6 30	40	1 9
3	14 15	9	5 49	50	48
4	11 39	10	5 16	70	21
5	9 46	15	3 32	90	0

We perceive from this table, that the refraction is nothing in the zenith, and is greatest in the horizon, where it amounts to about 35′.

87. *How the refraction may be determined by observation.*—The amount of refraction for different altitudes, may be determined by observation as follows: In latitudes greater than 45°, a star which

passes through the zenith of the place, may also be observed when it passes the meridian below the pole. Let the polar distance of such a star be measured both at the upper and lower culminations. In the former case there will be no refraction; the difference between the two observed polar distances will therefore be the amount of refraction for the altitude at the lower culmination; because if there were no refraction, the apparent diurnal path of the star would be a circle with the celestial pole for its centre. This method is strictly applicable only in latitudes greater than 45°, and by observations at one station we can only determine the refraction corresponding to a single altitude. Since, however, for zenith distances less than 45°, the amount of refraction is quite small, and is given with great accuracy by the Tables, we may safely extend the application of this method. We may therefore select any star within the circle of perpetual apparition, and observe its polar distance at the upper and lower culminations, and correct the former for refraction. The difference between this corrected value and the observed polar distance at the lower culmination, will be the refraction corresponding to the latter altitude.

88. *Second method of determining refraction.*—The following method is more general in its application, and requires no previous knowledge of the amount of refraction:

Observe the altitude of a star whose declination is known, and note the time by the clock. Observe also when the star crosses the meridian, and the difference of time between the observations will give the hour angle of the star from the meridian.

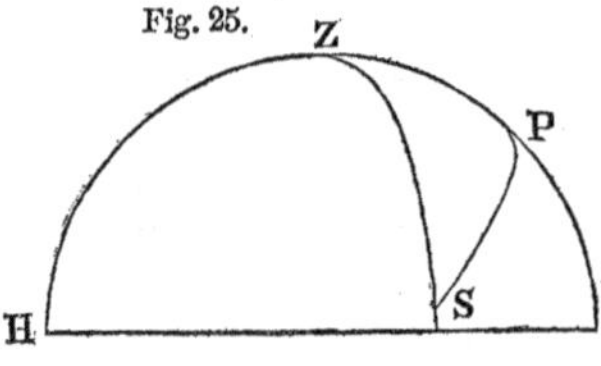

Let PZH be the meridian of the place of observation, P the pole, Z the zenith, and S the true place of the star. Let ZS be a vertical circle passing through the star, and PS an hour circle passing through the star. Then, in the triangle ZPS,

PZ=the complement of the latitude,

PS=the north polar distance of the star,

and ZPS=the angular distance of the star from the meridian.

In this triangle we know, therefore, two sides and the included angle, from which we can compute ZS, or the true zenith distance of the star. The difference between the computed value of ZS and its observed value, will be the refraction corresponding to this altitude.

If we commence our observations when the star is near the horizon, and continue them at short intervals until it reaches the meridian, we may, by a proper selection of stars, determine the amount of refraction for all altitudes from zero to 90°.

89. *Corrections for temperature and pressure.*—The amount of refraction at a given altitude is not constant, but depends upon the temperature, and weight of the air. Tables have been constructed, partly from observation and partly from theory, by which we may at once obtain the mean refraction for any altitude; and rules are given by which a correction may be made for the state of the barometer and thermometer.

90. *Effect of refraction upon the time of sunrise.*—Since refraction increases the altitudes of the heavenly bodies, it must accelerate their rising and retard their setting, and thus render them longer visible. The amount of refraction at the horizon is about 35′, which being a little more than the apparent diameters of the sun and moon, it follows that these bodies, at the moment of rising and setting, are visible above the horizon, when in reality they are wholly below it.

91. *Effect of refraction upon the figure of the sun's disc.*—When the sun is near the horizon, the lower limb, being nearest the horizon, is most affected by refraction, and therefore more elevated than the upper limb, the effect of which is to bring the two limbs apparently closer together by the difference between the two refractions. The apparent diminution of the vertical diameter sometimes amounts at the horizon to one fifth of the whole diameter. The disc thus assumes the form of an ellipse, of which the major axis is horizontal.

92. *Enlargement of the sun near the horizon.*—The apparent enlargement of the sun and moon near the horizon is an *optical illusion.* If we measure the apparent diameters of these bodies with any suitable instrument, we shall find that they subtend a *less* angle near the horizon, than they do when near the zenith. It is, then, wholly owing to an error of judgment that they seem to us larger near the horizon.

Our judgment of the absolute magnitude of a body is based upon our judgment of its distance. If two objects at unequal distances subtend the same angle, the more distant one must be the larger. Now the sun and moon, when near the horizon, appear to us *more distant* than when they are high in the heavens. They seem more distant in the former position, partly from the number of intervening objects, and partly from diminished brightness. When the moon is near the horizon, a variety of intervening objects shows us that the distance of the moon must be considerable; but when the moon is on the meridian no such objects intervene, and the moon appears quite near. For the same reason, the vault of heaven does not present the appearance of a hemisphere, but appears flattened at the zenith, and spread out at the horizon.

Our estimate of the distance of objects is also affected by their *brightness.* Thus, a distant mountain, seen through a perfectly clear atmosphere, appears much nearer than when seen through a hazy atmosphere.

93. *Cause of twilight.*—The sun continues to illumine the clouds and the upper strata of the air, after it has set, in the same manner as it shines on the summits of mountains after it has set to the inhabitants of the adjacent plains. The air and clouds thus illumined reflect light to the earth below them, and produce *twilight.* As the sun continues to descend below the horizon, a less part of the visible atmosphere receives his direct light; less light is transmitted by reflection to the surface of the earth; until, at length, all reflection ceases, and night begins. This takes place when the sun is about 18° below the horizon.

Before sunrise in the morning, the same phenomena are exhibited in the reverse order. If there were no atmosphere, none of the sun's rays could reach us after his actual setting, or before his rising.

Fig. 26.

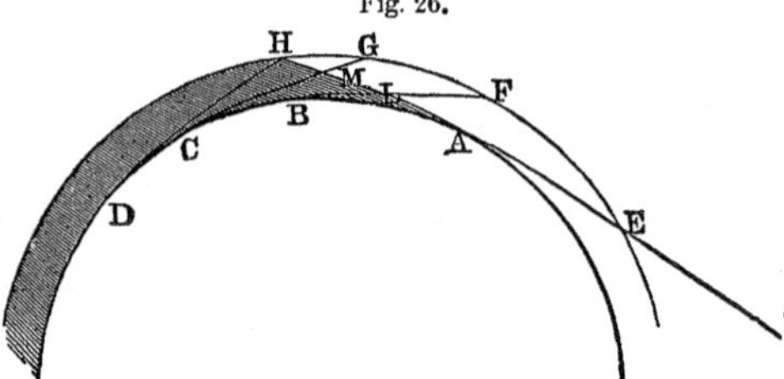

Let ABCD represent a portion of the earth, A a point on its surface where the sun, S, is in the act of setting, and let SAH be a ray of light just grazing the earth at

A, and leaving the atmosphere at the point H. The point A is illuminated by the whole reflective atmosphere HGFE. The point B, to which the sun has set, receives no direct solar light, nor any reflected from that part of the atmosphere which is below ALH, but it receives a twilight from the portion HLF, which lies above the visible horizon BF. The point C receives a twilight only from the small portion of the atmosphere HMG, while at D the twilight has ceased altogether.

94. *Duration of twilight at the equator.*—The duration of twilight varies with the season of the year, and with our position upon the earth's surface. At the equator, where the circles of daily rotation are perpendicular to the horizon, when the sun is in the celestial equator, it descends through 18° in an hour and twelve minutes ($\frac{18}{15}=1\frac{1}{5}$ hours); that is, twilight lasts 1h. and 12m. When the sun is not in the equator, the duration of twilight is somewhat increased.

95. *Duration of twilight at the poles.*—At the north pole there is night as long as the sun is south of the equator; but whenever it is not more than 18° south, the sun is never more than 18° below the horizon. About the close of September, the sun sinks below the horizon, and there is continual twilight until November 12th, when it attains a distance of 18° from the equator. From this date there is no twilight until January 29th, from which time there is continual twilight until about the middle of March, when the sun rises above the horizon, and continues above the horizon uninterruptedly for six months.

96. *Duration of twilight in middle latitudes.*—At intermediate points of the earth, the duration of twilight may vary from 1h. 12m. to several weeks. In latitude 40°, during the months of March and September, twilight lasts about an hour and a half, while in midsummer it lasts a little over two hours.

In latitude 50°, where the north pole is elevated 50° above the horizon, the point which is on the meridian 18° below the north point of the horizon, is 68° distant from the north pole, and therefore 22° distant from the equator. Now, during the entire month of June, the distance of the sun from the equator exceeds 22°; that is, in latitude 50° there is continual twilight from sunset to sunrise, during a period of more than a month.

At places nearer to the pole, the period of the year during which twilight lasts through the entire night, is still longer.

97. *Consequences if there were no atmosphere.*—If there were no atmosphere, the darkness of midnight would instantly succeed the setting of the sun, and it would continue thus until the instant of the sun's rising. During the day the illumination would also be much less than it is at present, for the sun's light could only penetrate apartments which were directly accessible to his rays, or into which it was reflected from the surface of natural objects. On the summits of mountains, where the atmosphere is very rare, the sky assumes the color of the deepest blue, approaching to blackness, and stars become visible in the daytime.

CHAPTER IV.

THE EARTH'S ANNUAL MOTION.—SIDEREAL AND SOLAR TIME.—THE EQUATION OF TIME.—THE CALENDAR.—THE CELESTIAL GLOBE.

98. *Sun's apparent motion in right ascension.*—If we observe the exact position of the sun with reference to the stars, from day to day through the year, we shall find that it has an apparent motion among them along a great circle of the celestial sphere, whose plane makes an angle of 23° 27′ with the plane of the celestial equator. This motion may be determined by observations with the transit instrument and mural circle.

If the sun's transit be observed daily, and its right ascension be determined, it will be found that the right ascension increases each day about four minutes of time, or one degree, so that in a year the sun makes a complete circuit round the heavens, moving constantly among the stars from west to east. This daily motion in right ascension is not uniform, but varies from 215s. to 266s., the mean being about 236s., or 3m. 56s.

99. *Sun's apparent motion in declination.*—If the point at which the sun's centre crosses the meridian be observed daily with the mural circle, it will be found to change from day to day. Its declination is zero on the 20th of March, from which time its north

declination increases until it becomes 23° 27′ on the 21st of June. It then decreases until the 22d of September, when the sun's centre is again upon the equator. Its south declination then increases until it becomes 23° 27′ on the 21st of December, after which it decreases until the sun's centre returns to the equator on the 20th of March.

If we trace upon a celestial globe the course of the sun from day to day, we shall find its path to be a great circle of the heavens, inclined to the equator at an angle of 23° 27′. This circle is called the *ecliptic,* because solar and lunar eclipses can only take place when the moon is very near this plane.

100. *The equinoxes and solstices.*—The ecliptic intersects the celestial equator at two points diametrically opposite to each other. These are called the *equinoctial points;* because, when the sun is at these points, it is for an equal time above and below the horizon, and the days and nights are therefore equal.

The point at which the sun passes from the south to the north side of the celestial equator, is called the *vernal* equinoctial point, and the other is called the *autumnal* equinoctial point. The *times* at which the sun's centre is found at these points are called the *vernal* and *autumnal equinoxes.* The vernal equinox, therefore, takes place on the 20th of March, and the autumnal on the 22d of September.

Those points of the ecliptic which are midway between the equinoctial points are the most distant from the celestial equator, and are called the solstitial points; and the times at which the sun's centre passes those points are called the *solstices.* The summer solstice takes place on the 21st of June, and the winter solstice on the 21st of December.

101. The *equinoctial colure* is the hour circle which passes through the equinoctial points. The *solstitial colure* is the hour circle which passes through the solstitial points. The solstitial colure is at right angles both to the ecliptic and to the equator, for it cuts both these circles 90 degrees from their common intersection; that is, from the equinoctial points.

The distance of either solstitial point from the celestial equator is 23° 27′. The more distant the sun is from the celestial equator, the more unequal will be the days and nights; and, therefore, the longest day of the year will be the day of the summer solstice, and the shortest that of the winter solstice. In southern latitudes the seasons will be reversed.

102. The *zodiac* is a zone of the heavens extending eight degrees each side of the ecliptic. The sun, the moon, and all the principal planets, have their motions within the limits of the zodiac.

The zodiac is divided into twelve equal parts, called *signs*, each of which contains 30 degrees. Beginning with the vernal equinox, they are as follows:

Sign.	Symbol.	Sign.	Symbol.
I. Aries.	♈	VII. Libra.	♎
II. Taurus.	♉	VIII. Scorpio.	♏
III. Gemini.	♊	IX. Sagittarius.	♐
IV. Cancer.	♋	X. Capricornus.	♑
V. Leo.	♌	XI. Aquarius.	♒
VI. Virgo.	♍	XII. Pisces.	♓

The vernal equinox is at the first point of Aries, and the autumnal equinox at the first of Libra. The summer solstice is at the first of Cancer, and the winter solstice at the first of Capricorn.

103. The *tropics* are two small circles parallel to the equator, and passing through the solstices. That on the north of the equator is called the Tropic of Cancer, and that on the south the Tropic of Capricorn.

The *Polar circles* are two small circles parallel to the equator, and distant 23° 27′ from the poles. One is called the *Arctic* and the other the *Antarctic* circle.

A great circle of the celestial sphere passing through the poles of the ecliptic, is called *a circle of latitude.*

The *latitude* of a star is the distance of the star from the ecliptic, measured on a circle of latitude. It may be north or south, and is counted from zero to 90 degrees.

The *longitude* of a star is the distance from the vernal equinox to the circle of latitude passing through the star, measured on the ecliptic in the order of the signs. Longitude is counted from zero to 360 degrees.

104. *Appearances produced by the earth's annual motion.*—The

apparent annual motion of the sun may be explained either by supposing a real revolution of the sun around the earth, or a revolution of the earth around the sun. But it follows from the principles of Mechanics that the earth and sun must both revolve around their common centre of gravity, and this point is very near the centre of the sun.

If the earth could be observed by a spectator upon the sun, it would appear among the fixed stars in the point of the sky opposite to that in which the sun appears as viewed from the earth.

Thus, in Fig. 27, let S represent the sun, and ABPD the earth's orbit: a spectator upon the earth will see the sun projected among the fixed stars in the point of the sky opposite to that occupied by the earth; and, as the earth moves from A to B and P, the sun will appear to move among the stars from P to D and A, and in the course of the year will appear to trace out in the sky the plane of the ecliptic. When the earth is in Libra we see the sun in the opposite sign Aries; and as the earth moves from Libra to Scorpio, the sun appears to move from Aries to Taurus, and so on through the ecliptic.

Fig. 27.

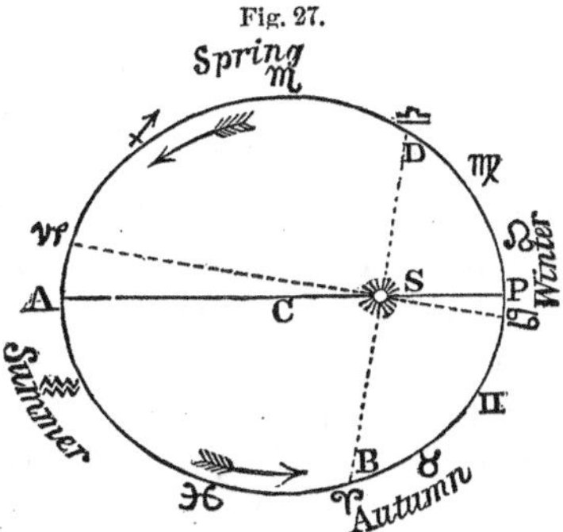

105. *Phenomena within the arctic circle.*—At the summer solstice, on the arctic circle, the sun's distance from the north pole is just equal to the latitude of the place, and the sun's diurnal path just touches the horizon at the north point. Within the arctic circle, there will be several days during which the sun never sinks below the horizon. So, also, near the winter solstice, within the arctic circle, there will be several days during which the sun does not rise above the horizon.

106. *Division of the earth into zones.*—The earth is naturally divided into five zones, depending on the appearance of the diurnal path of the sun.

These zones are,

1st. The two *frigid* zones, included within the polar circles. Within these zones there are several days of the year during which the sun does not rise above the horizon, and other days during which the sun does not sink below the horizon.

2d. The *torrid* zone, extending from the Tropic of Cancer to the Tropic of Capricorn. Throughout this zone, the sun every year passes through the zenith of the observer, when the sun's declination is equal to the latitude of the place.

3d. The north and south *temperate* zones, extending from the tropics to the polar circles. Within these zones the sun is never seen in the zenith, and it rises and sets every day.

107. *Cause of the change of seasons.*—While the earth revolves annually round the sun, it has a motion of rotation upon an axis which is inclined 23° 27′ from a perpendicular to the ecliptic; and *this axis continually points in the same direction.* Hence result the alternations of day and night, and the succession of seasons.

In June, when the north pole of the earth inclines toward the sun, the greater portion of the northern hemisphere is enlightened, and the greater portion of the southern hemisphere is dark. The days are, therefore, longer than the nights in the northern hemi-

Fig. 28.

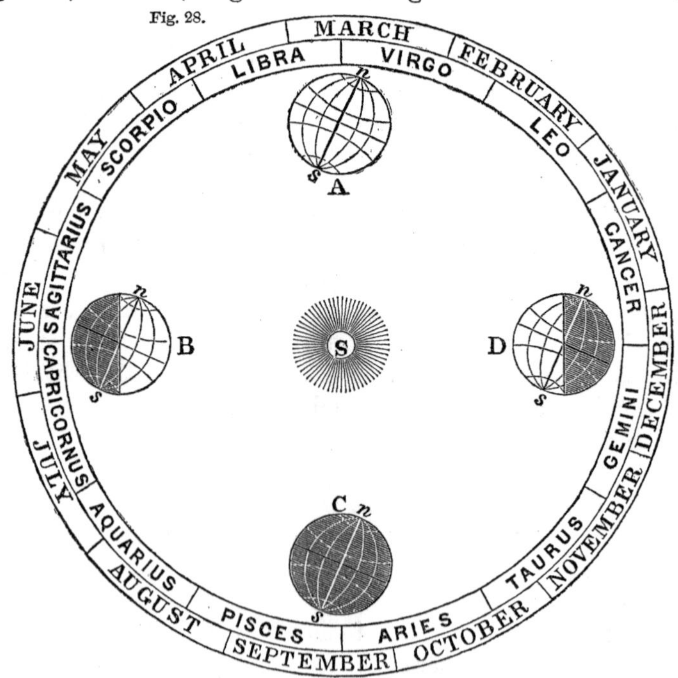

sphere. The reverse is true in the southern hemisphere; but on the equator, the days and nights are equal. In December, when the south pole inclines toward the sun, the days are longer than the nights in the southern hemisphere.

In March and September, when the earth's axis is perpendicular to the direction of the sun, the circle which separates the enlightened from the unenlightened hemisphere, passes through the poles, and the days and nights are equal all over the globe.

These different cases are illustrated by Fig. 28. Let S represent the position of the sun, and ABCD different positions of the earth in its orbit, the axis *ns* always pointing toward the same fixed star. At A and C the sun illumines from *n* to *s*, and as the globe turns upon its axis, the sun will appear to describe the equator, and the days and nights will be equal in all parts of the globe. When the earth is at B, the sun illumines 23½° beyond the north pole *n*, and falls the same distance short of the south pole *s*. When the earth is at D, the sun illumines 23½° beyond the south pole *s*, and falls the same distance short of the north pole *n*.

108. *Under what circumstances would there have been no change of seasons?*—If the earth's axis had been perpendicular to the plane of its orbit, the equator would have coincided with the ecliptic; day and night would have been of equal duration throughout the year, and there would have been no diversity of seasons.

109. *In what case would the change of seasons have been greater than it now is?*—If the inclination of the equator to the ecliptic had been greater than it is, the sun would have receded farther from the equator on the north side in summer, and on the south side in winter; and the heat of summer, as well as the cold of winter, would have been more intense; that is, the diversity of the seasons would have been greater than it is at present. If the equator had been at right angles to the ecliptic, the poles of the equator would have been situated in the ecliptic; and at the summer solstice the sun would have appeared at the north pole of the celestial sphere, and at the winter solstice it would have been at the south pole of the celestial sphere. To an observer in the middle latitudes, the sun would therefore, for a considerable part of summer, be within the circle of perpetual apparition, and for several weeks be constantly above the horizon. So, also, for a con-

siderable part of winter, he would be within the circle of perpetual occultation, and for several weeks be constantly below the horizon. The great vicissitudes of heat and cold resulting from such a movement of the sun, would be extremely unfavorable to both animal and vegetable life.

110. *To determine the obliquity of the ecliptic.*—The inclination of the equator to the ecliptic, or the obliquity of the ecliptic, is equal to the sun's greatest declination. It may therefore be ascertained by measuring, by means of the mural circle, the sun's declination at the summer, or at the winter solstice. The greatest declination of the sun is found to be 23° 27′ 25″, both north and south of the equator. This arc is, however, diminishing at the rate of about half a second annually.

111. *Form of the earth's orbit.*—The path of the earth around the sun is nearly, but not exactly, a circle. The relative distances of the sun from the earth may be found by observing the changes in the sun's apparent diameter. The apparent diameter of the sun, at different distances from the spectator, varies inversely as the distance. Thus, in Fig. 29,

Fig. 29.

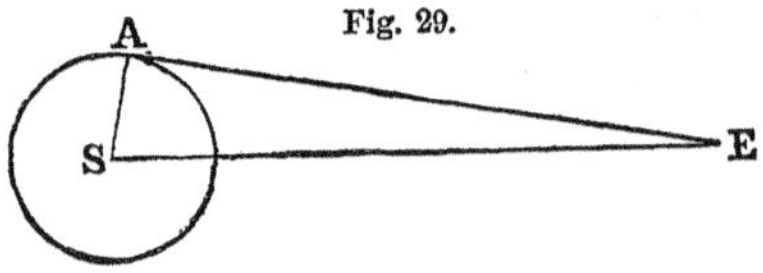

$$R : \sin. E :: ES : AS.$$

$$\therefore \sin. E = \frac{AS}{ES}, \text{ or varies as } \frac{1}{ES}.$$

Since the sines of small angles are nearly proportional to the angles, E varies as $\frac{1}{ES}$, very nearly.

By measuring, therefore, the sun's apparent diameter from day to day throughout the year, we have the means of determining the relative distances of the sun from the earth.

Ex. 1. On the 1st of January, 1864, the sun's apparent diameter was 32′ 36″.4, and on the 1st of July, 1864, his diameter was 31′ 31″.8. Find the relative distances of the sun at these two periods. *Ans.* 0.96698.

Ex. 2. On the 1st of April, 1864, the apparent diameter of the sun was 32′ 3″.4. Find the ratio of its distance to the distances in July and January. *Ans.* 0.98357 and 1.01716.

112. *The earth's orbit is an ellipse.*—By observations of the sun's apparent diameter continued throughout the year, we find that

the true form of the earth's orbit is an ellipse, having the sun in one of the foci. The sun's apparent diameter is least on the 1st of July, and greatest on the 1st of January. We may then construct a figure showing the form of the orbit, by setting off lines, SA, SB, SC, etc., corresponding to the sun's distances, and making angles with each other equal to the sun's angular motion between the times of observation. The figure thus formed is found to be an ellipse, with the sun occupying one of the foci, as S.

Fig. 30.

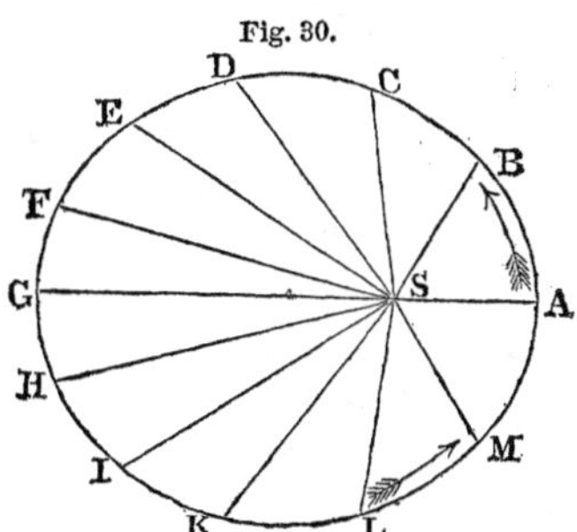

113. *To find the eccentricity of the earth's orbit.*—The point A of the orbit where the earth is nearest the sun, is called the *perihelion*, and this happens on the 1st of January. The point G most distant from the sun is called its *aphelion*, and this happens on the 1st of July; that is, the earth is more distant from the sun in summer than in winter.

The distance from the centre of the ellipse to the focus, divided by the semi-major axis, is called the *eccentricity* of the ellipse, and its value may be determined as follows:

If a denote the semi-major axis, and e the eccentricity of the earth's orbit, then

$$\text{the earth's aphelion distance} = a(1+e);$$
$$\text{the earth's perihelion distance} = a(1-e).$$

If we represent the aphelion distance by A, and the perihelion distance by P, we have

$$\frac{A}{P}=\frac{1+e}{1-e}.$$

Solving this equation, we obtain

$$e=\frac{A-P}{A+P}=\frac{1-\frac{P}{A}}{1+\frac{P}{A}}.$$

But
$$\frac{P}{A}=\frac{31'\ 31''.8}{32'\ 36''.4}=0.96698.$$

Hence
$$e=0.01678,$$

which is about $\frac{1}{60}$th.

This eccentricity is subject to a diminution of 0.000042 in one hundred years. If this change were to continue indefinitely, the earth's orbit must eventually become circular; but Le Verrier has proved that the diminution is not to continue beyond 24,000 years, when the eccentricity will be equal to .0033, and after that time the eccentricity will increase.

114. *Law of the earth's motion in its orbit.*—The radius vector of the earth's orbit describes equal areas in equal times. Let A and B be the positions of the earth in its orbit on two successive days; let θ represent the angle ASB, and R represent AS. Draw AC perpendicular to SB. Then $AC = AS \sin. ASB = R \sin. \theta$; and the area $ASB = \frac{1}{2}R^2 \sin. \theta$. But since the earth's diurnal motion in the ecliptic is small, we may assume that the arc θ is equal to its sine, and hence the area $= \frac{1}{2}R^2\theta$.

Fig. 31.

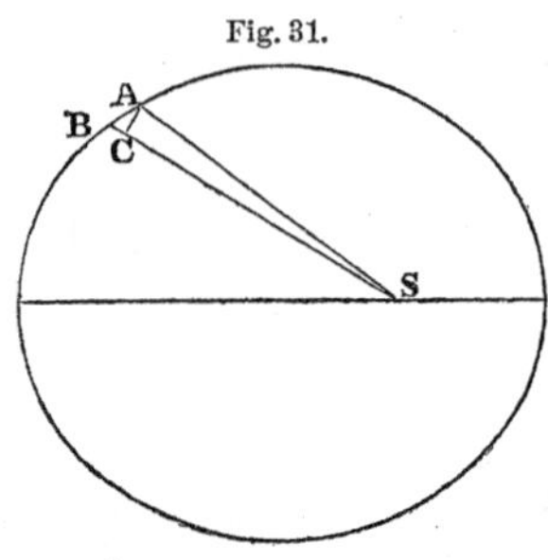

If this area described by the radius vector in one day, is a constant quantity, then $R^2\theta$ will evidently be a constant quantity. But R varies inversely as the apparent diameter of the sun. Hence, putting D for the sun's apparent diameter, $\frac{\theta}{D^2}$ must be a constant quantity; or

$$\theta : \theta' :: D^2 : D'^2;$$

that is, the sun's diurnal motion in different parts of its orbit, must vary as the square of its apparent diameter.

Now we find this supposition verified by observation. Thus:

From noon of January 1st to noon of January 2d, 1864, the sun moved through 1° 1′ 9″.9 of the ecliptic; and his apparent diameter at the same time was 32′ 36″.4.

From noon of July 1st to noon of July 2d, 1864, the sun moved through 57′ 12″.9; and his apparent diameter at the same time was 31′ 31″.8.

Reducing these values to seconds, we have

$$3669.9 : 3432.9 :: 1956.4^2 : 1891.8^2.$$

We find the same law to hold true in other parts of the orbit, and hence it is considered as established by observation, that *the radius vector of the earth's orbit describes equal areas in equal times.*

115. *Why the greatest heat and cold do not occur at the solstices.*—The influence of the sun in heating a portion of the earth's surface depends upon its altitude above the horizon, and upon the length of time during which it continues above the horizon. The greater the altitude, the less obliquely will the rays strike the surface of the earth at noon, and the greater will be their heating power. Both these causes conspire to produce the increased heat of summer, and the diminished heat of winter. It might be inferred that the hottest day ought to occur on the 21st of June, when the sun rises highest, and the days are the longest. Such, however, is not the case, for the following reason: As midsummer approaches, the quantity of heat imparted by the sun during the day is greater than the quantity lost during the night, and hence each day there is an increase of heat. On the 21st of June this daily augmentation reaches its maximum; but there is still each day an accession of heat, until the heat lost during the night is just equal to that imparted during the day, which happens, at most places in the northern hemisphere, some time in July or August.

For the same reason, the greatest cold does not occur on the 21st of December, but some time in January or February.

Sidereal and Solar Time.

116. *Sidereal Time.*—The interval between two successive returns of the vernal equinox to the same meridian, is called a *sidereal* day. This interval represents the time of the rotation of the earth upon its axis, and is not only invariable from one month to another, but has not changed so much as the hundredth part of a second, in two thousand years.

117. *Solar Time.*—The interval between two successive returns of the sun to the same meridian, is called a *solar* day.

The sun passes through 360 degrees of longitude in one year, or 365 days 5 hours 48 minutes and 47.8 seconds; so that the sun's mean daily motion in longitude is found by the proportion

$$\text{one year} : \text{one day} :: 360° : \text{daily motion} = 59' \ 8''.33.$$

This motion is not uniform, but is greatest when the sun is nearest the earth. Hence the solar days are unequal; and to avoid the inconvenience which would result from this fact, astronomers have recourse to a *mean* solar day, the length of which is equal to the mean or average of all the apparent solar days in a year.

118. *Sidereal and solar time compared.*—The length of the mean solar day is greater than that of the sidereal, because when the mean sun, in its diurnal motion, returns to a given meridian, it is 59′ 8″.3 eastward of its position on the preceding day.

An arc of the equator, equal to 360° 59′ 8″.3, passes the meridian in a mean solar day, while only 360° pass in a sidereal day. To find the excess of the solar day above the sidereal day, expressed in sidereal time, we have the proportion

360° : 59′ 8″.3 : : one day : 3m. 56.5s.

Hence 24 hours of mean solar time are equivalent to 24h. 3m. 56.5s. of sidereal time.

To find the excess of the solar day above the sidereal day, expressed in solar time, we have the proportion

360° 59′ 8″.3 : 59′ 8″.3 :: one day : 3m. 55.9s.

Hence 24 hours of sidereal time are equivalent to 23h. 56m. 4.1s. of mean solar time.

119. *Civil day, and astronomical day.*—The civil day begins at midnight, and consists of two periods of 12 hours each; but modern astronomers commence their day at noon, because this is a date which is marked by a phenomenon which can be accurately observed, viz., the passage of the sun over the meridian; and because observations being chiefly made at night, it is inconvenient to have a change of date at midnight. The astronomical day commences 12 hours later than the civil day, and the hours are numbered continuously up to 24. Thus July 4th, 9 A.M. civil time, corresponds to July 3d, 21 hours of astronomical time.

120. *Apparent time, and mean time.*—The interval between two successive returns of the sun to the same meridian, is an *apparent* solar day; and *apparent time* is time reckoned in apparent solar days, while *mean time* is time reckoned in mean solar days.

The difference between apparent solar time and mean solar time, is called the *equation of time.*

If a clock were required to keep apparent solar time, it would be necessary that its rate should change from day to day according to a complicated law. It has been found in practice impossible to accomplish this, and hence clocks are now regulated to indicate mean solar time. A clock, therefore, should not indicate 12h. when the sun is on the meridian, but should sometimes indicate more than 12h. and sometimes less than 12h., the difference being equal to the equation of time.

121. *Cause of the inequality of the solar days.*—The inequality of the solar days depends on two causes, the unequal motion of the earth in its orbit, and the inclination of the equator to the ecliptic.

While the earth is revolving round the sun in an elliptical orbit, its motion is greatest when it is nearest the sun, and slowest when it is most distant. Let ADGK represent the elliptic orbit of the earth, with the sun in one of its foci at S, and let the direction of motion be from A toward E.

Fig. 32.

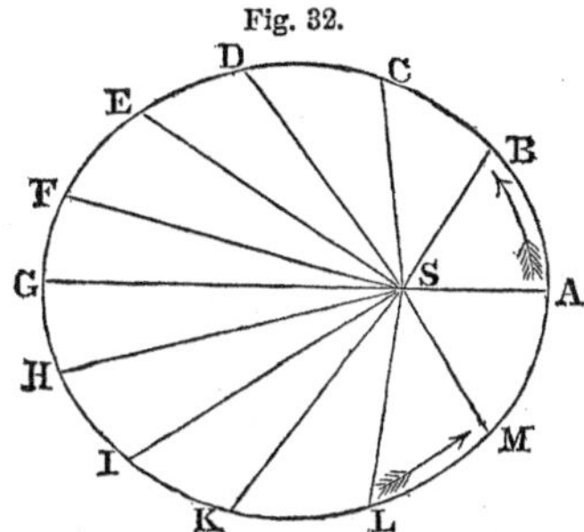

We have found that the sun's mean daily motion as seen from the earth, or the earth's mean daily motion as seen from the sun, is 59′ 8″.3. But when the earth is nearest the sun its daily motion is 61′ 10″. In passing from A toward E its daily motion diminishes, and at G it is only 57′ 12″. While moving, therefore, from A through E to G, the earth will be in advance of its mean place, while at G, having completed a half revolution, the true and the mean places will coincide. For a like reason, in going from G to A, the earth will be behind its mean place; but at A the mean and true places will again coincide. This point A in the diagram, corresponds to about the 1st of January.

Now the apparent direction of the sun from the earth, is exactly opposite to that of the earth from the sun. Hence, when the earth is nearest to the sun, the apparent solar day will be longer than the mean solar day. If, then, we conceive a fictitious sun to move uniformly through the heavens, describing 59′ 8″ per day, and that the true and fictitious suns are together on the 1st of January, it is evident that on the 2d of January the fictitious sun will come to the meridian a few seconds before the true sun; on the 3d of January the fictitious sun will be still more in advance of the true sun, and this difference will go on increasing for about three months, when it amounts to a little more than 8 minutes. From this time the difference will diminish until about the 1st of July, when the positions of the true and fictitious suns will coincide. But on the 2d of July the fictitious sun will come to the meridian a few seconds later than the true sun; on the 3d of July it will have fallen still more behind the true sun, and this difference will go on increasing for about three months, when it amounts to a little more than 8 minutes. From this time the difference will diminish until the 1st of January, when the positions of the true and fictitious suns will again coincide.

So far, then, as it depends upon the unequal motion of the earth in its orbit, the equation of time is positive for six months, and then negative for six months, and its greatest value is 8m. 24s.

122. *Second cause for the inequality of the solar days.*—Even if the earth's motion in its orbit were perfectly uniform, the apparent solar days would be unequal, because the ecliptic is inclined to the equator. Let A*g*N represent the equator, and AGN the northern half of the ecliptic. Let the ecliptic be divided into equal portions, AB, BC, CD, etc., supposed to be described by the sun in equal portions of time; and through the points B, C, D, etc., let hour circles be made to pass, cutting the equator in the points *b*, *c*, *d*, etc. The arc AGN is equal to the arc A*g*N, for all great circles bisect each other; also AG is equal to A*g*, since the former is one half of AGN, and the latter of A*g*N. Now, since AB*b* is a right-angled triangle, AB is greater than A*b*; for the same reason, AC is greater than A*c*; AD is greater than A*d*, and so on. But AG is equal to A*g*; therefore A*g* is divided into *unequal* portions at the points *b*, *c*, *d*, etc. Now B and *b* come to the meridian at the same instant; so also C and *c*, D and *d*, and so on.

Fig. 33.

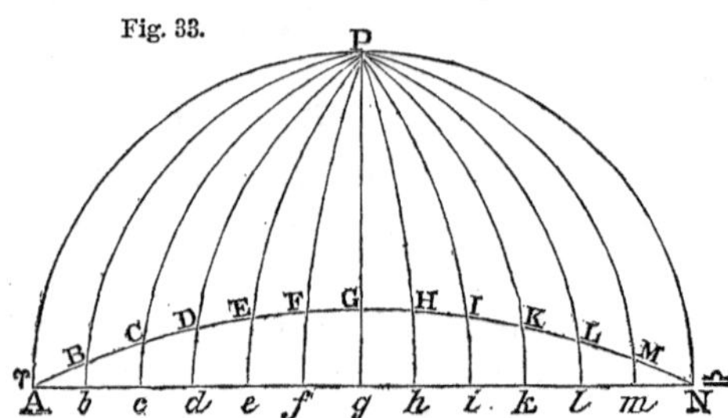

Suppose now that a fictitious sun moves in the equator at the rate of 59′ 8″ per day, while the real sun moves in the ecliptic at the same rate, and let them start together from A at noon on the 20th of March. On the 21st of March, at noon, the real sun will have advanced toward B 59′ 8″, which distance projected on the equator will be less than 59′ 8″, while the fictitious sun will have advanced toward *b* 59′ 8″; that is, the fictitious sun will be eastward of the real sun, and the real sun will come to the meridian sooner than the fictitious one. The same will happen during the motion of the sun through the entire quadrant AG. The two suns will reach the points G and g on the 21st of June, and then they will both come to the meridian at the same instant.

During the motion of the sun through the second quadrant, the real sun will come to the meridian later than the fictitious one, but both will reach the point N on the 22d of September at the same instant. During the motion through the third quadrant, the real sun will come to the meridian sooner than the fictitious one, until the 21st of December, when they will be found 180° from the points G and g. During the motion through the last quadrant, the real sun will come to the meridian later than the fictitious one, but both will reach the point A at the same instant on the 21st of March. Thus we see that, so far as it depends upon the obliquity of the ecliptic, the equation of time is positive for three months; then negative for three months; then positive for three months; and then negative for another three months.

The amount of the equation of time due to this cause, may be computed as follows: Suppose the sun to have advanced 45° from A; then, in the right-angled triangle ADd, the angle at A is 23° 27′, and the hypothenuse is 45°. Ad is then computed from the equation,

$$\text{tang. A}d = \text{cos. A tang. AD},$$

whence Ad is found to be 42° 31′ 47″.

The difference between AD and Ad is 2° 28′ 13″, or 9m. 52.8s. in time; and this is about the greatest amount of the equation of time, due to the obliquity of the ecliptic.

123. *Resulting values of the equation of time.*—The influence of each of these causes upon the equation of time, is artificially represented in the following figure, where AE is supposed to represent a year divided into twelve equal parts to represent the months; and the ordinates of the curve ABCDE, measured from the line AE as an axis, represent the values of the equation of time, so far as it depends upon the unequal motion of the earth in its orbit; and the ordinates of the curve FGHI represent the values of the equation of time, so far as it depends upon the inclination of the equator to the ecliptic. The actual equation of time will be found by taking the algebraic sum of the effects due to these two separate causes. The result is the curve MNOPQR,

Fig. 34.

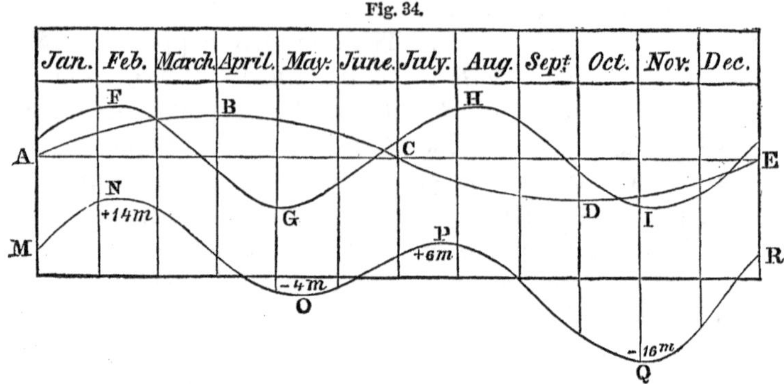

the ordinates being measured from the lower horizontal line in the figure. From this we see that the equation of time has two annual maxima and two annual minima, and there are four periods when the equation is zero. These dates and the corresponding values of the equation of time are as follows:

February 11,	+14m.	32s.	July 26,	+ 6m.	12s.
April 15,	0	0	September 1,	0	0
May 14,	− 3	55	November 2,	−16	18
June 14,	0	0	December 24,	0	0

These dates and the values of the equation of time change slightly from one year to another, so that, where great accuracy is required, a table of the equation of time is required for each year. Such a table is annually published in the Nautical Almanac.

The Calendar.

124. *The Julian Calendar.*—The interval between two successive returns of the sun to the vernal equinox, is called a *tropical year.* Its average length expressed in mean solar time is 365d. 5h. 48m. 47.8s. But in reckoning time for the common purposes of life, it is most convenient to have the year contain a certain number of *whole* days. In the calendar established by Julius Cæsar, and hence called the Julian Calendar, three successive years were made to consist of 365 days each, and the fourth of 366 days. The year which contained 366 days was called a *bissextile* year, because the 6th of the Kalends of March was twice counted. It is also frequently called leap-year. The others are called common years. The odd day inserted in a bissextile year is called the *intercalary* day.

The reckoning by the Julian calendar supposes the length of

the year to be 365¼ days. A Julian year, therefore, exceeds the tropical year by 11m. 12s. This difference amounts to a little more than 3 days in the course of 400 years.

125. *The Gregorian Calendar.*—At the time of the Council of Nice, in the year 325, the Julian calendar was introduced into the Church, and at that time the vernal equinox fell on the 21st of March; but in the year 1582 the error of the Julian calendar had accumulated to nearly 10 days, and the vernal equinox fell on the 11th of March. If this erroneous reckoning had continued, in the course of time spring would have commenced in September, and summer in December. It was therefore resolved to reform the calendar, which was done by Pope Gregory XIII., and the first step was to correct the loss of the ten days, by counting the day after the 4th of October, 1582, not the 5th, but the 15th of the month. In order to keep the vernal equinox to the 21st of March in future, it was concluded that three intercalary days should be omitted every four hundred years. It was also agreed that the omission of the intercalary days should take place in those years which were not divisible by 400. Thus the years 1700, 1800, and 1900, which, according to the Julian calendar, would be bissextile, would, according to the reformed calendar, be common years.

The calendar thus reformed is called the *Gregorian* Calendar. The error of this calendar amounts to less than one day in 4000 years.

126. *Adoption of the Gregorian Calendar.*—The Gregorian calendar was immediately adopted at Rome, and soon afterward in all Catholic countries. In Protestant countries the reform was not so readily adopted, and in England and her colonies it was not introduced till the year 1752. At this time there was a difference of 11 days between the Julian and Gregorian calendars, in consequence of the suppression in the latter, of the intercalary day in 1700. It was therefore enacted by Parliament that 11 days should be left out of the month of September in the year 1752, by calling the day following the 2d of the month, the 14th instead of the 3d.

The Gregorian calendar is now used in all Christian countries except Russia. The Julian and Gregorian calendars are frequently designated by the terms *old style* and *new style.* In consequence of the intercalary days omitted in the years 1700 and 1800, there is now 12 days difference between the two calendars.

127. *When does the year begin?*—In the different countries of Europe, the year has not always been regarded as commencing at the same date. In certain countries, the year has been regarded as commencing at Christmas, on the 25th of December; in others, on the 1st of January; in others, on the 1st of March; in others, on the 25th of March; and in others at Easter, which may correspond to any date between March 22d and April 25th. In England, previous to the year 1752, the legal year commenced on the 25th of March; but the same act that introduced the Gregorian calendar established the 1st of January as the commencement of the year. In this manner the year 1751 lost its month of January, its month of February, and the first 24 days of March. This change in the calendar explains the double date which is frequently found in English books. For example, Feb. 15, $\frac{1751}{1752}$, means the 15th of February, 1751, according to the old mode of counting the years from the 25th of March, and 1752 according to the new method prescribed by Parliament. In order to distinguish the one mode of reckoning from the other, it was for a long time customary to attach to each date the letters O. S. for old style, or N. S. for new style. Thus the date of General Washington's birth was either written Feb. 11, 1731, O. S., or Feb. 22, 1732, N. S.

128. *First and last days of the year.*—Since a common year consists of 365 days, or 52 weeks and 1 day, the last day of each common year must fall on the same day of the week as the first; that is, if the year begins on Sunday it will end on Sunday. But if leap-year begins on Sunday it will end on Monday, and the following year will begin on Tuesday.

PROBLEMS ON THE CELESTIAL GLOBE.

129. *To find the right ascension and declination of a star.*

Bring the star to the brass meridian; the degree of the meridian over the star will be its declination, and the degree of the equinoctial under the meridian will be its right ascension. Right ascension is sometimes expressed in hours and minutes of time, and sometimes in degrees and minutes of arc.

Ex. Required the right ascension and declination of Arcturus.

130. *The right ascension and declination of a star being given, to find the star upon the globe.*

Bring the degree of the equator which marks the right ascension to the brass meridian; then under the given declination marked on the meridian will be the star required.

Ex. Required the star whose right ascension is 10h. 1m. 7s., and declination 12° 37′ N.

131. *To set the celestial globe in a position similar to that of the heavens, at a given place, at a given day and hour.*

Set the brass meridian to coincide with the meridian of the place; elevate the pole to the latitude of the place; bring the sun's place in the ecliptic to the meridian, and set the hour index at 12; then turn the globe westward until the index points to the given hour. The constellations would then have the same appearance to an eye situated at the centre of the globe, as they have at that moment in the heavens.

Ex. Required the appearance of the heavens at New Haven, Lat. 41° 18′, June 20th, at 10 o'clock P.M.

132. *To determine the time of rising, setting, and culmination of a star for any given day and place.*

Elevate the pole to the latitude of the place; bring the sun's place in the ecliptic for the given day to the meridian, and set the hour index to 12. Turn the globe until the star comes to the eastern horizon, and the hour shown by the index will be the time of the star's rising. Bring the star to the brass meridian, and the index will show the time of the star's culmination. Turn the globe until the star comes to the western horizon, and the index will show the time of the star's setting.

Ex. Required the time when Aldebaran rises, culminates, and sets at Cincinnati, October 10th.

133. *To determine the position of the planets in the heavens at any given time and place.*

Find the right ascension and declination of the planets for the given day from the Nautical Almanac, and mark their places upon the globe; then adjust the globe as in Art. 131, and the position of the planets upon the globe will correspond to their position in the heavens. We may then determine the time of their rising and setting as in Art. 132. The time of rising and setting of a comet may be determined in the same manner.

CHAPTER V.

PARALLAX.—ASTRONOMICAL PROBLEMS.

134. *Diurnal parallax defined.*—The direction in which a celestial body would be seen if viewed from the centre of the earth, is called its *true place;* and the direction in which it is seen from any point on the surface, is called its *apparent place.* The arc of the heavens intercepted between the true and apparent places—that is, the apparent displacement which would be produced by the transfer of the observer from the centre to the surface, is called the *diurnal parallax.*

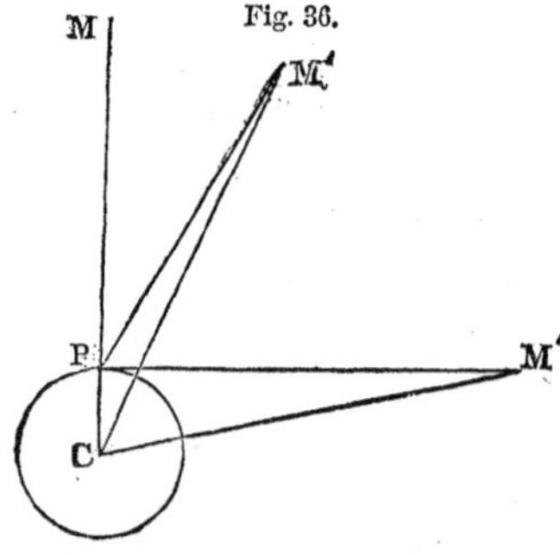

Fig. 36.

Let C denote the centre of the earth; P the place of the observer on its surface; M an object seen in the zenith at P; M′ the same object seen at the zenith distance MPM′; and M″ the same object seen in the horizon.

It is evident that M will appear in the same direction whether it be viewed from P or C. Hence, in the zenith, there is no diurnal parallax, and there the apparent place of an object is its true place.

If the object be at M′, its apparent direction is PM′, while its true direction is CM′, and the parallax corresponding to the zenith distance MPM′ will be PM′C.

As the object is more remote from the zenith, the parallax increases; and when the object is in the horizon, as at M″, the diurnal parallax becomes greatest, and is called the *horizontal parallax.* It is the angle PM″C which the radius of the earth subtends at the object.

It is evident that parallax *increases* the zenith distance, and consequently diminishes the altitude. Hence, to obtain the true zenith distance from the apparent, the parallax must be *subtracted;* and to obtain the true altitude from the apparent, the parallax must be *added.* The azimuth of a heavenly body is not affected by parallax.

135. *To deduce the parallax at any altitude from the horizontal parallax.*—In the triangle CPM′ we have

$$\text{CM}' : \text{CP} :: \sin.\ \text{CPM}'(=\sin.\ \text{MPM}') : \sin.\ \text{CM}'\text{P}. \qquad (1)$$

Also, in the triangle CPM″, we have

$$\text{CM}'' : \text{CP} :: 1 : \sin.\ \text{CM}''\text{P}. \qquad (2)$$

Hence $\quad 1 : \sin.\ \text{CM}''\text{P} :: \sin.\ \text{MPM}' : \sin.\ \text{CM}'\text{P},$

or $\quad \sin.\ \text{CM}'\text{P} = \sin.\ \text{CM}''\text{P} \times \sin.\ \text{MPM}'$;

that is, *the sine of the parallax at any altitude, is equal to the product of the sine of the horizontal parallax, by the sine of the apparent zenith distance.*

The parallax of the sun and planets is so small that we may, without sensible error, employ the parallax itself instead of its sine; that is, *the parallax at any altitude is equal to the product of the horizontal parallax, by the sine of the apparent zenith distance.*

136. *Relation of the parallax of a heavenly body to its distance.*

Let us put z=the zenith distance MPM′;
p=the parallax CM′P;
r=CP, the radius of the earth;
R=CM′, the distance of the heavenly body.

Then, by equation (1),

$$\text{R} : r :: \sin.\ z : \sin.\ p,$$

or

$$\sin.\ p = \frac{r}{\text{R}} \sin.\ z,$$

or

$$p = \frac{r}{\text{R}} \sin.\ z, \text{ very nearly.}$$

The parallax at any given altitude varies, therefore, inversely as the distance, very nearly.

When the zenith distance becomes 90°, sin. z becomes unity; and if we denote the horizontal parallax by P, we shall have

$$\sin.\ \text{P} = \frac{r}{\text{R}},$$

or

$$\text{P} = \frac{r}{\text{R}}, \text{ very nearly.}$$

137. *To determine the parallax of the moon by observation.*—Let A, A′ be two places on the earth situated under the same merid-

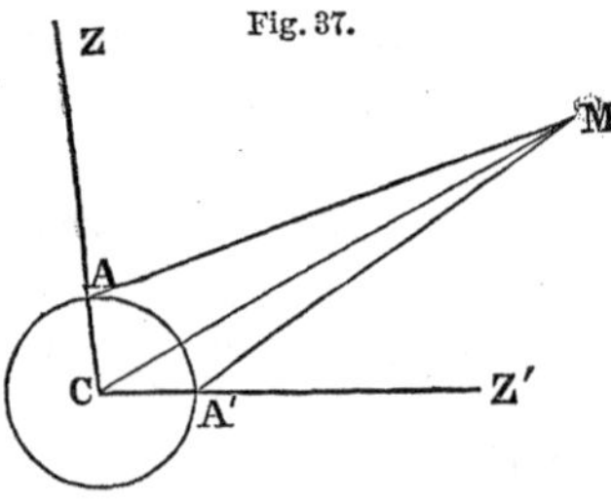

Fig. 37.

ian, and at a great distance from each other; let C be the centre of the earth, and M the moon.

Let AC be denoted by r, and CM by R, and let ZAM, Z'A'M, which are the moon's zenith distances as measured at the two observatories, be denoted by z, and z'. Then the moon's parallax, AMC, at the station A, will be

$$p=\frac{r}{\mathrm{R}}\sin. z,$$

and the parallax A'MC at the station A' will be

$$p'=\frac{r}{\mathrm{R}}\sin. z'.$$

Adding these equations together, we find

$$p+p'=\frac{r}{\mathrm{R}}(\sin. z+\sin. z').$$

But the angle $p+p'$, or AMA', is equal to the difference between ZCZ' and the sum of the angles z and z'; and since, if the places be situated one north and the other south of the equator, we have ZCZ' equal to the sum of the latitudes of the stations $l+l'$, we obtain $\qquad p+p'=z+z'-l-l'.$

Substituting this value in the preceding equation, we find

$$z+z'-l-l'=\frac{r}{\mathrm{R}}(\sin. z+\sin. z'),$$

or

$$\frac{r}{\mathrm{R}}=\frac{z+z'-l-l'}{\sin. z+\sin. z'}.$$

But $\frac{r}{\mathrm{R}}$ is the horizontal parallax of the moon, which was required to be found.

138. *Stations of observation.*—It is not essential that the two observers should be exactly on the same meridian; for if the meridian zenith distances of the moon be observed on several consecutive days, its change of meridian zenith distance in a given time will be known. Then, if the difference of longitude of the two places is known, the zenith distance of the moon as observed at one of the meridians, may be reduced to what it would have been found to be, if the observations had been made in the same latitude at the other meridian.

139. *Results obtained by this method.*—There is an observatory at the Cape of Good Hope, in Lat. 33° 56′ S., where the moon's meridian altitude has been observed daily for many years, whenever the weather would permit; and similar observations are regularly made at Greenwich Observatory, in Lat. 51° 28′ N., as also at numerous other observatories in Europe. By combining these observations, the moon's parallax has been ascertained with great precision. It is found that the parallax varies considerably from one day to another. The equatorial parallax, when greatest, is about 61′ 32″, and when least, 53′ 48″. Its average value is 57′ 2″.

By the preceding method the sun's parallax may be ascertained to be about 9″. It can, however, be found more accurately by observations of the transits of Venus, as will be explained hereafter.

The parallax of the planets can also be determined in the same manner as that of the moon; but in the case of the nearest planet the parallax never exceeds 32″, and that of the remoter planets never amounts to 1″; and there are other methods by which these quantities can be more accurately determined.

140. *To compute the distance of a heavenly body.*—When we know the earth's radius and the horizontal parallax of a heavenly body, we can compute its distance. For (Fig. 36)

sin. PM″C : PC :: radius : CM″,

or the distance of the object equals the radius of the earth, divided by the sine of the horizontal parallax.

141. *Effect of the ellipticity of the earth upon parallax.*—The horizontal parallax of the moon is the angle which the earth's radius would subtend to an observer at the moon. On account of the spheroidal figure of the earth, this horizontal parallax is not the same for all places on the earth, but varies with the earth's radius, being greatest at the equator, and diminishing as we proceed toward either pole. It is necessary, therefore, always to compute the earth's radius for the place of the observer, and this may be done from the known properties of an ellipse. The moon's horizontal parallax for any given latitude is equal to the horizontal parallax at the equator multiplied by the radius of the earth at the given latitude, the radius of the equator being considered as unity.

It is this corrected value of the equatorial parallax which should be employed in all computations which involve the parallax of a particular place.

ASTRONOMICAL PROBLEMS.

142. *To find the latitude of any place.*—The latitude of a place may be determined by measuring the altitude of any circumpolar star, both at its upper and lower culminations, as explained in Art. 76. It may also be determined by measuring a single meridian altitude of any celestial body whose declination is known.

Fig. 38.

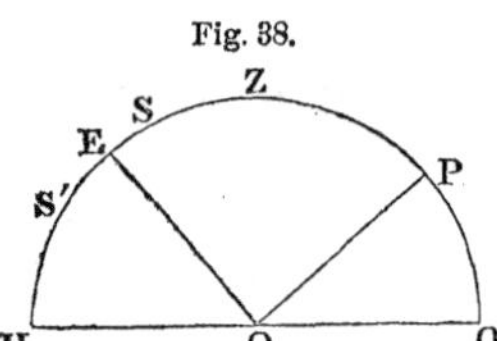

Let S or S′ be a star on the meridian; SE or S′E its declination. Measure SH, the altitude of the star S, and correct it for refraction. Then

$$EH=SH-SE=S'H+S'E.$$

But EH is the complement of PO, which is the latitude sought. The declinations of all the brighter stars have been determined with great accuracy, and are recorded in catalogues of the stars.

143. *To find the latitude at sea.*—At sea the latitude is usually determined by observing with the sextant the greatest altitude of the sun's lower limb above the sea horizon at noon. The observations are commenced about half an hour before noon, and the altitude of the sun is repeatedly measured until the altitude ceases to increase. This greatest altitude is considered to be the altitude on the meridian. To this altitude we must add the sun's semi-diameter in order to obtain the altitude of the sun's centre, and this result must be corrected for refraction. To this result we must add the sun's declination if south of the equator, or subtract it if north, and we shall obtain the elevation of the equator, which is the complement of the latitude. The Nautical Almanac furnishes the sun's declination for every day of the year.

144. *To find the time at any place.*—The time of apparent noon is the time of the sun's meridian passage, and is most conveniently found by means of a transit instrument adjusted to the meridian. Mean time may be derived from apparent time by applying the equation of time with its proper sign.

The time of apparent noon may also be found by noting the

times when the sun has equal altitudes before and after passing the meridian, and bisecting the interval between them. When great accuracy is required, the result obtained by this method requires a slight correction, since the sun's declination changes between morning and evening.

145. *To find the time by a single altitude of the sun.*—The time may also be computed from an altitude of the sun measured at any hour of the day, provided we know the sun's declination and the latitude of the place.

Let PZH be the meridian of the place of observation, P the pole, Z the zenith, and S the place of the sun. Measure the zenith distance, ZS, and correct it for refraction. Then, in the spherical triangle ZPS, we know the three sides, viz., PZ, the complement of the latitude, PS, the distance of the sun from the north pole, and ZS, the sun's zenith distance. In this triangle we can compute (Trigonometry, Art. 223) the angle ZPS, which, if expressed in time, will be the interval between the moment of observation and noon. This observation can be made at sea with a sextant, and this is the method of determining time which is commonly practiced by navigators.

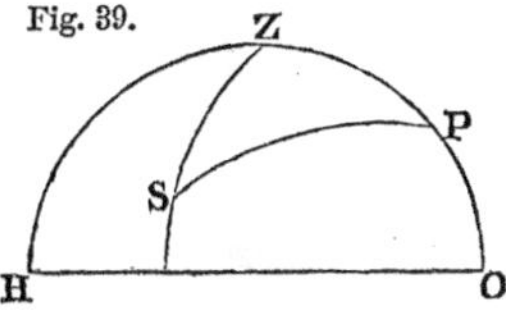

Fig. 39.

146. *A meridian mark, and sun-dial.*—If, upon a horizontal plane, we trace a meridian line, and at the south extremity of this line erect a vertical rod freely exposed to the sun, we may determine the time of apparent noon by the passage of the shadow of the rod over the meridian line. Or, if we set up a straight rod in a position parallel to the axis of the earth, its shadow, as cast upon a horizontal plane, will have the same direction at any given hour, at all seasons of the year. If, then, we graduate this horizontal plane in a suitable manner, and mark the lines with the hours of the day, we may determine the apparent time whenever the sun shines upon the rod. Such an instrument is called a *sun-dial*, and it may be constructed with sufficient precision to answer the ordinary purposes of society. This instrument will always indicate *apparent* time; but *mean* time may be deduced from it by applying the equation of time.

147. *To compute the longitude, right ascension, and declination of the sun, any one of these quantities, together with the obliquity of the ecliptic, being given.*

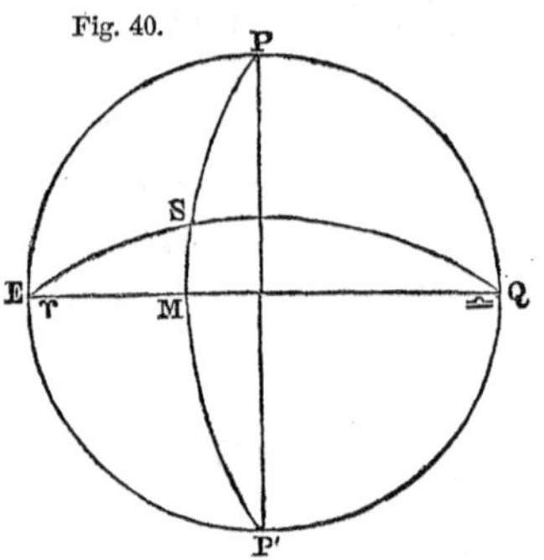

Let EPQP′ represent the equinoctial colure, EMQ the equator, ESQ the ecliptic, E the first point of Aries, S the place of the sun, PSP′ an hour circle passing through the sun; then EM is the sun's right ascension, SM his declination, ES his longitude, and MES the obliquity of the ecliptic. Then, in the triangle ESM, we have, by Napier's rule,

$$\text{R cos. E} = \text{tang. ME cot. SE};$$

that is, representing the obliquity by ω, and the right ascension by R. A.

$$\text{tang. R. A.} = \text{tang. Long. cos. } \omega, \qquad (1)$$

and

$$\text{tang. Long.} = \frac{\text{tang. R. A.}}{\text{cos. } \omega}. \qquad (2)$$

Also,

$$\text{R sin. ME} = \text{tang. MS cot. E};$$

that is,

$$\text{sin. R. A.} = \text{tang. Dec. cot. } \omega, \qquad (3)$$

and

$$\text{tang. Dec.} = \text{sin. R. A. tang. } \omega. \qquad (4)$$

Also,

$$\text{R sin. MS} = \text{sin. E sin. ES};$$

that is,

$$\text{sin. Dec.} = \text{sin. } \omega \text{ sin. Long.}, \qquad (5)$$

and

$$\text{sin. Long.} = \frac{\text{sin. Dec.}}{\text{sin. } \omega}. \qquad (6)$$

Also,

$$\text{R cos. ES} = \text{cos. ME cos. MS};$$

that is,

$$\text{cos. Long.} = \text{cos. R. A. cos. Dec.}, \qquad (7)$$

and

$$\text{cos. R. A.} = \frac{\text{cos. Long.}}{\text{cos. Dec.}}. \qquad (8)$$

Ex. 1. On the 1st of June, 1864, at Greenwich mean noon, the sun's right ascension was 4h. 38m. 27.75s., and his declination 22° 7′ 55″.2 N.; required his longitude. *Ans.* 71° 10′ 35″.9.

Ex. 2. On the 1st of January, 1864, the sun's longitude was 280° 23′ 52″.3, and his declination 23° 2′ 52″.2 S.; required his right ascension. *Ans.* 18h. 45m. 14.70s.

Ex. 3. On the 20th of May, 1864, the sun's longitude was 59° 40′ 1″.6, and the obliquity of the ecliptic 23° 27′ 18″.5; required his right ascension and declination.

Ans. R. A. 3h. 49m. 52.62s.
Dec. 20° 5′ 33″.9 N.

Ex. 4. On the 27th of October, 1864, the sun's right ascension was 14h. 8m. 19.06s., and the obliquity of the ecliptic 23° 27′ 17″.8; required his longitude and declination.

Ans. Long. 214° 20′ 34″.7.
Dec. 12° 58′ 34″.4 S.

Ex. 5. On the 8th of August, 1864, the sun's declination was 16° 0′ 56″.4 N., and the obliquity of the ecliptic 23° 27′ 18″.2; required his right ascension and longitude.

Ans. R. A. 9h. 14m. 19.20s.
Long. 136° 7′ 6″.5.

148. *Given the latitude of a place and the sun's declination, to find the time of his rising or setting.*

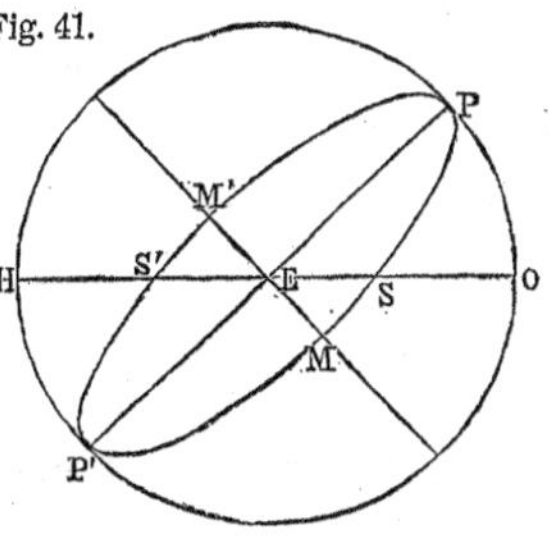

Fig. 41.

Let PEP′ represent the hour circle, which is six hours from the meridian, and which intersects the horizon in the east point, E. Let S or S′ be the position of the sun in the horizon, and through S draw the hour circle PSP′; also through S′ draw the hour circle PS′P′. Then, in the right-angled spherical triangle EMS, or EM′S′,

EM or EM′ = the distance of the sun from the six o'clock hour circle.

MS or M′S′ = the sun's declination, which we will represent by δ.

MES = M′ES′ = the complement of the latitude.

Now, by Napier's rule,

$$\text{R sin. EM} = \text{tang. MS cot. MES.}$$

Representing the latitude by ϕ,

$$\text{sin. EM} = \text{tang. } \delta \text{ tang. } \phi.$$

The time from the sun's rising to his passing the meridian = 6 hours ± EM.

Ex. 1. Required the time of sunrise at New York, Lat. 40° 42′, on the 10th of May, when the sun's declination is 17° 49′ N.

Ans. 4h. 56m.

Ex. 2. Required the time of sunset at Cincinnati, Lat. 39° 6′, on the 5th of November, when the sun's declination is 15° 56′ S.

Ans. 5h. 6m. apparent time.

149. *To find the time when the sun's upper limb rises, allowance being made for refraction.*—The preceding method gives the time when the sun's centre would rise if there were no refraction. The effect of refraction is to cause the sun to be seen above the sensible horizon sooner in the morning, and later in the afternoon, than he actually is; and moreover, when the sun's upper limb coincides with the horizon, the centre is about 16′ below. At the instant, therefore, of sunrise or sunset, his centre is 90° 50′ from the zenith; the semi-diameter being about 16′, and the horizontal refraction 34′. In order, therefore, to compute the apparent time of rising of the sun's upper limb, we must compute when the sun's centre is 90° 50′ from the zenith. This may be done as follows:

Fig. 42.

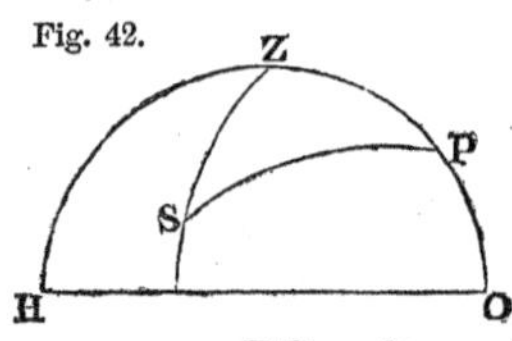

Let PZH be the meridian of the place of observation, P the pole, Z the zenith, and S the place of the sun. In the spherical triangle ZPS, the three sides are known, viz.,

PZ = the co-latitude $=\psi$;
ZS = the zenith distance $=z$;
PS = the north polar distance of the sun $=d$.

In this triangle we can compute ZPS, which is the angular distance of the sun from the meridian.

By Trigonometry,

$$\sin.\tfrac{1}{2}A=\sqrt{\frac{\sin.(S-b)\sin.(S-c)}{\sin.b\sin.c}}.$$

Put $$2S=z+d+\psi;$$

then $$\sin.\tfrac{1}{2}P=\sqrt{\frac{\sin.(S-\psi)\sin.(S-d)}{\sin.\psi\sin.d}}.$$

Ex. 1. Required the time of sunset at New York, Lat. 40° 42′, on the 10th of May, when the sun's declination is 17° 49′ N.

Here			
$\psi=$ 49° 18′		$\sin.(S-\psi)=9.922892$
$d=$ 72 11		$\sin.(S-d)=9.747281$
$z=$ 90 50		$\text{cosec.}\psi=0.120254$
S = 106 9½		$\text{cosec.}d=0.021345$
$S-\psi=$ 56 51½		2)9.811772
$S-d=$ 33 58½	½P = 53° 37½′ sin.	= 9.905886
	P = 107° 15′ = 7h. 9m.	

Hence the sun sets at 7h. 9m. apparent time; or, subtracting 4m. for equation of time, we have 7h. 5m. mean time.

Ex. 2. Required the mean time of sunrise at Boston, Lat. 42° 21′, on the 15th of October, when the sun's declination is 8° 47′ S., mean time being 14 minutes slow of apparent time.

Ans. 6h. 14m.

150. *To find the time of beginning or end of twilight.*—At the beginning or end of twilight, the sun is 18° below the horizon; that is, his zenith distance is 108°. Hence this problem can be solved by the formula of the last article.

Ex. 1. Required the time of the commencement of twilight at Washington, Lat. 38° 53′, on the 1st of June, when the sun's declination is 22° 10′ N., mean time being 2 minutes slow of apparent time. *Ans.* 2h. 41m. mean time.

Ex. 2. Required the time of ending of twilight at New Orleans, Lat. 29° 57′, on the 19th of February, when the sun's declination is 11° 19′ S., mean time being 14 minutes fast of apparent time.

Ans. 7h. 12m. mean time.

151. *To compute the distance between two stars whose right ascensions and declinations are known.*

Fig. 43.

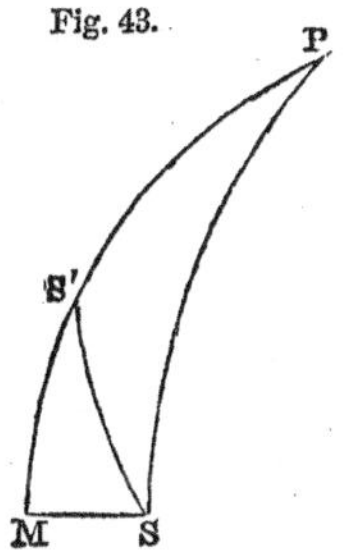

Let P be the pole, and S and S′ two stars whose places are known. Then PS and PS′ will represent their polar distances, and SPS′ will be the difference of their right ascensions. Draw SM perpendicular to PS′ produced. Then

R cos. P = tang. PM cot. PS.

Therefore, tang. PM = cos. P tang. PS.

Also, S′M = PM − PS′.

And cos. PM : cos. S′M :: cos. PS : cos. S′S.

Ex. 1. Required the distance from Aldebaran, R. A. 4h. 27m. 25.9s., polar distance 73° 47′ 33″, to Sirius, R. A. 6h. 38m. 37.6s., polar distance 106° 31′ 2″.

P = 2h. 11m. 11.7s. =	32°	47′	55″	cos. = 9.924579
PS =	106	31	2	tang. = 0.527916
PM =	109	25	55	tang. = 0.452495
PS′ =	73	47	33	
S′M =	35	38	22	cos. = 9.909930
PS =	106	31	2	cos. = 9.453782
PM =	109	25	55	sec. = 0.477964
SS′ =	46	0	44	cos. = 9.841676

Ex. 2. Required the distance from Regulus, R. A. 10h. 0m. 29.1s., polar distance 77° 18′ 41″, to Antares, R. A. 16h. 20m. 20.3s., polar distance 116° 5′ 55″. *Ans.* 99° 55′ 45″.

152. *Distance between two stars on the same parallel of declination.*—If two stars have the same declination, their distance can be computed as follows:

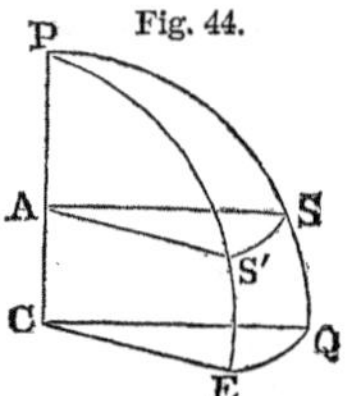

Fig. 44.

Let P be the pole, EQ a portion of the equator, and SS′ a portion of any parallel of declination, and PCE, PCQ two meridians passing through S and S′.

Then, by Geometry,

arc EQ : *arc* SS′ :: CQ : AS :: 1 : cos. Dec.

Therefore SS′=EQ cos. Dec.=EPQ cos. Dec.

That is, the distance between the two stars is equal to their difference of right ascension, multiplied by the cosine of their declination. This distance is, however, not measured on an arc of a great circle, but on a parallel of declination.

153. *To find the longitude and latitude of a star, when its right ascension and declination are known.*

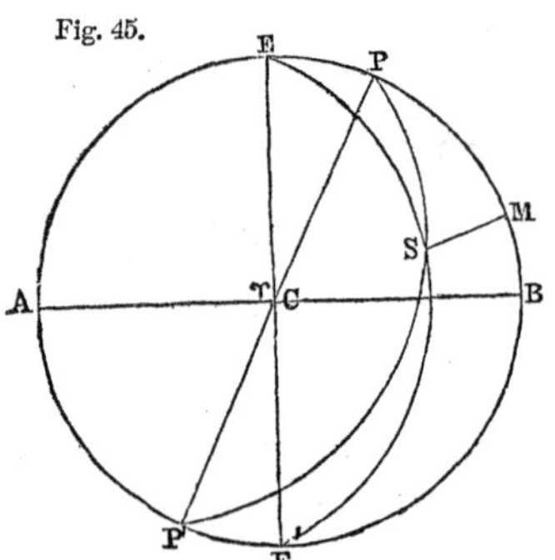

Fig. 45.

Let P represent the pole of the equator, E the pole of the ecliptic, C the first point of Aries, PSP′ an hour circle passing through the star S, and ESE′ a circle of latitude passing through the same star. Then AEBE′ represents the solstitial colure, EP represents the obliquity of the ecliptic, PS the polar distance of the star, ES its co-latitude; SPB is the complement of its right ascension, and SEB is the complement of its longitude. Draw SM perpendicular to PB. Represent PM by a; also represent the longitude of the star S by L, its latitude by l, and the obliquity of the ecliptic by ω.

Now, by Napier's rule, we have

R cos. SPM=tang. PM cot. PS;

that is, sin. R. A.=tang. a tang. Dec.,

or tang. a=sin. R. A. cot. Dec. (A)

Also, $\quad EM = EP + PM = a + \omega.$

Again, Trig., Art. 216, Cor. 3,

$$\sin. EM : \sin. PM :: \tan. SPM : \tan. SEM;$$

that is,

$$\sin.(a+\omega) : \sin. a :: \cot. R.A. : \cot. L :: \tan. L : \tan. R.A.,$$

or $$\tan. L = \frac{\tan. R.A. \times \sin.(a+\omega)}{\sin. a}. \qquad (1)$$

Also, $\quad R \cos. SEM = \tan. EM \cot. ES;$

that is, $$\tan. l = \cot.(a+\omega) \sin. L. \qquad (2)$$

Ex. 1. On the 1st of January, 1864, the R. A. of Capella was 5h. 6m. 42.01s., and its Dec. 45° 51′ 20″.1 N.; required its latitude and longitude, the obliquity of the ecliptic being 23° 27′ 19″.45.

By equation (A),

R. A. 76°	40′	30″.15		sin.	=9.988148
Dec. 45	51	20	.1	cot.	=9.987028
a=43	21	48	.2	tang.	=9.975176
ω=23	27	19	.45		
$a+\omega$=66	49	7	.65		

By equation (1),

tang. R. A.	=0.625527
sin. $(a+\omega)$	=9.963440
cosec. a	=0.163282
L=79° 58′ 3″.5 tang.	=0.752249

By equation (2),

cot. $(a+\omega)$	=9.631659
sin. L	=9.993308
l=22° 51′ 48″.3 tang.	=9.624967

Ex. 2. On the 1st of January, 1864, the R. A. of Regulus was 10h. 1m. 9.34s., and its Dec. 12° 37′ 36″.8 N.; required its latitude and longitude, the obliquity of the ecliptic being 23° 27′ 19″.45.

Ans. Latitude,

Longitude,

CHAPTER VI.

THE SUN—ITS PHYSICAL CONSTITUTION.

154. *Distance of the sun.*—The distance of the sun from the earth can be computed when we know its horizontal parallax, and the radius of the earth.

The mean value of the horizontal parallax of the sun has been found to be 8″.58, and the equatorial radius of the earth is 3963 miles.

Hence sin. 8″.58 : 3963 :: 1 : the sun's distance,
which is found to be 95,300,000 miles; or, in round numbers, 95 millions of miles.

155. *Velocity of the earth's motion in its orbit.*—Since the earth makes the entire circuit around the sun in one year, its daily motion may be found by dividing the circumference of its orbit by 365¼, and thence we may find the motion for one hour, minute, or second. The circumference of the earth's orbit is very nearly that of a circle whose radius is the sun's mean distance. We thus find the circumference of the orbit to be 598,800,000 miles; that the earth moves 1,639,000 miles per day; 68,300 miles per hour; 1138 miles per minute; and nearly 19 miles per second.

By the diurnal rotation, a point on the earth's equator is carried round at the rate of 1037 miles per hour, or 17 miles per minute. The motion in the orbit is, therefore, 66 times as rapid as the diurnal motion at the equator.

156. *The diameter of the sun.*—The sun's absolute diameter can be computed, when we know his distance and apparent diameter. The apparent diameter, as well as the distance, is variable, but the mean value of his apparent diameter is 32′ 3″.64. Hence we have the proportion

rad. : ES (95 millions) :: sin. 16′ 1″.8 : sun's radius,
which is found to be 444,406 miles; or his diameter is 888,812 miles.

The diameter of the sun is therefore 112 times that of the earth;

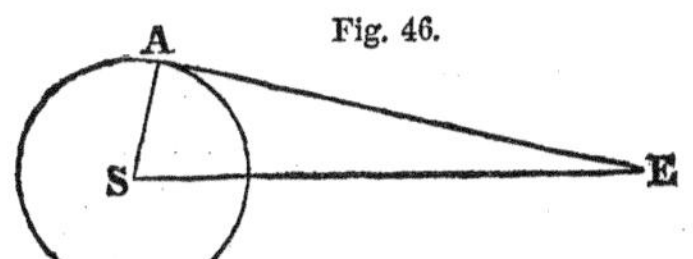

and, since spheres are as the cubes of their diameters, the volume of the sun is more than 1,400,000 times that of the earth.

The density of the sun is about one quarter that of the earth; and, therefore, his mass, which is equal to the product of his volume by his density, is found to be 355,000 times that of the earth.

157. *Figure of the sun's disc.*—Since the sun rotates upon an axis, as shown Art. 171, his figure can not be that of a perfect sphere. The oblateness of a heavenly body depends chiefly upon the ratio of the centrifugal force to the force of gravity upon its surface. Now, on account of its slow rotation, the centrifugal force of a point upon the sun's equator, is only about one sixth what it is upon the earth, while the force of gravity is nearly thirty times as great; hence the oblateness of the sun should be only about $\frac{1}{180}$th part of that of the earth. But the oblateness of the earth is about $\frac{1}{300}$th. Hence the oblateness of the sun should be only about $\frac{1}{54000}$, which corresponds to a difference of less than one twentieth part of a second between the equatorial and polar diameters. This quantity is too small to be detected by our observations; and although the sun's diameter has been measured many thousand times, still, with the exception of the effect due to refraction, explained in Art. 91, his disc is sensibly a perfect circle.

158. *Force of gravity on the sun.*—The attraction of a sphere being the same as if its whole mass were collected in its centre, will be proportional to the mass directly, and the square of the distance inversely; hence the force of gravity on the surface of the sun, will be to the force of gravity on the surface of the earth, as $\frac{355,000}{112^2}$ to unity, which is 27.9 to 1; that is, a pound of terrestrial matter at the sun's surface, would exert a pressure equal to what 27.9 such pounds would do at the surface of the earth. A body weighing 200 pounds on the earth, would produce a pressure of 5580 pounds on the sun.

At the surface of the earth, a body falls through $16\frac{1}{12}$th feet in one second; but a body on the sun would fall through $16\frac{1}{12} \times 27.9 = 448.7$ feet in one second.

PHYSICAL CONSTITUTION OF THE SUN.

159. *Solar spots.*—When we examine the sun with a good telescope, we frequently perceive upon his surface, black spots of irregular shape, sometimes extremely minute, and at other times of vast extent. They usually make their first appearance at the eastern limb of the sun; advance gradually toward the centre; pass beyond it, and disappear at the western limb, after an interval of about 14 days. They remain invisible about 14 days, and then sometimes reappear at the eastern limb in nearly the same position as at first, and again cross the sun's disc as before, having taken 27d. 7h. in the entire revolution.

The appearance of a solar spot is that of an intensely black, irregularly-shaped patch, called the *nucleus*, surrounded by a fringe which is less dark, and is called the *penumbra*. The form of this fringe is generally similar to that of the inclosed black spot; but this is not always the case, for several dark spots are occasionally included in a common penumbra.

Black spots have occasionally been seen without any penumbra; and sometimes we see a large penumbra without any central nucleus; but generally both the nucleus and penumbra are combined.

160. *Changes of the spots.*—These spots change their form from day to day, and sometimes from hour to hour. They usually commence from a point of insensible magnitude, grow very rapidly at first, and usually attain their full size in less than a day. Then they remain stationary, with a well-defined penumbra, and continue for ten, twenty, and some even for fifty days. Then the nucleus usually becomes divided by a narrow line of light; this line sends out numerous branches, which extend until the entire nucleus is covered by the penumbra.

Decided changes have been detected in the appearance of a spot within the interval of a single hour, indicating a motion upon the sun's surface of at least 1000 miles per hour.

The duration of the spots is very variable. A spot has appeared and vanished in less than 24 hours, while others have lasted for weeks, and even months. In 1840, a spot was identified for *nine* revolutions, which corresponds to a period of about eight months.

161. *Magnitude and number of the spots.*—Solar spots are sometimes of immense magnitude, so that they have repeatedly been visible to the naked eye. In June, 1843, a solar spot remained for a whole week visible to the naked eye. Its breadth measured 167″, which indicates an absolute diameter of 77,000 miles.

The number of spots seen on the sun's disc is very variable. Sometimes the disc is entirely free from them, and continues thus for weeks, or even months together; at other times a large portion of the sun's disc is covered with spots. Sometimes the spots are small, but numerous; and sometimes they appear in groups of vast extent. In a large group of spots which appeared in 1846, upward of 200 single spots and points were counted. In 1837 a cluster of spots covered an area of nearly 5 square minutes, or nearly 4000 millions of square miles.

162. *The black nucleus.*—It is not certain that the black nucleus of a spot is entirely destitute of light; for the most intense artificial light, when seen projected on the sun's disc, appears as dark as the spots themselves. Sir W. Herschel estimated that the light of the penumbra was less than one half that of the brighter part of the sun's surface, and the light of the nucleus less than one hundredth of the brighter surface.

163. *Upon what part of the sun do the spots appear?*—The spots do not appear with equal frequency upon every part of the sun's disc. With few exceptions, they are confined to a zone included between 30° of N. Latitude and 30° of S. Latitude, measured from the sun's equator. According to a series of observations, extending over a period of ten years, and comprehending 1700 spots, the distribution in zones is as follows:

	Per Cent.		Per Cent.
Beyond 30° N. Latitude	3	Beyond 30° S. Latitude	2
Between 20° and 30° N.	17	Between 20° and 30° S.	15
" 10 " 20	23	" 10 " 20	17
" 0 " 10	11	" 0 " 10	12

We thus perceive that 95 per cent. of all the spots are found within 30° of the sun's equator. There are only three cases on record, in which spots have been seen as far as 45° from the sun's equator.

164. *Appearance of the bright part of the sun's disc.*—Independently of the dark spots, the luminous part of the sun's disc is not uniformly bright. It exhibits a mottled appearance, like that which would be presented by a stratum of luminous clouds of irregular shape and variable depth. This mottled appearance is not confined, like the black spots, to a particular zone, but is seen on all parts of the surface, even near the poles of rotation.

Sometimes we observe upon the sun's disc curved lines, or streaks of light, more luminous than the rest of the surface. These are called *faculæ*, and they generally appear in the neighborhood of the black spots.

165. *Proof that the sun's outer envelope is not solid.*—The rapid changes which take place upon the surface of the sun, prove that his outer envelope is not solid. Admitting that the great mass of the sun is solid, that portion which we ordinarily see, must be either liquid or gaseous; and the rapid motion of 1000 miles per hour, which has been observed in solar spots, indicates that the luminous matter which envelops the sun must be gaseous, since liquid bodies could hardly be supposed to move with such velocity.

166. *The solar spots are not planetary bodies.*—It is evident that the solar spots are *at the surface* of the sun; for if they were bodies revolving around the sun at some distance from it, the time during which they would be seen on the sun's disc would be *less* than that occupied in the remainder of their revolution. Thus, let S represent the sun, E the earth, and suppose ABC to represent the path of an opaque body revolving about the sun. Then AB represents that part of the orbit in which the body would appear projected upon the sun's disc, and this is less than half the entire circumference; whereas the spot reappears on the opposite limb of the sun after an interval nearly equal to that required to pass across the disc.

Fig. 47.

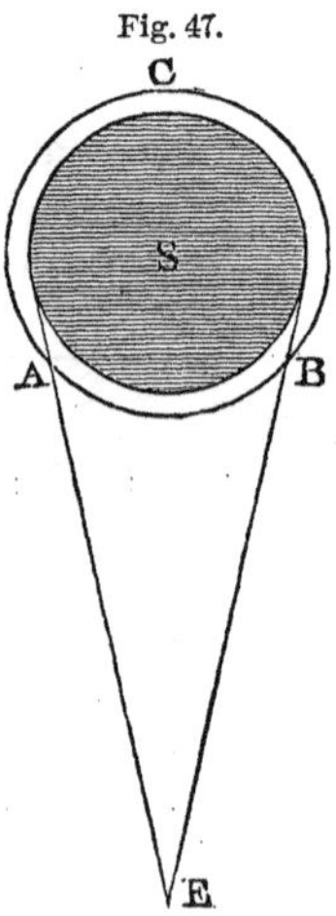

167. *The dark spots are depressions* in the luminous matter which envelops the sun. This was first proved by an observation made by Dr.

Wilson, of Glasgow, in November, 1769. He first noticed a spot November 22d, when it was not far from the sun's western limb; and he observed that the penumbra was about equally broad on every side of the nucleus. The next day the eastern portion of the penumbra had contracted in breadth, while the other parts remained nearly of their former dimensions. On the 24th the pe-

Fig. 48.

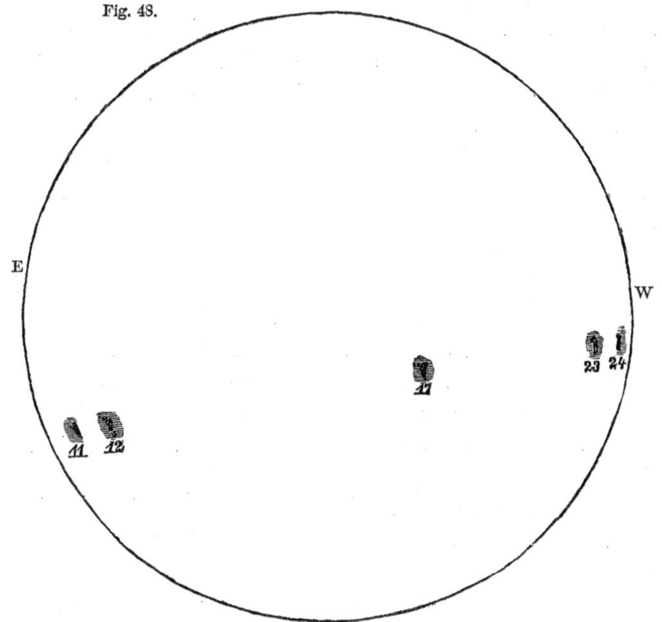

numbra had entirely *disappeared* from the eastern side, while it was still visible on the western side. On the 11th of December the spot reappeared on the sun's eastern limb, and now there was *no* penumbra on the western side of the spot, although it was distinctly seen on the remaining sides. The next day the penumbra came into view on the western side, though narrower than on the other sides. On the 17th the spot had passed the centre of the sun's disc, and now the penumbra appeared of equal extent on every side of the nucleus. From these observations, it is inferred that the penumbra is lower than the general level of the sun's bright surface, and the nucleus lower than the penumbra. Dr. Wilson computed that the depth of the spot just described was nearly 4000 miles.

Similar observations were repeatedly made by Sir W. Herschel. In 1794 he observed that, as a spot approached near the western limb of the sun, the black nucleus gradually contracted in breadth, while its length remained unchanged. It became reduced to a narrow black line, and then disappeared, *while the penumbra was still visible.* Similar observations have repeatedly been made by other astronomers.

In 1801 Sir W. Herschel observed that when a spot came near the western margin of the sun, he was able to distinguish the thickness of the stratum on the western border, but not on the eastern; and he hence computed that the depression of the penumbra below the bright surface of the sun was not less than 1800 miles. Similar observations have been made by M. Secchi at Rome.

168. *The bright streaks or faculæ are elevated ridges* rising above the general level of the sun's surface. This is proved by an observation made in 1859 by Mr. Dawes, of England. He had the good fortune to observe a bright streak of unusual size precisely at the edge of the sun's disc, and he perceived that it projected beyond the circular outline of the disc in the manner of a mountain ridge.

In 1862, as an uncommonly large spot was passing off the sun's disc, Mr. Howlett perceived a small notch in the sun's margin, precisely over the place where the great nucleus had previously been seen, and on either side of it the photosphere appeared to be heaped up above the general level of the sun's surface.

169. *To determine the time of the sun's rotation.*—It is found that a spot generally employs $27\frac{1}{4}$ days in passing from one limb of the sun around to the same limb again, and it is inferred that this apparent motion is caused by a rotation of the sun upon his axis. But the period above mentioned is not the time in which the sun performs one rotation about his axis; for, let AA′B represent the sun, and EE′D the orbit of the earth. When the earth is at E, the visible disc of the sun is AA′B; and if the earth was stationary at

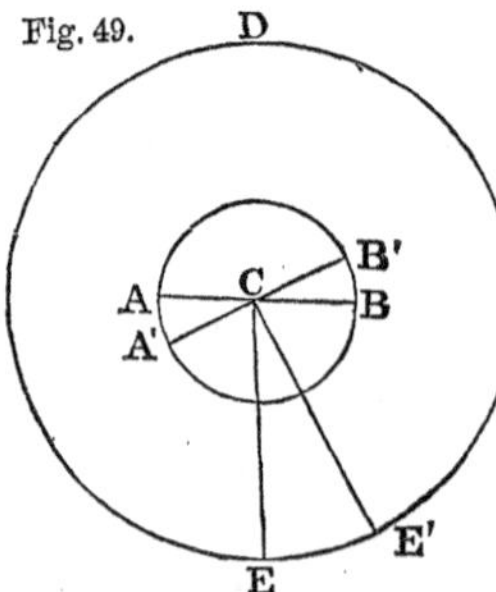

Fig. 49.

E, then the time required for a spot to move from the limb B round to the same point again would be the time of the sun's rotation. But while the spot has been performing its apparent revolution, the earth has advanced in her orbit from E to E′, and now the visible disc of the sun is A′B′, so that the spot has performed more than a complete revolution in the time it has taken to move from the western limb to the western limb again. Since an apparent rotation of the sun takes place in $27\frac{1}{4}$ days, the number of apparent rotations in a year will be $\frac{365\frac{1}{4}}{27\frac{1}{4}}$, or 13.4.

But, in consequence of the motion of the earth about the sun, if the sun had no real rotation, it would in one year make an *apparent* rotation in a direction contrary to the motion of the earth. Hence, in one year, there must be 14.4 real rotations of the sun, and the time of one real rotation is $\frac{365\frac{1}{4}}{14.4}$, or 25.3 days. Thus the time of a real rotation is found to be nearly two days less than that of an apparent rotation.

170. *Temperature of different parts of the sun's disc.*—By receiving the image of different portions of the sun upon a very sensitive thermometer, it has been discovered that the sun's disc has not throughout exactly the same temperature. The rays proceeding from the centre of the disc are hotter than those which proceed from the margin, and the black spots radiate less heat than the neighboring bright surface.

The luminous intensity of different portions of the sun's disc exhibits corresponding variations, the borders of the disc being found less luminous than the centre. This difference is quite noticeable in a photographic picture of the sun.

171. *Influence of solar spots upon terrestrial temperatures.*—It has been supposed that the presence of an unusual number of large spots on the sun's disc must influence the temperature of the earth, and there are some facts which favor this supposition. At Paris, out of 26 years of observations, the mean temperature of those years in which the spots were most numerous was half a degree lower than that of those years in which the spots were least frequent. But during the same years a slight effect of the opposite kind was observed upon the temperature of places in the

United States, so that we seem obliged to ascribe the differences in question to other causes than the solar spots.

172. *Position of the sun's equator.*—Besides the time of rotation, observations of the solar spots enable us to ascertain the position of the equator with reference to the ecliptic. The angle between the solar equator and the ecliptic has been determined to be about 7°. About the first weeks of June and December, the spots, in traversing the sun's disc, appear to us to describe straight lines, but at other times the apparent paths of the spots are somewhat elliptical, and they present the greatest curvature about the first weeks of March and September.

173. *Periodicity in the number of the solar spots.*—The number of the solar spots varies greatly in different years. Some years the sun's disc is *never* seen entirely free from spots, while in other years, for weeks and even months together, no spots of any kind can be perceived. From a continued series of observations, embracing a period of 38 years, it appears that the spots are subject to a certain periodicity. The number of the spots increases during 5 or 6 years, and then diminishes during about an equal period of time, the interval between two consecutive maxima being from 10 to 12 years.

As this period corresponds to the time of one revolution of Jupiter, it suggests the idea that possibly Jupiter may have the power of sensibly disturbing the sun's surface.

174. *The sun not a solid body.*—A comparison of the dark lines in the solar spectrum has led to the conclusion that the elements of which the sun is composed are to a great extent the same as those found upon the earth. The existence of iron, nickel, and several other well-known metals in the sun's atmosphere is considered as proved; and since the density of the sun is only one fourth that of the earth, while the force of gravity is 28 times its force upon the earth, we can not suppose that any large part of the sun's mass is in the condition of a solid or even a liquid body. The most refractory substances, iron and nickel, exist upon the sun in the state of elastic vapor. Hence the temperature of the sun's surface is extremely elevated, far beyond the heat of terrestrial volcanoes. It is possible that the centre of the sun consists

of matter in the liquid or even the solid state; but it is probable that the principal part of the sun's volume consists of matter in the gaseous condition.

175. *Nature of the sun's photosphere.*—The bright envelope of the sun, which we call its *photosphere*, consists of matter in a state analogous to that of aqueous vapor in terrestrial clouds; that is, in the condition of a precipitate suspended in a transparent atmosphere. This photosphere is not only intensely luminous, but intensely hot, and the thermoscope indicates that it radiates more heat than the solar spots; but this does not prove that the photosphere is really hotter than the nucleus of a solar spot, for gases radiate heat more feebly than solids of the same temperature. The matter of the photosphere probably consists of particles precipitated in consequence of their being cooled by radiation.

The sun's gaseous envelope extends far beyond the photosphere. During total eclipses we observe protuberances rising to a height of 80,000 miles above the surface of the sun, which requires us to admit the existence of bodies analogous to clouds floating at great elevations in an atmosphere; and if the extent of the solar atmosphere compared with the height of the visible clouds corresponds with what exists upon the earth, we must conclude that the solar atmosphere extends to at least a million of miles beyond his surface.

Fig. 50.

176. *Nature of the penumbra.*—The penumbra of a solar spot appears to be formed of filaments of photospheric light converging toward the centre of the nucleus, each of the filaments having the same light as the photosphere, and the sombre tint results from the dark interstices between the luminous streaks, as in a steel engraving shades are produced by

dark lines separated by white interstices. The convergence of the luminous streaks of the penumbra toward the centre of the spot indicates the existence of currents flowing toward the centre. These converging currents probably meet an ascending current of the heated atmosphere, by contact with which the matter of the photosphere is dissolved, and becomes non-luminous.

The faculæ are ascribed to commotions in the photosphere, by which the thickness of the phosphorescent stratum is rendered greater in some places than in others, and the surface appears brightest at those points where the luminous envelope is thickest.

177. *Motion of the solar spots.*—The spots are not stationary on the sun's disc, for the apparent time of revolution of some of the spots is much greater than that of others. In one instance, the time of the sun's rotation, as deduced from observations of a solar spot, was only 24d. 7h., while in another case it amounted to 26d. 6h. This difference can only be explained by admitting that the spots have a motion of their own relative to the sun's surface, just as our clouds have a motion relative to the earth's surface.

The motion of the solar spots in latitude is very small, and this motion is sometimes directed *toward* the equator, but generally *from* the equator. The motion of the spots in longitude is more decided. Spots near the equator have an apparent movement of rotation more rapid than those at a distance from the equator. While at the equator the daily angular velocity of rotation is 865′, in lat. 20° it is only 840′, and in lat. 30° it is 816′. Hence a point on the sun's equator makes a complete rotation in 25 days, but a point in lat. 30° makes one rotation in 26½ days.

178. *Cause of the movements of the solar spots.*—The heat of the sun must be continually dissipated by radiation. If this radiation is more obstructed in some regions than in others, heat must accumulate in such places. Now the phenomena observed during total eclipses indicate in the sun's atmosphere the existence of large masses analogous to terrestrial clouds. Wherever these clouds prevail, the free radiation of heat from the sun must be obstructed, and heat must rapidly accumulate. The solar atmosphere tends to move toward these heated centres, and this must be accompanied by an upward motion at the centre. The heated air thus ascending partly dissolves and partly divides the matter

of the photosphere, causing it to heap up in a ring around the opening, producing thus around the margin of the penumbra the appearance of a border of light more intense than the general photosphere.

A general movement of the atmosphere toward one point must create a tendency to revolve around this centre, for the same reason that terrestrial storms sometimes rotate about a vertical axis. Such a motion of the solar spots has been repeatedly indicated by observation. Moreover, solar spots have sometimes exhibited a spiral structure such as might be supposed to result from rotation about a vertical axis. Fig. 51 represents such a spot observed by M. Secchi at Rome in 1857. The nucleus exhibited two centres perfectly black, while the penumbra showed numerous dark lines extending spirally from these centres, and a large spiral filament, in the form of an eagle's beak, extended far within the nucleus.

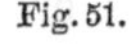

Fig. 51.

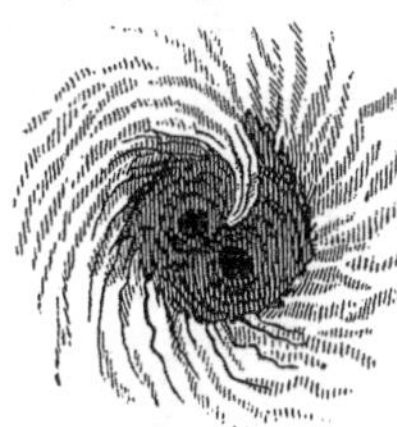

179. *Zodiacal light.*—The zodiacal light is a faint light, somewhat resembling that of the Milky Way, or more nearly that of the tail of a comet, and is seen at certain seasons of the year in the west after the close of twilight in the evening, or in the east before its commencement in the morning. Its apparent form is nearly that of a cone with its base toward the sun, and its axis is situated nearly in the plane of the ecliptic. The season most favorable for observing this phenomenon, is when its direction, or the direction of the ecliptic, is most nearly perpendicular to the horizon. For places near the latitude of New York, this occurs about the 1st of March for the evening, and about the 10th of October for the morning.

Fig. 53.

The distance to which the zodiacal light extends from the sun, varies from 20° or 30° to 80° or 90°, and sometimes even more than 90°. Its breadth at its base perpendicularly to its length, varies from 8° to

30°. It is brightest in the parts nearest the sun, and in its upper part its light fades away by insensible gradations, so that different observers at the same time and place assign to it different limits. Under favorable circumstances, it has been seen to extend entirely across the heavens.

It is probable that the zodiacal light is an envelope of very rare matter surrounding the sun, and extending beyond the orbits of Mercury and Venus, and at times even beyond the orbit of the earth. If the sun could be viewed from one of the other stars, it would probably appear to be surrounded by a nebulosity, similar to that in which some of the fixed stars appear to be enveloped, as seen from the earth.

CHAPTER VII.

PRECESSION OF THE EQUINOXES.—NUTATION.—ABERRATION.—LINE OF THE APSIDES.

179. *Fixed position of the ecliptic.*—By comparing catalogues of stars formed in different centuries, we find that the *latitudes* of the stars continue always nearly the same. Hence the position of the ecliptic among the stars must be well-nigh invariable.

180. *Precession of the equinoxes.*—It is found that the *longitudes* of the stars are continually increasing, at the rate of about 50″ in a year. Since this increase of longitude is common to all the stars, and is nearly the same for each star, we can not ascribe it to motions in the stars themselves. We hence conclude that the vernal equinox, the point from which longitude is reckoned, has a backward or retrograde motion along the ecliptic, amounting to 50″ in a year, while the inclination of the equator to the ecliptic remains nearly the same. This motion is called *the precession of the equinoxes*, because the place of the equinox among the stars each year *precedes* (with reference to the diurnal motion) that which it had the previous year.

The amount of precession is 50″.2 annually. In order to determine how many years will be required for a complete revolution of the equinoctial points, we divide 1,296,000, the number of seconds in the circumference of a circle, by 50″.2, and obtain 25,800 years.

181. *The pole of the equator revolves round the pole of the ecliptic.*—Since the position of the ecliptic is fixed, or nearly so, it is evident that the equator must change its position, otherwise there could be no motion in the equinoctial points; and a motion of the equator implies a motion of the poles of the equator. Since the obliquity of the ecliptic remains nearly constant, the distance from the pole of the equator to the pole of the ecliptic must remain nearly constant; and we may conceive the phenomena of precession to arise from the revolution of the pole of the celestial equator around the pole of the ecliptic, in the period of 25,800 years, at a constant distance of about 23½ degrees.

182. *The signs of the zodiac and the constellations of the zodiac.*—At the time of the formation of the first catalogue of stars, 140 years before Christ, the signs of the ecliptic corresponded very nearly to the constellations of the zodiac bearing the same names. But in the interval of 2000 years since that period, the vernal equinox has retrograded about 28°; so that the sign Taurus now corresponds nearly with the constellation Aries, the sign Gemini with the constellation Taurus, and so for the others.

183. *The pole star varies from age to age.*—The pole of the equator in its revolution about the pole of the ecliptic, must pass in succession by different stars. At the time the first catalogue of the stars was formed, the north pole was nearly 12° distant from the present pole star, while its distance from it is now less than 1½ degrees. The pole will continue to approach this star till the distance between them is about half a degree, and will then recede from it. After a lapse of about 12,000 years, the pole will have arrived within about 5° of α Lyræ, the brightest star in the northern hemisphere.

184. *Cause of the precession of the equinoxes.*—The earth may be considered as a sphere surrounded by a spheroidal shell, thickest at the equator, Art. 45. The matter of this shell may be regarded as forming a ring round the earth, in the plane of the equator. Now the tendency of the sun's action on this ring, except at the time of the equinoxes, is always to make it turn round the intersection of the equator with the ecliptic, toward the plane of this latter circle.

The solar force exerted on the part of this ring that is above the ecliptic, may be resolved into two forces, one of which is in the plane of the equator, and the other perpendicular to it. The latter force tends to impress on the ring a motion round its intersection with the ecliptic. So, also, the solar force exerted on the part of the ring that is below the ecliptic, may be resolved into two, one in the plane of the equator, and the other perpendicular to it. The sun's attraction upon the nearest half of the ring, tends to bring the plane of the ring nearer to the plane of the ecliptic; while its attraction upon the remoter half of the ring produces an opposite effect. But on account of the greater distance, the latter effect is less than the former; so that, besides the motion of translation produced by the various other forces, the ring would turn slowly around its intersection with the ecliptic, and the two planes would ultimately coincide.

185. *How to find the resultant of two rotary motions.*—While the equatorial ring has this tendency to turn about the line of the equinoxes, it also rotates on an axis perpendicular to its plane in twenty-four hours; that is, it has a tendency to rotate simultaneously about two different axes. The result is a tendency to rotate about an intermediate axis, whose position is determined by the following theorem:

Fig. 54.

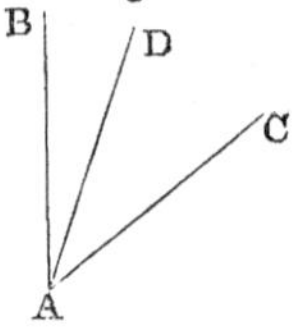

If a body is revolving freely round the axis AB, with the angular velocity V, and if a force be impressed upon it which would make it revolve about the axis AC with an angular velocity V′, then the body will not revolve about either of the axes AB, AC, but about a third axis AD, situated in the plane BAC, and the angle BAC will be divided so that

sin. BAD : sin. CAD : : V′ : V.

Let PP′ represent the axis of diurnal rotation of the equatorial ring, and AB the line of the equinoxes, about which it also tends slowly to revolve.

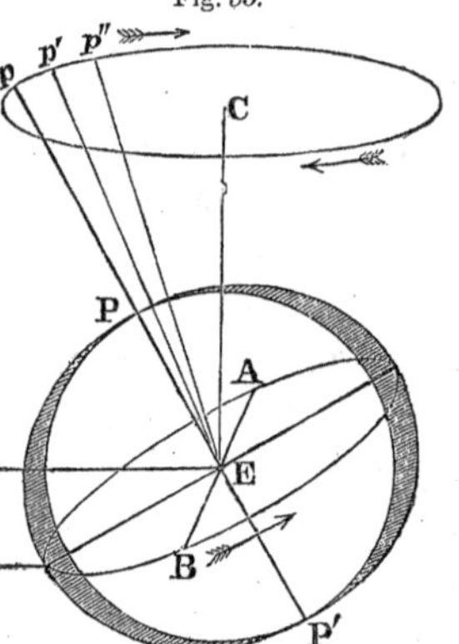

Fig. 55.

The new axis of rotation, Ep', will be situated in the plane pEA, and the sine of its angular distance from each of the former axes will be in the inverse ratio of the angular velocity round that axis. Repeating the same construction for the following instant, we shall find the new position of the axis will be Ep'', and so on; that is, the point p will be made to describe a curve around C, the pole of the ecliptic.

186. *Illustration from the Gyroscope.*—This motion of the earth's equatorial ring may be very closely imitated by a modified form of the gyroscope. Let AB represent a brass ring, supported by wires AD, BD, which are connected with an axis, DC, whose extremity is a little above the centre of gravity of the ring AB, and rests upon a support, CE. When the ring AB is at rest, its axis DC will have a vertical position. If, however, the axis be inclined from the vertical, and be made to rotate by twirling it with the fingers, the plane of the ring will turn slowly round in azimuth, preserving, however, a nearly constant inclination to the horizon; that is, the axis of the ring will describe the surface of a cone, or the point F will describe the circumference of a circle about the point G.

Fig. 56.

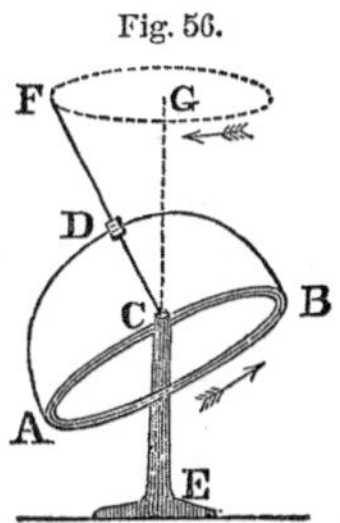

187. *Why the precession is so slow.*—If the earth were a perfect sphere, the solar forces acting on the opposite hemispheres would exactly balance one another, and could produce no motion in the earth or its axis. If now we conceive the equatorial ring already described, to be attached to the spherical part of the earth, which is far heavier than the ring, it is evident that the ring, having to drag around with it this great inert mass, will have its velocity of retrogradation proportionally diminished. Thus, then, the entire globe must have a motion similar to that ascribed to the ring, but the motion will be extremely slow.

The moon produces a similar retrogradation in the intersection of the equator with the plane of the lunar orbit, but, on account of its nearness to the earth, its effect is more than double that of the sun. The planets also, by their attraction, exert a small influence upon the position of the equatorial ring, but the result is slightly to diminish the amount of precession. The whole effect of the sun and moon is 50″.37, and that of the planets 0″.16, leaving the actual amount of precession 50″.21 annually.

Nutation.

188. The effect of the action of the sun and moon upon the earth's equatorial ring, depends upon their position with regard to the equator. When either body is in the plane of the equator, its action can have no tendency to change the position of this plane, and consequently none to change the positions of the equinoctial points. Its effect in producing these changes, increases with the distance of the body from the equator, and is greatest when that distance is greatest. Twice a year, therefore, viz., at the equinoxes, the effect of the sun to produce precession is nothing, while at the solstices the effect of the sun is a maximum. On this account, the precession of the equinoxes, as well as the obliquity of the ecliptic, is subject to a semi-annual variation, which is called the *solar nutation.* There is also an inequality depending upon the position of the moon which is called *lunar nutation.* The maximum value of the lunar nutation in longitude is 17″.2, and that of the solar nutation 1″.2.

In consequence of this oscillatory motion of the equator, its pole, in revolving about the pole of the ecliptic, does not move strictly in a circle, but in a waving curve, which passes alternately within and without the circle, somewhat similar to that in Fig. 57.

Fig. 57.

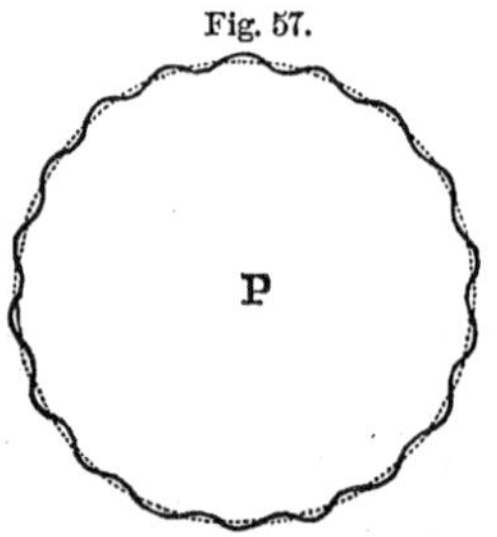

189. *Tropical and sidereal years.* — The time occupied by the sun in moving from the vernal equinox to the vernal equinox again, is called *a tropical year.*

The time occupied by the sun in moving from one fixed star to the same fixed star again, is called *a sidereal year.*

On account of the precession of the equinoxes, the tropical year is less than the sidereal year, the vernal equinox having gone westward so as to meet the sun. The tropical year is less than the sidereal year, by the time that the sun takes to move over 50″.2 of his orbit. This amounts to 20m. 22s.

The mean length of a tropical year expressed in mean solar time is 365d. 5h. 48m. 48s. The length of the sidereal year is therefore 365d. 6h. 9m. 10s.

Aberration.

190. The annual motion of the earth, combined with the motion of light, causes the stars to appear in a direction different from their true direction. This displacement is called *aberration.* The nature of this effect may be understood from the following illustration:

If we suppose a shower of rain to fall during a dead calm in vertical lines, if the observer be at rest the rain will *appear* to fall vertically; and if the observer hold in his hand a tube in a vertical position, a drop of rain may descend through the tube without touching the sides; but if the observer move forward, the rain will strike against his face; and, in order that a drop of rain may descend through the tube without touching the sides, the tube *must be inclined* forward. Suppose, while a rain-drop is falling from E to D with a uniform velocity, the spectator moves from C to D, and carries the tube inclined in the direction EC. A drop of rain entering the tube at E, when the tube has the position EC, would reach the ground at D when the tube has come into the position FD; that is, the drop of rain will *appear* to follow the direction EC.

Fig. 58.

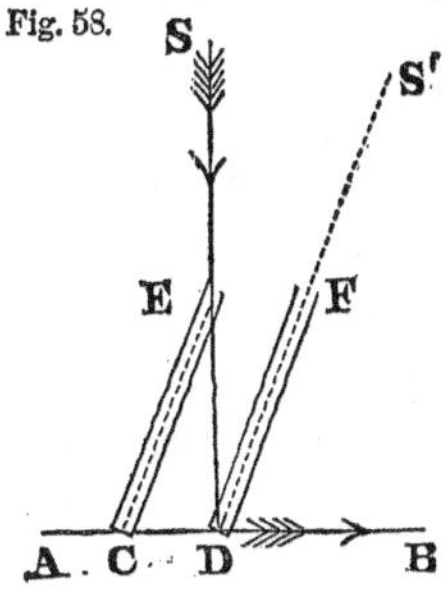

Now $$CD = ED \times \text{tang. } CED;$$
that is, the velocity of the observer = velocity of the rain × tangent of the apparent deflection of the rain-drop.

191. *To determine the amount of aberration.*—The aberration of light is explained in a similar manner. Let AB be a small portion of the earth's orbit, and S the position of a star. Let CD be the distance through which the observer is carried in 1s., and ED the distance through which light moves in 1s. If a straight tube be conceived to be directed from the eye at C to the light at E, so that the light shall be in the centre of its opening, and if the tube moves with the eye from C to D, remaining constantly parallel to itself, the light, in moving from E to D, will pass along the axis of the tube, and will arrive at D when the earth reaches the same point. It is evident that the star will appear in the direction of the axis of the tube; that is, the star *appears* in the direc-

tion S'D instead of SD. The velocity of the earth in its orbit is 19 miles per second; the velocity of light is 192,000 miles per second.

In the triangle ECD, we have

$$\text{tang. CED} = \frac{\text{CD}}{\text{ED}} = \frac{19}{192,000}.$$

Hence $\text{CED} = 20''$;

that is, the aberration of a star which is 90° from the path in which the earth is moving, amounts to 20''.

192. *Effect of aberration upon a star situated at the pole of the ecliptic.*—It is obvious that the aberration is always in the direction in which the earth is moving. Its effect, therefore, upon the apparent position of a star, will vary with the season of the year.

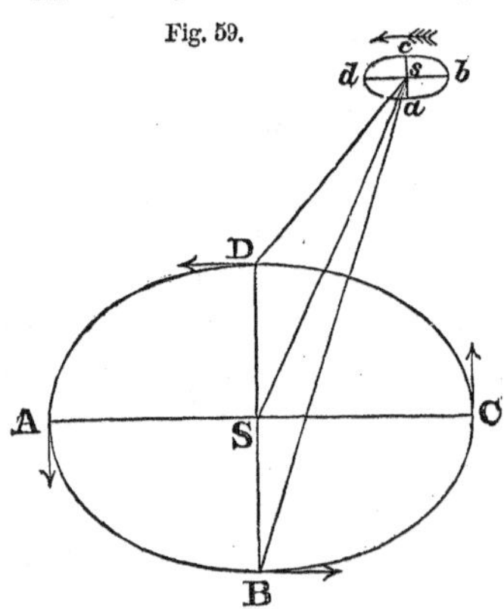

Fig. 59.

Let ABCD represent the annual path of the earth around the sun; let S be the place of the sun, and *s* the place of a star so situated that the line S*s* is perpendicular to the plane of the ecliptic.

When the earth is at the point A, moving toward B, the aberration will be in the direction *sa*; that is, the star appears at the point *a*.

When the earth has arrived at B, the aberration will be in the direction *sb*; that is, the star appears at the point *b*.

When the earth has arrived at C, the star appears at the point *c*; and when the earth has arrived at D, the star appears at the point *d*. But *sa*, *sb*, *sc*, *sd* are each 20'', and therefore the star will appear annually to describe a small circle in the heavens, 40'' in diameter.

193. *Effect upon a star situated in the plane of the ecliptic.*—If the star were situated in the plane of the ecliptic, in the direction of the line AC produced, then, when the earth is at C, the aberration will be 20'', as before; but when the earth is at D, the aberration will be nothing, because the earth and the light of the star are moving in the same direction. When the earth is at A, the ab-

erration will again be 20″, but in a direction opposite to what it was at C; and when the earth is at B, the aberration will again be nothing. Hence we see that if a star be in the plane of the ecliptic, it will appear to oscillate to and fro along a straight line, 20″ on each side of the true position of the star, and this line will be situated in the plane of the ecliptic.

A star situated between the ecliptic and its poles, will appear annually to describe an ellipse whose major axis is 40″, but its minor axis will increase with its distance from the plane of the ecliptic.

194. *The apsides of the earth's orbit.*—The points of perihelion and aphelion of the earth's orbit, are called by the common name of *apsides*. The major axis of the earth's orbit is therefore called the *line of the apsides.*

By comparing very distant observations, it is found that the line of the apsides has a progressive motion, or a motion eastward amounting to about 12″ annually. Since the equinox from which longitude is reckoned moves in the opposite direction 50″ annually, the longitude of the perihelion increases about 62″ annually.

At this rate, the line of the apsides would complete a sidereal revolution in 108,000 years, or a tropical revolution in 20,900 years. For the cause of this motion, see Arts. 279 478.

195. *Changes in the position of the line of the apsides.*—The line of the apsides, thus continually moving round, must at one period have coincided with the line of the equinoxes. The longitude of the perihelion in 1864 was 100° 16′, which point the earth passed on the 1st of January. The time required to move over an arc of 100¼° at the rate of 62″ annually, is about 5818 years, which extends back nearly 4000 years before the Christian era—a period remarkable for being that to which chronologists refer the creation of the world. At this time the winter and spring were equal, and longer than the summer and autumn, which were also equal.

196. *Mean place and true place; mean anomaly and true anomaly.*—The *mean* place of a body revolving in an orbit, is the place where the body *would* have been if its angular velocity had been

uniform; the *true* place of a body is the place where the body actually is at any time. *Equations* are corrections which are applied to the mean place of a body, in order to get its true place.

The angular distance of a planet from its perihelion, as seen from the sun, is called its *anomaly.*

If an imaginary planet be supposed to move from perihelion to aphelion with a uniform angular motion round the sun, in the same time that the real planet moves between the same points with a variable angular motion, the angular distance of this imaginary planet from perihelion is called its *mean anomaly*, while its actual distance at the same moment in its orbit is called its *true anomaly.*

197. *Equation of the centre.*—The difference between the mean and the true anomaly is called the *equation of the centre.*

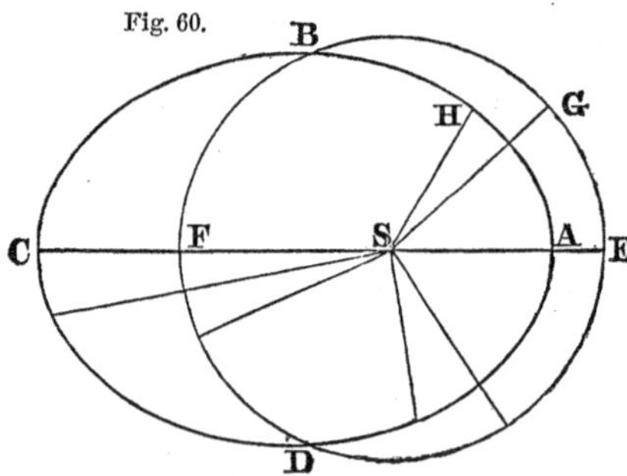

Fig. 60.

Let ABCD be the orbit of a planet having the sun in one of the foci at S. With the centre S, and a radius equal to the square root of the product of the semi-axes of the ellipse, describe the circle EBFD; the area of this circle will be equal to that of the ellipse. At the same time that a planet departs from A, the perihelion, to describe the orbit ABCD, let an imaginary planet start from E, and describe the circle EBFD with a uniform motion, and perform a whole revolution in the same period that the planet describes the ellipse. The imaginary planet will describe around S, sectors of circles which are proportional to the times, and equal to the elliptic areas described in the same time by the planet. Suppose the imaginary planet to be at G; then take the sector ASH=ESG, and H will be the place of the planet in the ellipse. The angle ESG is called the *mean anomaly;* ASH is the *true anomaly;* and GSH is the *equation of the centre.*

If we consider the mean and the true anomaly as agreeing at A, the angles ESG and ASH must increase unequally, and the true anomaly must exceed the mean. The equation of the centre increases till the planet reaches the point B. From B to C the mean anomaly gains upon the true, until at C they coincide —that is, the equation of the centre is nothing. Proceeding from C, the mean anomaly must exceed the true, and the equation of the centre increases until the planet reaches the point D. From D to A the true anomaly gains upon the mean, until at A they coincide again. At the points B and D the equation of the centre is the greatest possible.

The greatest value of the equation of the centre for the sun is 1° 55′ 27″.

198. *The anomalistic year.*—The time occupied by the earth in moving from the perihelion to the perihelion again, is called the *anomalistic year.* This period must be a little longer than the sidereal year, since the earth must describe a further arc of 11″.8 before reaching the perihelion; and the difference will be equal to the time necessary for the earth to describe 11″.8 of its orbit, or 4m. 35s., which gives 365d. 6h. 13m. 45s. for the length of the anomalistic year. This period is occasionally used in astronomical investigations, but mankind are generally more concerned in the *tropical* year, on which the return of the seasons depends.

CHAPTER VIII.

THE MOON—ITS MOTION—PHASES—TELESCOPIC APPEARANCE.

199. *Distance of the moon.*—The distance of the moon can be computed when we know its horizontal parallax. This parallax varies considerably during a revolution of the moon round the earth. The equatorial parallax, when least, is 53′ 48″, and when greatest, 61′ 32″. The mean horizontal parallax of the moon at the equator is 57′ 2″.3. Hence the mean distance will be found by the proportion

sin. 57′ 2″.3 : 3963.35 :: 1 : the moon's distance,

which is found to be 238,883 miles.

In the same manner, the moon's greatest distance is found to be 253,263 miles, and its least distance 221,436 miles.

200. *Diameter of the moon.*—The absolute diameter of the moon can be computed when we know its apparent diameter, and its

distance from the earth. The apparent diameter varies according to its distance from the earth. When nearest to us, it is 33′ 31″.1; but at its greatest distance it is only 29′ 21″.9. At its mean distance the apparent diameter is 31′ 7″.0. Hence the absolute diameter will be found by the proportion

1 : 238,883 :: sin. 15′ 33″.5 : the moon's semi-diameter,

which is found to be 1081.1 miles. Hence the moon's diameter is 2162 miles.

Since spheres are as the cubes of their diameters, the *volume* of the moon is $\frac{1}{49}$th that of the earth. Its *density* is about $\frac{3}{5}$ths (.615) the density of the earth, and its *mass* $(=\frac{1}{49}\times.615)$ is about $\frac{1}{80}$th of the mass of the earth.

201. *Definitions.*—A body is said to be in *conjunction* with the sun when its longitude is the same as that of the sun; it is said to be in *opposition* to the sun when their longitudes differ 180°; and to be in *quadrature* when their longitudes differ 90° or 270°. The term *syzygy* is used to denote either conjunction or opposition.

The *octants* are the four points midway between the syzygies and quadratures.

The two points in which the orbit of the moon or a planet is cut by the plane of the ecliptic are called *nodes.* That node at which the body passes from the south to the north side of the ecliptic is called the *ascending node*, and the other the *descending node.*

202. *Revolution of the moon.*—If the situations of the moon be observed on successive nights, it will be found that it changes its position among the stars, moving among them from west to east; that is, in a direction *contrary* to that of the diurnal motion. By this motion it makes a complete circuit of the heavens in about 27 days. Hence either the moon revolves round the earth, or the earth round the moon. Strictly speaking, the earth and moon both revolve about their common centre of gravity. This is a point in the line joining their centres, situated at an average distance of 2690 miles from the centre of the earth, or about 1270 miles beneath the surface of the earth.

203. *Sidereal and synodic revolutions.*—The interval of time oc-

cupied by the moon in performing one *sidereal* revolution round the earth, or the time which elapses between her leaving a fixed star until she again returns to it, is 27d. 7h. 43m. 11s.

The moon's mean daily motion is found by dividing 360° by the number of days in one revolution. The mean daily motion is thus found to be 13°.1764, or about $13\frac{1}{6}$ degrees.

The *synodical* revolution of the moon is the interval between two consecutive conjunctions or oppositions.

The synodical revolution of the moon is longer than the sidereal by 2d. 5h. 0m. 51s., which is the time required by that body to describe with its mean angular velocity of $13\frac{1}{6}$ degrees per day the arc traversed by the sun since the previous conjunction. Hence we find the duration of the synodical period to be 29d. 12h. 44m. 2s.

204. *How the synodical period is determined.*—The mean synodical period may be determined with great accuracy by observations of eclipses of the moon. The middle of an eclipse is very near the instant of opposition, and from the observations of the eclipse the exact time of opposition may be easily computed. Now eclipses have been very long observed, and the time of the occurrence of some has been recorded even before the Christian era. By comparing an eclipse observed by the Chaldeans, 720 B.C., with recent observations, the duration of the mean synodic period has been ascertained with great accuracy.

205. *How the sidereal period is derived from the synodical.*—The sidereal period may be deduced from the synodical as follows:

Let P = the length of the sidereal year,
p = the sidereal revolution of the moon,
T = the synodical period of the moon.

Then the arc which the moon describes in order to come into conjunction with the sun, exceeds 360° by the space which the sun has passed over since the preceding conjunction. This excess is found by the proportion

$$P : T :: 360° : \frac{360T}{P}.$$

Then, as the whole distance the moon must move from the sun to reach it again, is to one circumference, so is the time of describing the former, to the time of describing the latter; that is,

$$360+\frac{360\text{T}}{\text{P}}:360::\text{T}:p;$$

or

$$1+\frac{\text{T}}{\text{P}}:1::\text{T}:p.$$

Whence $$p=\frac{\text{PT}}{\text{P}+\text{T}}=\frac{365.25\times 29.53}{365.25+29.53}=27.32 \text{ days};$$

and this is the most accurate mode of determining the sidereal period of the moon.

206. *Moon's path.*—The moon's observed right ascension and declination enable us to determine her latitude and longitude. By observing the moon from day to day when she passes the meridian, we find that her path does not coincide with the ecliptic, but is inclined to it at an angle of 5° 8′ 48″, and intersects the ecliptic in two opposite points, which are called the moon's nodes.

207. *Form of the moon's orbit.*—It can be proved in a manner similar to that given for the sun, Arts. 111 and 114, that the moon in her orbit round the earth obeys the following laws:

1st. The moon's path is an ellipse, of which the earth occupies a focus.

2d. The radius vector of the moon describes equal areas in equal times.

The point in the moon's orbit nearest the earth is called her *perigee*, and the point farthest from the earth her *apogee*. The line joining the apogee and perigee is called the line of the *apsides*.

208. *Eccentricity of the moon's orbit.*—The eccentricity of the lunar orbit may be found by observing the greatest and least apparent diameters of the moon, in the same manner as was done in the case of the sun, Art. 113.

Example. In the month of October, 1862, the greatest apparent diameter of the moon was 33′ 0″.6, and the least was 29′ 34″.0. From these data determine the eccentricity of the lunar orbit during that month.

The ratio of A to P is 0.89569.

Hence, by the formula $$e=\frac{\text{A}-\text{P}}{\text{A}+\text{P}},$$

we find $$e=0.0550, \text{ or about } \tfrac{1}{18}.$$

209. *Interval of moon's transits.*—The moon's mean daily motion in right ascension is 13°.17, or 12°.19 greater than that of the sun. Hence, if on any given day we suppose the moon to be on the meridian at the same instant with the sun, on the next day she will not arrive at the meridian till 51m. after the sun; that is, the interval between two successive meridian passages of the moon is, on the average, 24h. 51m.

In consequence of the inequalities in the moon's motion in right ascension, this interval varies from 24h. 38m. to 25h. 6m.

210. *Moon's meridian altitude.*—The moon's altitude when it crosses the meridian is very variable. The meridian altitude of the sun at the summer solstice is 46° 54′ (twice the obliquity of the ecliptic) greater than it is at the winter solstice. Now, since the moon's orbit is inclined 5° 9′ to the plane of the ecliptic, the moon will sometimes be distant from the ecliptic by this quantity on the north side, and at other times by the same quantity on the south side; hence the greatest meridian altitude of the moon will exceed its least by 46° 54′+10° 18′, or 57° 12′. In latitude 41° 18′, the greatest meridian altitude of the moon is 77° 18′, and its least 20° 6′.

211. *The moon's phases.*—The different forms which the moon's visible disc presents during a synodic revolution are called *phases.*

The moon's phases are completely accounted for by assuming her to be an opaque globular body, rendered visible by reflecting light received from the sun.

Let E be the earth, and ABCDH the orbit of the moon, the sun being supposed to be at a great distance in the direction AS. When the moon is in conjunction at A, the enlightened half is turned directly from the earth, and she must then be invisible. It is then said to be *new moon.*

About 7½ days after new moon, when she is in quadrature at C, one half of her illumined surface is turned toward the earth, and her enlightened disc appears as a semicircle. She is then said to be in her *first quarter.*

About 15 days after new moon, when she is in opposition at F, the whole of her illumined surface is turned toward the earth, and she appears as a full circle of light. It is then said to be *full moon.*

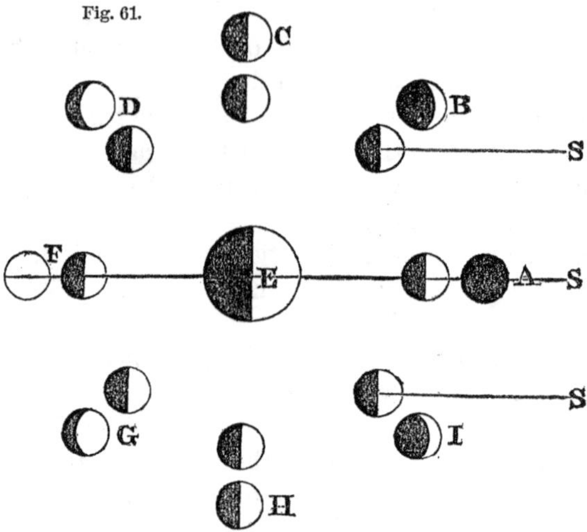

About 7½ days after full moon, when she is again in quadrature at H, one half of her illumined surface being turned toward the earth, she again appears as a semicircle. She is then said to be at her *last quarter*.

From new moon to first quarter, and from last quarter to new moon, her enlightened disc is called a *crescent*. This phase is represented at B and I. The two extremities of the crescent are called *cusps*, or *horns*. From first quarter to full moon, and from full moon to last quarter, the form of her enlightened disc is said to be *gibbous*. This phase is represented at D and G. These phases prove conclusively that the moon shines by light borrowed from the sun.

The interval from one new moon to the next new moon is called a *lunation*, or *lunar month*. It is evidently the same as a synodical revolution of the moon.

212. *Obscure part of the moon's disc.*—When the moon is just visible after new moon, the whole of her disc is quite perceptible, the part not fully illumined appearing with a faint light. As the moon advances, the obscure part becomes more and more faint, and it entirely disappears before full moon. This phenomenon depends on light reflected from the earth to the moon, and from the moon back to the earth.

When the moon is near to A, she receives light from nearly

the whole of the earth's illumined surface, and this light, being in part reflected back, renders visible that portion of the disc that is not directly illumined by the sun. As the moon advances toward opposition at F, the quantity of light she receives from the illumined surface of the earth decreases; and its effect in rendering the obscure part visible, is farther diminished by the increased light of the part which is directly illuminated by the sun's rays.

It is obvious that, to an observer at the moon, the earth must appear as a splendid moon, presenting all the phases of the moon as seen from the earth, and having more than three times its apparent diameter.

213. *Daily retardation of the moon's rising or setting.*—The average daily retardation of the moon's rising or setting is the same as that of her passage over the meridian; but the actual retardation, being affected by the moon's changes in declination, as well as by the inequalities of her motion in right ascension, is subject to greater variation. In the latitude of New York, the least daily retardation is 23 minutes, and the greatest is 1h. 17m.

214. *Harvest Moon.*—The less or greater retardation of the moon's rising attracts most attention when it occurs at the time of full moon. When the retardation has its least value near the time of full moon, the moon rises soon after sunset on several successive evenings; whereas, when the retardation is greatest, the moon ceases in two or three days to be seen in the early part of the evening.

When the moon is in that part of her orbit which makes the least angle with the horizon, 13 degrees of her orbit (which is her average progress in a day) rises above the horizon at New York in less than 30 minutes. This happens for the full moon near the time of the autumnal equinox. As this is about the period of the English harvest, this moon is hence called the *Harvest Moon.*

215. *Effect of altitude on the moon's apparent diameter.*—The apparent diameter of the moon is not the same at the same instant for all points of the earth, on account of their different distances from the moon. As the moon rises above the horizon (if we sup-

pose its distance from the centre of the earth to remain constant), its distance from the place of observation must diminish, while its altitude increases, and, consequently, its apparent diameter must increase. This effect attains its maximum when the moon is in the zenith of the spectator.

The distance AB is about equal to CB or CG, and exceeds AG by AC, the radius of the earth, which is about one sixtieth of the moon's distance. Hence the angle GAH, which the moon's radius subtends when in the zenith, exceeds the angle BAD, which the moon's radius subtends when in the horizon, by about one sixtieth of the whole quantity; that is, the augmentation of the moon's diameter on account of her apparent altitude may amount to more than half a minute.

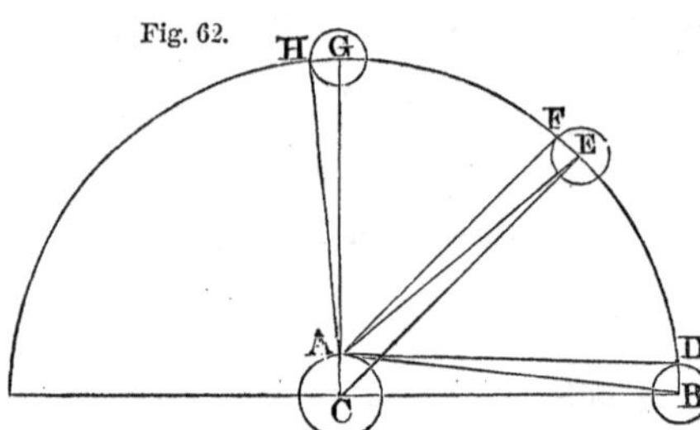

Fig. 62.

The apparent enlargement of the moon near the horizon is an optical illusion, as explained Art. 92.

216. *Has the moon an atmosphere?*—There is no considerable atmosphere surrounding the moon. This is proved by the absence of twilight. Upon the earth, twilight continues until the sun is 18° below the horizon; that is, day and night are separated by a belt 1200 miles in breadth, in which the transition from light to darkness is not sudden, but gradual—the light fading away into the darkness by imperceptible gradations. This twilight results from the refraction and reflection of light by our atmosphere; and if the moon had an atmosphere, we should notice, in like manner, a gradual transition from the bright to the dark portions of the moon's surface. Such, however, is not the case. The boundary between the light and darkness, though irregular, is perfectly well defined and sudden. Close to this boundary, the unillumined portion of the moon appears just as dark as any portion of the moon's unillumined surface.

217. *Argument from the absence of refraction.*—The absence of an atmosphere is also proved by the absence of refraction when the moon passes between us and the distant stars. Let AB represent

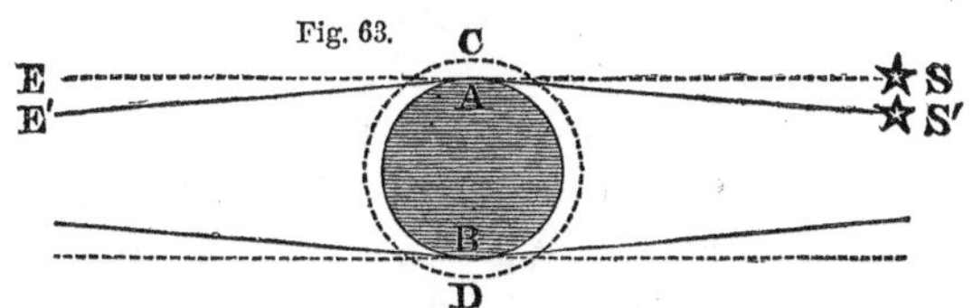

Fig. 63.

the disc of the moon, and CD an atmosphere supposed to surround it. Let SAE represent a straight line touching the moon at A, and proceeding toward the earth, and let S be a star situated in the direction of this line. If the moon had no atmosphere, this star would appear to touch the edge of the moon at A; but if the moon had an atmosphere, this atmosphere would refract light; and a star behind the edge of the moon in the position S′ would be visible at the earth, for the ray S′A would be bent by the atmosphere into the direction AE′. So, also, near the opposite limb of the moon, a star might be seen at the earth, although really behind the edge of the moon. Hence we see that if the moon had an atmosphere, the time during which a star would be concealed by the moon would be less than if it had no atmosphere; and the amount of this effect must be proportional to the density of the atmosphere.

Many thousand occultations of stars by the moon have been observed, and no appreciable effect of refraction has ever been detected. This species of observation is susceptible of such accuracy, that if the refraction amounted to 4″ of arc, it is believed that it could not fail to be detected in the mean of a large number of observations. Now the earth's atmosphere changes the direction of a ray of light more than half a degree when it enters the atmosphere, and the same when it leaves it, making a total deflection of over 4000″. Hence we conclude that if the moon have an atmosphere, its density can not exceed one thousandth part of the density of our own. Such an atmosphere is more rare than that which remains under the receiver of the best air-pump when it has reached its limit of exhaustion.

218. *Light of the full moon.*—The light received from the full moon was compared by Bouguer with the light received from the sun, by comparing each with the light of a candle. The light of the sun being admitted into a dark room through a small aperture, he placed in front of the operator a concave lens, to diminish the intensity of the sun's rays by causing them to diverge. He then placed a candle at such a distance that its light received

upon a screen was exactly equal to that of the sun received upon the same screen.

Repeating this experiment at night with the full moon, he compared the light of the moon with that of the candle. By several experiments of this kind, he arrived at the conclusion that the sun illumines the earth 300,000 times more than the full moon.

Professor G. P. Bond compared the light of the moon with that of the sun by placing in the sun's light a glass globe with a silvered surface, and comparing the brightness of the reflected image of the sun with an artificial light, and afterward comparing the light of the full moon with the same standard. He hence inferred that the light of the sun was 470,000 times that of the full moon.

219. *Heat of the moon.*—Until recently, the most delicate experiments had failed to detect any heat in the light of the moon. The light of the full moon has been collected into the focus of a concave mirror of such a magnitude as, if exposed to the sun's light, would have been sufficient to evaporate platinum; yet no sensible effect was produced upon the bulb of a differential thermometer so delicate as to show a change of temperature amounting to the 500th part of a degree. This experiment, if reliable, would indicate that the moon reflects a less proportion of the heating rays than of the luminous rays of the sun.

In 1846 Melloni repeated this experiment on the top of Mount Vesuvius with a lens of three feet diameter, and found feeble indications of heat when the light of the moon was concentrated upon a delicate thermo-multiplier.

In the summer of 1856, Professor Smyth repeated this experiment on the summit of Teneriffe, over 10,000 feet above the sea, and found that the heat of the full moon was equal to one third that of an ordinary candle placed at a distance of 15 feet.

Even this small amount of heat appears to be absorbed by the atmosphere before reaching the earth; and near the earth's surface, the moon's heat is inappreciable by the most delicate means of observation hitherto employed.

220. *Telescopic appearance of the moon.*—If with a telescope we examine the bounding line between the illumined and dark portions of the moon's surface, especially about the time of the first

quarter, we shall find it to be very broken and irregular. At some distance from the generally illumined surface we may notice bright spots, often entirely surrounded by a dark ground; and we also find dark spots entirely surrounded by an illumined surface. These appearances change sensibly in a few hours. As the light of the sun advances upon the moon, the dark spots become bright; and at full moon they all disappear, and we only notice that certain regions appear more *dusky* than others. The moon's surface is therefore uneven; and, by observing the passage of the sun's light over these spots, we may form a judgment of their dimensions and figure.

The most favorable time for observing these inequalities is near the first or third quarter, because then the shadows of the mountains appear of their greatest length, and are not shortened by being seen obliquely. See Plate II., Fig. 2, which gives a representation of a small portion of the moon's surface as seen through a powerful telescope.

221. *Particular phenomena described.*—Near the bounding line of the moon's illumined surface we frequently observe the following phenomena: A bright ring nearly circular; within it, on the side next the sun, a black circular segment; and without it, on the side opposite to the sun, a black region with a boundary more or less jagged. Near the centre of the circle we sometimes notice a bright spot, and a black stripe extending from it opposite to the sun. After a few hours, the black portions are found to have contracted in extent, and in a day or two entirely disappear.

After about two weeks these dark portions reappear, but on the side opposite to that on which they were before seen; and they increase in length until they pass entirely within the dark portion of the moon. These appearances can only be explained by admitting the existence of a circular wall, rising above the general level of the moon's surface, and inclosing a large basin, from the middle of which rises a conical peak.

222. *Height of the lunar mountains.*—If the distance of the illuminated summit of a mountain from the enlightened part of the disc be measured with a micrometer, and the positions of the sun and moon at the time be obtained by observation or computation, the height of the mountain may be computed.

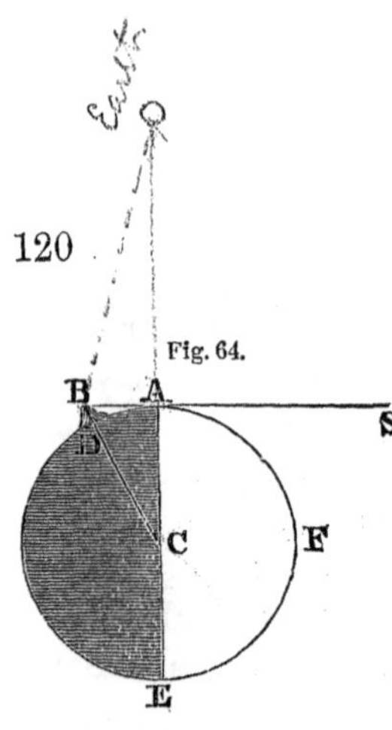

Fig. 64.

Let AFE be the illuminated hemisphere of the moon, SA a ray of the sun touching the moon at A, and let BD be a mountain so elevated that its summit just reaches to the ray SAB, and is illumined while the intervening space AB is dark. Suppose now the earth to be in the direction of the diameter AE produced. Let the angle which AB subtends at the earth be measured with a micrometer; then, since the distance of the moon from the earth is known, the absolute length of AB can be computed. Then, in the right-angled triangle ABC, AC, the radius of the moon, is known, whence BC can be computed; and subtracting AC from BC, gives BD, the height of the mountain.

If the earth is so situated that the line AB is not seen perpendicularly, since we know the relative positions of the sun and moon, we can determine the inclination at which AB is seen, and hence the absolute length of AB.

The height of a mountain may also be computed from the measured length of the shadow it casts.

The greatest elevation of any lunar mountain which has been observed is 23,800 feet. The altitudes of the higher mountains in the moon are probably as accurately known as those of the highest mountains on the earth.

223. *Circular craters.*—Mountain ranges, approaching nearly to the form of circles, are very common on the moon's surface. They sometimes have a diameter of over 50 miles, and a height of 2 or 3 miles. Tycho, Kepler, and Copernicus are among the most remarkable of these mountain ranges. See Plate II., Fig. 1. Tycho, No. 1, is near the moon's southern limb; Kepler, No. 2, near the eastern limb; and Copernicus, No. 3, a little west of Kepler. These circular mountains bear an obvious analogy to the volcanic craters upon the earth.

224. *The crater of Kilauea*, on one of the Sandwich Islands, is a vast basin, more than three miles in its longer diameter, and nearly 1000 feet deep. From the bottom of the basin rise numerous little cones, from which smoke is almost constantly emitted, and sometimes melted lava. The craters of most volcanoes exhibit

an irregular circular wall of considerable height, sometimes 2 or

Fig. 65.

3 miles, and within this wall rise one or two cones formed by the occasional overflowing of the lava.

225. *Lunar volcanoes compared with terrestrial.*—The lunar volcanoes differ from the terrestrial in their enormous dimensions and immense number. This may be due, in some degree, to the feeble attraction of the moon, since objects on the moon's surface weigh only *one sixth* what they would on the earth.

226. It is certain that most of the lunar volcanoes are *entirely extinct;* and it is doubted whether any signs of eruption have ever been noticed. The spot called Aristarchus, marked 4 on Fig. 1, Plate II., is so brilliant that some have concluded it to be an active volcano. Herschel observed on the dark portion of the moon three bright points, which he ascribed to volcanic fires; but the same lights may be seen every month, and they are probably to be ascribed to mountain peaks which have an unusual power of reflecting the feeble light which is emitted by the earth. It is believed that all the inequalities of brightness observed on the moon's surface (with the exception of the shadows described in Arts. 220–1) result from a difference in the nature of the reflecting materials. Two distinguished astronomers, Beer and Mädler, who have studied the moon's surface with greater care than any one else, assert that they have never seen any thing that could au-

thorize the conclusion that there are in the moon volcanoes now in a state of ignition.

227. *Streaks of light from Tycho.*—Very remarkable streaks of light are seen diverging from several of the lunar craters. These are quite conspicuous about Tycho, Kepler, Copernicus, and Aristarchus. One of these streaks of light diverging from Tycho can be traced 1700 miles. These streaks cross ridges and valleys without interruption; and some of them have been noticed to cast shadows. They are thought to have resulted from some violent volcanic eruption, by which enormous crevices were opened in the moon's surface. These crevices are supposed to have been filled with melted lava, which congealed into a glassy rock, having a more brilliant reflecting surface than the general disc of the moon. Similar phenomena, but upon a far less extensive scale, have taken place on the earth's surface.

228. There is *no water* on the moon's surface. The dusky regions, which were once supposed to be seas, are regions comparatively level; but upon which, with a good telescope, we can detect black shadows, indicating the existence of permanent inequalities, which could not exist on a fluid surface. Moreover, if there were any water on the moon's surface, a portion of it would rise in vapor, and form an atmosphere which would refract light to an extent far beyond what we actually observe.

229. *Can volcanoes exist without air or water?*—It may be objected that volcanoes could not exist without air or water. It is not certain that the presence of air is necessary to the activity of a volcano. Volcanoes may be ascribed to the primitive heat of the globe, or to galvanic action on a large scale. A commotion of the melted lava would be instantly produced by the introduction of water, which would suddenly generate large quantities of steam; and it might also be produced by the presence of various other bodies; as, for example, sulphur, which almost invariably accompanies volcanic eruptions. Some similar substance might cause an eruption of a lunar volcano without the agency of water.

230. *Can animal life exist upon the moon?*—Air and water are necessary to the support of both animal and vegetable life. It is

doubtful, therefore, whether even the humblest form of life with which we are acquainted could exist upon the moon. Nothing has ever been discovered upon the moon's surface to indicate the agency of human beings, or the presence of any form of animal or vegetable life. The extremes of temperature upon the moon's surface must be far more violent than they are upon the earth. For 14 successive days the sun shines uninterruptedly upon the same portion of the moon, and for the next 14 days his light is entirely withdrawn. During the first period, the moon's surface must become intensely heated; and during the next fortnight the cold must be equally severe, since there is no atmosphere or clouds to obstruct the radiation of heat.

While, then, we are compelled to say that Infinite wisdom and power can create beings to live in such a world, we can safely assert that no varieties of animal or vegetable life with which we are acquainted can exist in the moon.

231. *Does the moon influence the weather?*—The impression is almost universal (especially among uneducated men) that the moon exerts a sensible influence upon the weather. Various and even opposite effects have been ascribed to the moon's action; but most people are confident that the moon exerts a powerful influence of some sort. One opinion, which has been defended even by some men of science, is that the full moon has a decided influence in dissipating the clouds. But, from a comparison of seven years' observations at Greenwich, I have found that at full moon the average cloudiness of the sky is precisely the same as at new moon.

It has also been claimed that the amount of rain is affected by the moon's age; and some observations appear to indicate that from the first quarter to the full, the amount of rain exceeds the average; but the results obtained at different stations are diverse and often contradictory, indicating that these differences are due mainly to other causes than the moon.

232. *Does the moon influence the pressure of the air?*—Many have imagined that inasmuch as the moon elevates the water of the ocean, its disturbing influence ought to be much greater upon a fluid of such mobility as our atmosphere. The moon does indeed influence the pressure of the air, but its disturbing force is extremely small. At Singapore, under the equator, when the moon

is on the meridian, the barometer is higher by $\frac{6}{1000}$th of an inch than when the moon is six hours from the meridian; at St. Helena, in Lat. 15° 55′, this difference amounts to $\frac{4}{1000}$th of an inch; and in our latitude the difference should be still less. This effect is so minute that it can only be detected by the most accurate observations, continued for a period of several years. Indeed, it has never been shown that the moon exerts any influence upon the weather, except that which is of the feeblest kind, and which is only appreciable after a very long series of the best observations.

233. *Moon's rotation upon an axis.*—The various spots on the moon always occupy nearly the same positions upon the disc, from which it follows that nearly the same surface is always turned toward the earth. Hence we conclude that the moon rotates upon an axis in the same time that she makes a revolution in her orbit. If the moon had no motion of rotation, then in opposite parts of her orbit she would present opposite sides to the earth. In order that a globe which revolves in a circle around a centre should turn continually the same hemisphere toward that centre, it is necessary that it should make one rotation upon its axis in the time it takes to revolve about the centre.

234. *Librations of the moon.*—Although it is true that nearly the same hemisphere of the moon is always turned toward the earth, yet the moon has apparently a slight oscillatory motion, which allows us to see a portion of the opposite hemisphere. This oscillatory motion is called *libration.*

Libration in longitude.—While the moon's angular velocity on its axis is rigorously uniform throughout the month, its angular velocity in its orbit is not uniform, being most rapid when nearest the earth. Hence we see at one time a little more of the eastern or western edge of the moon than we do at another time. This is called the libration in longitude.

Libration in latitude.—The axis of the moon is not quite perpendicular to the plane of her orbit, but makes an angle with it of 83½ degrees. On account of this inclination, the northern and southern poles of the moon incline alternately 6½° to and from the earth. When the north pole leans toward the earth, we see a little more of that region, and a little less when it leans the contrary way. This variation is called the libration in latitude.

Diurnal libration.—By the diurnal motion of the earth, we are carried with it round its axis; and if the moon presented exactly the same hemisphere toward the earth's centre, the hemisphere visible to us when the moon rises, would be different from that which would be visible to us when the moon sets. This is another cause of a variation in the edges of the moon's disc, and is called the diurnal libration.

In consequence of all these librations, we can see somewhat more than half of the surface of the moon; yet there remains about $\frac{3}{7}$ths of its surface which is always hidden from our view.

235. *Lunar day.*—The rotation of the moon upon its axis, being equal to that of its revolution in its orbit, is $27\frac{1}{4}$ days. The intervals of light and darkness to the inhabitants of the moon, if there were any, would be altogether different from those upon the earth. There would be about 328 hours of continued light, alternating with 328 hours of continued darkness. The heavens would be perpetually serene and cloudless. The stars and planets would shine with extraordinary splendor as well in the day as in the night. The inclination of her axis being small, there would be no sensible change of seasons. The inhabitants of one hemisphere could never see the earth; while the inhabitants of the other would have it constantly in their firmament by day and by night, and always nearly in the same position. To those who inhabit the central part of the hemisphere presented to us, the earth would appear stationary in the zenith, with the exception of the small effect due to libration.

The earth illumined by the sun would appear as the moon does to us, but with a superficial magnitude about fourteen times as great. Its phases would also be similar to those which we see in the moon.

236. *Equality of the periods of rotation and revolution.*—That the moon should rotate on an axis in exactly the same time that is required for a revolution around the earth, can not be supposed to be accidental.

We are forced, then, to seek for some physical cause to explain this coincidence. If we admit that originally these two motions were *nearly* equal, the *exact* equality may be explained as follows: The moon, like the earth, was probably once in a plastic condition. The earth would then act upon the moon as the moon acts upon the earth in raising the tides, only with much greater power; that is, it would give the moon an elongated figure, its major axis pointing toward the centre of the earth. If the moon has such an elongated figure, the earth must act upon it as upon a pendulum. When a pendulum is deflected from the vertical position, the earth's attraction brings it back again, causing it to oscillate to and fro. So, also, if the longer axis of the moon were deflected from pointing toward the earth, the earth's attraction would tend to bring it back to this position, thus tending to establish a rigorous equality between the times of rotation and revolution of the moon.

237. *Position of the moon's centre of gravity.*—From a careful study of the moon's motions, Hansen concludes that the centre of gravity of the moon does not coincide with its centre of figure, and that the centre of figure is nearer to us by 33 miles than the centre of gravity; in other words, the hemisphere which is turned toward the earth is lighter than the opposite hemisphere, and may be regarded as an enormous mountain, rising 33 miles above the mean level of the moon. This lightness may be the result of volcanic energy, upheaving the crust, and leaving large cavities beneath; and these cavities must be mainly on the side of the moon which is turned toward the earth. This cause may have contributed to produce that elongated figure of the moon which enables us to explain the exact equality between the time of rotation upon its axis and of revolution about the earth. This conclusion of Hansen is not accepted by all astronomers.

238. *Path of the moon in its motion about the sun.*—While the moon revolves about the earth, it also accompanies the earth in its motion about the sun. The actual path described by the moon will then be an undulating line, alternately within and without the orbit of the earth. The undulations are, however, so small, in comparison with the dimensions of the earth's orbit, that the path of the moon is always concave toward the sun. The distance, AB, passed over by the earth in a fortnight, is about 24 millions of miles. If we draw a chord connecting these points, this chord, at its middle point, will fall about 700,000 miles within the orbit of the earth, while the greatest distance of the moon from

Fig. 66.

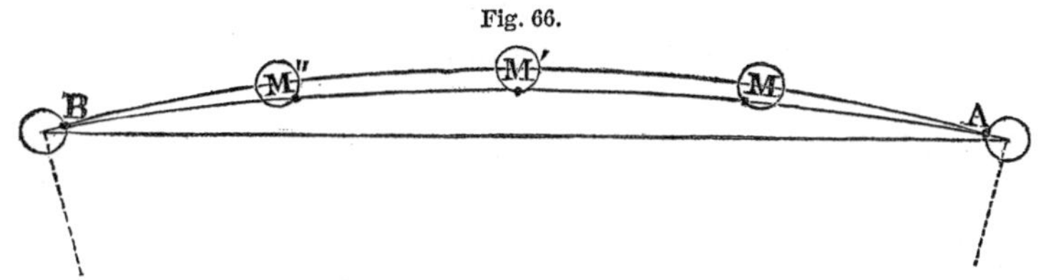

the earth is only 253,000 miles. The moon's path, therefore, approaches so near to that of the earth as to be always concave toward the sun.

239. *Changes of the moon's orbit.*—The elliptic path described by the moon, changes gradually from month to month both in form and position. Its eccentricity varies within certain limits, being sometimes as great as 0.065, and sometimes as small as 0.049. Its mean value is 0.05484, or about $\frac{1}{18}$th.

The major axis of the moon's orbit is not fixed, but has a direct motion on the ecliptic at the rate of about 41° in a year, accomplishing a complete revolution in a little less than nine years; so that in $4\frac{1}{2}$ years the perigee arrives where the apogee was before. This motion of the line of the apsides is not equable throughout the whole of a lunar month; for when the moon is in syzygies, the line of apsides advances in the order of the signs, but is retrograde in quadratures. The direct motion is, however, greater than the retrograde.

240. *Motion of the line of the nodes.*—The line in which the plane of the moon's orbit cuts the ecliptic, is called the *line of the nodes.* The position of the nodes is found by observing the longitude of the moon when she has no latitude; and it appears, by a comparison of such observations, that the line of the nodes is not fixed, but has a slow retrograde motion at the rate of about 19° in a year. By this motion the nodes make a mean tropical revolution in 18 years and 224 days, nearly. It is not, however, an equable motion throughout the whole of the moon's revolution. The node is generally stationary when the moon is in quadrature, or in the ecliptic; in all other parts of the orbit it has a retrograde motion, which is greater the nearer the moon is to the syzygies, or the greater the distance from the ecliptic.

Thus we see that the path of the moon does not return into itself, but is a curve of the most complicated kind, whose form and position are both in a state of continual change.

241. *The lunar cycle.*—The lunar cycle consists of 235 synodical revolutions of the moon, which differ from 19 years of 365¼ days only by about an hour and a half.

For $29.5305887 \times 235 = 6939.688$ days.

And $365\frac{1}{4} \times 19 = 6939.75$ days.

If, then, full moon should happen on the 1st of January in the first year of the cycle, it will happen on that day (or within a very short time of its beginning or ending) again after a lapse of 19 years; and all the full moons in the interval will occur on the same days of the month as in the preceding cycle. This period of 19 years is sometimes called the *Metonic Cycle*, and the year of the Metonic cycle is called the *Golden Number*. This cycle of 19 years is used for finding *Easter*. Easter day is the first Sunday after the full moon which happens upon or next after the 21st day of March. The present lunar cycle began in 1862, when full moon occurred April 14th. Full moon also occurred on the same day of April in 1843, 1824, etc.

The following are the dates of the full moons next following the vernal equinox for several lunar cycles:

Year.	Year.	Year.	Year.	Full Moon.
1805	1824	1843	1862	April 14
1806	1825	1844	1863	April 3
1807	1826	1845	1864	March 22
etc.	etc.	etc.	etc.	

CHAPTER IX.

CENTRAL FORCES.—LAW OF GRAVITATION.—LUNAR IRREGULARITIES.

242. *Curvilinear motion.*—If a body at rest receive an impulse in any direction, it will, if entirely at liberty to obey that impulse, move in that direction, and with a uniform rate of motion. When a body moves in a curve line, there must then be some force which at every instant deflects it from the rectilinear course it tends to pursue in virtue of its inertia. We may then consider this motion in a curve line to arise from two forces: one a primitive im-

pulse given to the body, which alone would have caused it to describe a straight line; the other a deflecting force, which continually urges the body toward some point out of the original line of motion.

243. *Kepler's laws.*—Before Newton's discovery of the law of universal gravitation, the paths in which the planets revolve about the sun had been ascertained by observation; and the following laws, discovered by Kepler, and afterward called *Kepler's laws,* were known to be true:

1st. *The radius vector of every planet describes about the sun equal areas in equal times.*

2d. *The path of every planet is an ellipse, having the sun in one of its foci.*

3d. *The squares of the times of revolution are as the cubes of the mean distances from the sun, or as the cubes of the major axes of the orbits.*

From these facts, revealed by observation, we may deduce the law of attractive force upon which they depend.

244. *Theorem.*—*When a body moves in a curve, acted on by a force tending to a fixed point, the areas which it describes by radii drawn to the centre of force are in a constant plane, and are proportional to the times.*

Let S be the centre of attraction; let the time be divided into short and equal portions, and in the first portion let the body describe AB. In the second portion of time, if no new force were to act upon the body, it would proceed to *c* in the same straight line, describing B*c* equal to AB. But when the body has arrived at B, let a force tending to the centre S act on it by a single instantaneous impulse, and

Fig. 67.

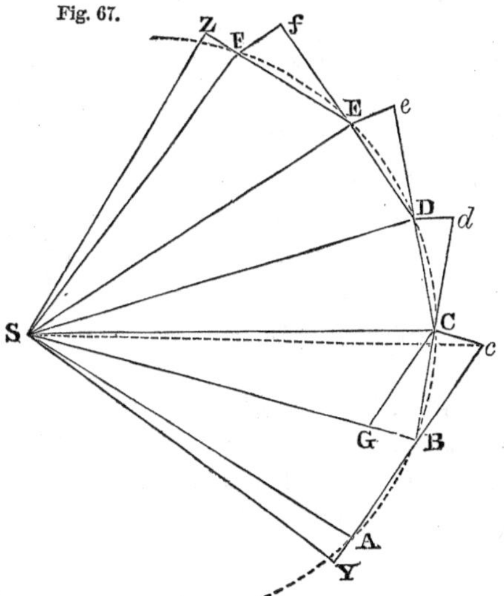

compel the body to continue its motion along the line BC. Draw C*c* parallel to BS, and at the end of the second portion of time, the body will be found in C, in the same plane with the triangle ASB. Join SC; and because SB and C*c* are parallel, the triangle SBC will be equal to the triangle SB*c*, and therefore also to the triangle SAB, because B*c* is equal to BA.

In like manner, if a centripetal force toward S act impulsively at C, D, E, etc., at the end of equal successive portions of time, causing the body to describe the straight lines CD, DE, EF, etc., these lines will all lie in the same plane, and the triangles SCD, SDE, SEF will each be equal to SAB and SBC. Therefore these triangles will be described in equal times, and will be in a con stant plane; and we shall have

polygon SADS : polygon SAFS :: time in AD : time in AF.

Let now the number of the portions of time in AD, AF be augmented, and their magnitude be diminished *in infinitum*, the perimeter ABCDEF ultimately becomes a curve line, and the force which acted impulsively at B, C, D, E, etc., becomes a force which acts continually at all points. Therefore, in this case also, we have

curvilinear area SADS : curvilinear area SAFS
:: time in AD : time in AF.

245. *Theorem.—The velocity of a body moving in a curve and attracted to a fixed centre, is inversely as the perpendicular from the fixed centre upon the tangent to the curve.*

For the velocities in the polygon at two points, A, E, are as AB, EF, because these lines are described in equal portions of time. But if SY, SZ be drawn perpendicular to these lines, SY.AB = SZ.EF, because the triangles SAB, SEF are equal. Therefore velocity at A : velocity at E :: SZ : SY.

And ultimately, the velocity in the polygon becomes the velocity in the curve, and the lines AY, EZ are the tangents to the curve at A and E.

246. *Theorem.—If a body moves in a curve line in a constant plane, and by a radius drawn to a fixed point, describes areas about that point proportional to the times, it is urged by a central force tending to that point.*

Every body which moves in a curve line is deflected from a straight line by some force acting upon it. If the body were to describe the polygon ABCDEF, describing the equal triangles SAB, SBC, etc., in equal times, it must at B be acted on by a force directed toward S. For in AB produced, take B*c* equal to AB. Then the triangle ASB=BS*c*. But, by supposition, ASB=BSC. Therefore, BSC=BS*c*; and, consequently, C*c* is parallel to SB. Now BC may be regarded as the resultant of two forces, one the impulse in the direction of AB produced, and the other a deflecting force C*c*, which is parallel to SB; that is, the deflecting force at B is directed toward the sun. But ultimately the motion in the polygon will coincide with the motion in the curve, and the force in the polygon will be the same as the force in the curve. Therefore in the curvilinear motion the proposition is true.

Now since the planets describe about the sun equal areas in equal times, it follows that the force which deflects them from a straight line is directed toward the centre of the sun.

247. *Theorem.—When bodies describe different circles with uniform motions, the forces tend to the centres of the circles, and are as the squares of the velocities divided by the radii of the circles.*

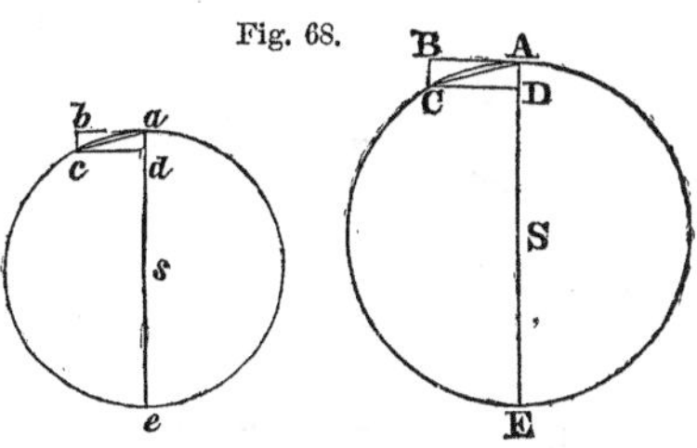

Fig. 68.

By Art. 246 the forces tend to the centres of the circles. Let AC, *ac* be arcs described in two different circles in equal times. Draw the tangents AB, *ab*; draw BC, *bc* perpendicular to the tangents, and CD, *cd* parallel to them. Draw also the chords AC, *ac*. Then BC, *bc*, or AD, *ad*, are the spaces through which the bodies are deflected from the tangents by the action of the forces to S and *s*. Then

$$AD : AC :: AC : AE; \text{ whence } AD=\frac{AC^2}{2AS}.$$

Also

$$ad=\frac{ac^2}{2as}.$$

Now when the arc is taken indefinitely small, we shall have the centripetal force at A : centripetal force at *a* :: the square of the arc AC divided by the radius AS : the square of the arc *ac* divided by the radius *as*. But the arcs AC, *ac*, described in equal times, are as the velocities; hence in circles, if F represent the centripetal force, V the velocity, and R the radius of the circle, we shall have $$F : f :: \frac{V^2}{R} : \frac{v^2}{r},$$

or $$F \text{ varies as } \frac{V^2}{R}.$$

248. *Theorem.—When bodies describe different circles with uniform motions, the central force is as the radius of the circle divided by the square of the time of one revolution.*

Let R be the radius of the circle, V the velocity, and T the time of describing the whole circle. The circumference of the circle will be represented by $2\pi R$, which equals VT; hence $V = \frac{2\pi R}{T}$.

But by Art. 247, F varies as $\frac{V^2}{R}$, or $\frac{4\pi^2R^2}{T^2R}$, which varies as $\frac{R}{T^2}$;

that is, $$F : f :: \frac{R}{T^2} : \frac{r}{t^2}.$$

249. *Theorem.—If a body describes an ellipse, being continually urged by a force directed toward the focus, that force must vary inversely as the square of the distance.*

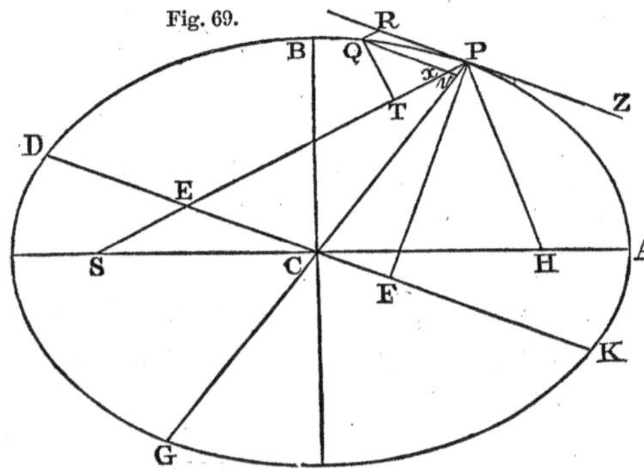

Fig. 69.

Let APB represent the elliptic orbit of a planet, and S the focus occupied by the sun. Let PQ be an arc described by the planet in an indefinitely short time, t. Draw the diameter PG; also the ordinate Qv parallel to the tangent at P; and let DK be the diameter which is conjugate to PG. Draw the radius vector SP, cutting the diameter DK in E, and the ordinate Qv in x, and complete the parallelogram $QxPR$. Also draw QT perpendicular to SP, and PF perpendicular to DK.

If the arc PQ be taken indefinitely small, it may be considered as a straight line described by the joint action of the force which is directed toward S, and of the projectile force which acts in the direction of PR. That is, the force PQ may be resolved into the two forces Qx and Px. *During the time t, the deflecting force, if it acted alone, would cause the body to describe* Px. Hence, denoting the intensity of this force by F, we have, by Mechanics,

$$Px = \tfrac{1}{2}Ft^2;$$

and taking t for the unit of time, we have

$$F = 2Px.$$

By similar triangles, $Px : Pv :: PE : PC$.

By Ellipse, Prop. XIX., $Gv . Pv : Qv^2 :: PC^2 : CD^2$.

Compounding these two proportions, we have

$$Gv . Px : Qv^2 :: PC . AC : CD^2,$$

since $PE = AC$, Geom., Ellipse, Prop. VII.

But when the arc PQ is taken indefinitely small, $Qv = Qx$, and $Gv = 2PC$.

Hence $$Qx^2 = \frac{Gv . Px . CD^2}{PC . AC} = \frac{2Px . CD^2}{AC}. \quad (1)$$

Again, by similar triangles,

$$Qx : QT :: PE(=CA) : PF.$$

Also (Ellipse, Prop. XVI.),

$$CD . PF = CA . CB, \text{ or } CA : PF :: CD : CB.$$

Hence $$Qx : QT :: CD : CB;$$

$$\therefore Qx = \frac{QT . CD}{CB}, \text{ and } Qx^2 = \frac{QT^2 . CD^2}{CB^2}. \quad (2)$$

From equations (1) and (2) we have

$$\frac{2Px . CD^2}{AC} = \frac{QT^2 . CD^2}{CB^2}.$$

Represent half the major axis of the ellipse by a, and half the minor axis by b; then

$$2Px = \frac{QT^2 . AC}{CB^2} = \frac{QT^2 . a}{b^2}. \quad (3)$$

If now we denote the area of the elliptical sector SQP by k, we have $$k = \tfrac{1}{2}SP . QT;$$

and hence $$QT = \frac{2k}{SP}, \text{ and } QT^2 = \frac{4k^2}{SP^2}.$$

Substituting this value in equation (3), we have

$$F = 2Px = \frac{4k^2}{SP^2} \cdot \frac{a}{b^2}.$$

If we consider the action of the deflecting force at some other point of the ellipse, as P′, and denote the intensity of the force by F′, we shall have $$F' = \frac{4k^2}{SP'^2} \cdot \frac{a}{b^2}.$$

But by Kepler's first law, k is a constant quantity; hence we have

$$F : F' :: SP'^2 : SP^2;$$

or the deflecting force varies inversely as the square of the distance of the planet from the sun.

250. *Theorem.—When several bodies revolve in ellipses about the same centre of force, varying inversely as the square of the distance, the squares of the periodic times will vary as the cubes of the major axes.*

Let T denote the periodic time of a planet, expressed in seconds, and k the area described by the radius vector in one second; then the entire area of the ellipse will be represented by Tk. But this area is also represented by πab (Ellipse, Prop. XXI.). Hence

$$Tk = \pi ab, \text{ or } k = \frac{\pi ab}{T}.$$

Represent the distance of the planet from the sun by R; then, by the last article,

$$F = \frac{4\pi^2 a^2 b^2}{T^2} \cdot \frac{a}{b^2 R^2} = \frac{4\pi^2 a^3}{T^2 R^2}.$$

If we represent by f the value of the deflecting force F at the distance of unity, then, by hypothesis,

$$f : F :: \frac{1}{1^2} : \frac{1}{R^2}; \text{ that is, } F = \frac{f}{R^2}.$$

Hence
$$\frac{f}{R^2} = \frac{4\pi^2 a^3}{T^2 R^2};$$

that is
$$T^2 = \frac{4\pi^2 a^3}{f}, \text{ or } T = \frac{2\pi a^{\frac{3}{2}}}{\sqrt{f}}.$$

If T′ denote the periodic time of a second planet, and a' half the major axis of its orbit, we shall have

$$T' = \frac{2\pi a'^{\frac{3}{2}}}{\sqrt{f}};$$

whence
$$T : T' :: a^{\frac{3}{2}} : a'^{\frac{3}{2}}, \text{ or } T^2 : T'^2 :: a^3 : a'^3.$$

Thus we perceive that Kepler's first law would hold true, whatever might be the law by which the deflecting force depended upon the distance; but the second and third laws prove that in the solar system this deflecting force varies inversely as the square of the distance.

251. *Modification of Kepler's third law.*—Kepler's third law is strictly true only in the case of planets whose quantity of matter is inappreciable in comparison with that of the central body. In considering the motion of a planet, for instance Jupiter, round the sun, it is necessary to remember that while the sun attracts Jupiter, Jupiter also attracts the sun. The motion which the attraction of Jupiter produces in the sun, is less than the motion which the attraction of the sun produces in Jupiter, in the same ratio in which Jupiter is smaller than the sun. If the sun and Jupiter were allowed to approach one another, their rate of approach would be the sum of the motions of the sun and Jupiter, and would therefore be greater than their rate of approach if the sun were not movable, in the same ratio in which the sum of the masses of the sun and Jupiter is greater than the sun's mass. Consequently, in comparing the orbits described by different planets round the sun, we must suppose the central force to be the attraction of a mass equal to the sum of the sun and planet.

If we regard the mass of the sun as unity, and represent the masses of two planets by m and m', then we shall have

$$T^2 : T'^2 :: \frac{a^3}{1+m} : \frac{a'^3}{1+m'};$$

and this proportion is rigorously true.

252. *The force that retains the moon in her orbit is the same as that which causes bodies to fall near the earth's surface, the force being diminished in proportion to the square of the distance from the earth's centre.*

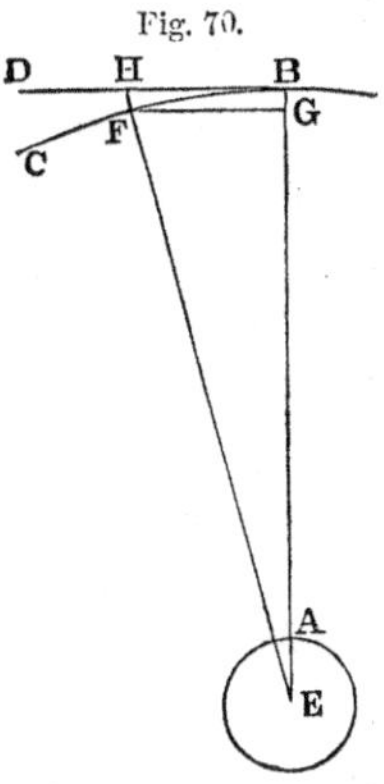

Fig. 70.

Let E be the centre of the earth, A a point on its surface, and BC a part of the moon's orbit assumed to be circular. When the moon is at any point, B, in her orbit, she would move on in the direction of the line BD, a tangent to the orbit at B, if she was not acted upon by some deflecting force. Let F be her place in her orbit one second of time after she was at B, and let FG be drawn parallel to BD, and FH parallel to EB. The line FH, or its equal BG, is the distance the moon has been drawn, during one second, from the tangent toward the earth at E. If we divide the circumference of the moon's orbit by the number of seconds in the time of one revolution, we shall have the length of the arc BF. Now, by Geometry,

2BE : BF :: BF : BG.

But the chord BF does not differ sensibly from the arc BF, already obtained. BG is thus found, by computation, to be 0.0534 inch.

At the equator, a body falls through 192½ inches in the first second. At the distance of the moon, the force of gravity (if it diminishes in proportion to the square of the distance from the earth's centre) will be found by the proportion

$59.964^2 : 1^2 :: 192\frac{1}{2} : 0.0535$ inch,

which agrees very nearly with the distance above computed.

The space through which the moon actually falls toward the earth in one second is a little less than that computed from the force of gravity at the earth's surface, because (as we shall see hereafter) the action of the sun diminishes by a small quantity the moon's gravity toward the earth.

253. *Deductions from Kepler's laws.*—We are thus led, by the laws of Kepler, to consider the centre of the sun as the focus of an attractive force, which extends infinitely in every direction, decreasing in the ratio of the square of the distance. The law of the proportionality of the areas described by the radius vector to the times of description, shows that the principal force acting on the planets and comets is always directed toward the centre of the sun. The ellipticity of the planetary orbits, and the almost parabolic orbits of the comets, prove that, for each planet and comet, this force is inversely proportional to the square of the distance of the body from the sun; and from the law of the proportionality of the squares of the times of revolution to the cubes of the major axes of the orbits, it follows that this force is the *same* for all the planets and comets, placed at equal distances from the sun; so that, in this case, these bodies fall toward it with the same velocity.

254. *Motion of the satellites.*—If from the planets we pass to the satellites, we shall find that as the laws of Kepler are very nearly observed in the motions of the satellites about their primary planets, they ought to gravitate toward the centres of these planets in the inverse ratio of the square of their distances from those cen-

tres; the satellites ought likewise to gravitate toward the sun in nearly the same manner as their planets, in order that the relative motions about their primary planets may be very nearly the same as if these planets were at rest. For each system of satellites, the squares of the times of their revolutions are as the cubes of their mean distances from the centre of the planet. The satellites are, therefore, attracted toward the planets, and toward the sun, by forces inversely proportional to the squares of the distances.

255. *The law of gravitation extends to all the bodies of the solar system.*—Hence it follows that the sun, and the planets which have satellites, are endowed with an attractive force, extending indefinitely, decreasing inversely as the square of the distance, and including all bodies in the sphere of their activity. Moreover, since a body can not act on another without experiencing an equal and contrary reaction, and since the planets and comets are attracted toward the sun, they must, in like manner, attract that body. For the same reason, the satellites attract their planets; this attractive property is therefore common to the planets, comets, and satellites; and, consequently, we may consider the gravitation of the heavenly bodies toward each other as a general law of the universe. The law of gravitation in the inverse ratio of the square of the distance, represents with the greatest precision all the known inequalities of the motions of the heavenly bodies; and this accordance, taken in connection with the simplicity of the law, authorizes the belief that it is rigorously the law of nature.

256. *Gravitation is also proportional to the masses;* for if the planets and comets are supposed to be at equal distances from the sun, they would fall freely toward it through equal spaces in equal times; consequently, their gravities would be proportional to their masses. The motions of the satellites about their primary planets prove that the satellites gravitate, like the planets, toward the sun in the ratio of their masses. Hence we see that the comets, planets, and satellites, placed at the same distance from the sun, would gravitate toward it in the ratio of their masses; and since action and reaction are equal and contrary, it follows that they attract the sun in the same ratio; consequently, their actions on the sun are proportional to their masses, divided by the square of their distances from its centre.

This universal gravitation is the cause of various perturbations of the motions of the heavenly bodies. For the planets and comets in obeying their mutual attractions must vary a little from the elliptical motion which they would exactly follow if they were attracted only by the sun. The satellites, disturbed in their motions about their planets by their mutual attractions, and by that of the sun, vary also from these laws.

257. *The heavenly bodies all move in conic sections.*—It was demonstrated by Newton that if a body (a planet, for instance) is impelled by a projectile force, and is continually attracted toward the sun's centre by a force varying inversely as the square of the distance, and no other forces act upon the body, the body will move in one of the following curves—a circle, an ellipse, a parabola, or an hyperbola; that is, it will move in one of the *conic sections.* The form of the orbit will depend upon the direction and intensity of the projectile force.

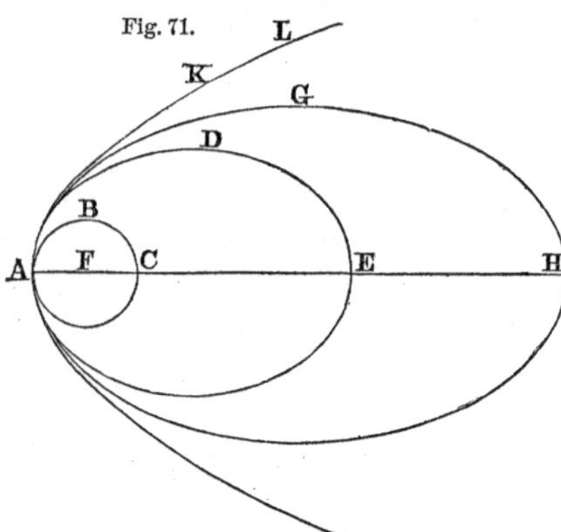

Fig. 71.

If we conceive F to be the centre of an attractive force, and a body at A to be projected in a direction at right angles to the line AF, then there is a certain velocity of projection which would cause the body to describe the circle ABC; a greater velocity would cause it to describe the ellipse ADE, or the more eccentric ellipse AGH; and if the velocity of projection be sufficient, the body will describe the semi-parabola AKL. If the velocity of projection be still greater, the body will describe an hyperbola. The curve can not be a circle unless the body be projected in a direction perpendicular to AF, and, moreover, unless the velocity with which the planet is projected is neither greater nor less than one particular velocity, determined by the length of FA and the mass of the central body. If it differs little from this particular velocity (either greater or less), the body will move in an ellipse; but if it is much greater, the body will move in a parabola or an hyperbola.

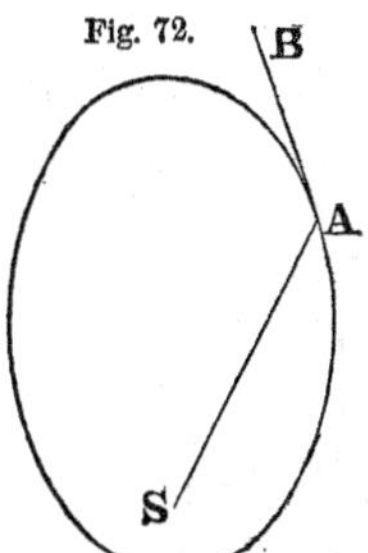

If the body be projected in a direction AB oblique to SA, and the velocity of projection is small, the body will move in an ellipse; but if the velocity is great, it may move in a parabola or hyperbola, but not in a circle.

If a body describe a circle, the sun is in the centre of the circle. If the body describe an ellipse, the sun is not in the centre of the ellipse, but in one focus. If the body describe a parabola or an hyperbola, the sun is in the focus. The planets describe ellipses which differ little from circles. A few of the comets describe very long ellipses; and nearly all the others that have been observed are found to move in curves which can not be distinguished from parabolas. There is reason to think that two or three comets which have been observed move in hyperbolas.

258. *Motions of projectiles.*—The motions of projectiles are governed by the same laws as the motions of the planets. If a body be projected in a horizontal direction from the top of a mountain, it is deflected by the attraction of the earth from the rectilinear path which it would otherwise have pursued, and made to describe a curve line which at length brings it to the earth's surface; and the greater the velocity of projection, the farther it will go before it reaches the earth's surface. We may therefore suppose the velocity to be so increased that it shall pass entirely round the earth without touching it.

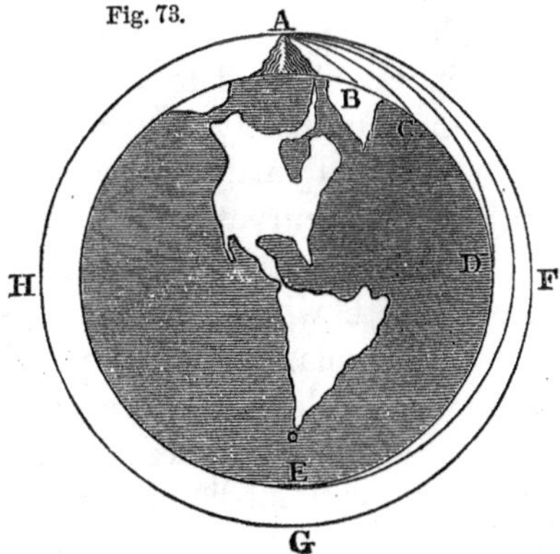

Let BCD represent the surface of the earth; AB, AC, AD the curve lines which a body would describe if projected horizontally from the top of a high mountain, with successively greater and greater velocities. Supposing there were no air to offer resistance, and the velocity were sufficiently great, the body would pass entirely round the earth, and return to the point from which it was projected.

259. *Time of revolution near the earth's surface.*—By means of Kepler's third law, we are able to compute the time required to complete a revolution in such an orbit near the earth's surface. We may regard such a body as a satellite revolving round the earth's centre in an orbit whose radius is equal to the radius of the earth, while the moon completes one revolution in 27.32 days in an orbit whose radius is 59.96 times the radius of the earth. If we put T to represent the periodic time of such a satellite, we shall have the proportion

$$59.96^3 : 1^3 :: 27.32^2 : T^2;$$

from which we find $T = 0.0588$ days, or 1h. 24m. 35s.

If the velocity of projection were too small to carry it entirely round the earth, and the impenetrability of the earth did not prevent, it would describe an ellipse, of which the earth's centre would occupy the lower focus, and it would return again to the point from which it started. This conclusion is easily reconciled with the doctrine of Mechanics, that the path of a projectile is a parabola, for it is there assumed that gravity acts in parallel directions, and that it is a constant accelerating force. These principles are sensibly true for small distances, but they are not true when great distances are considered.

Problem.—How much faster than at present must the earth rotate upon its axis, in order that bodies on its surface at the equator may lose all their gravity? *Ans.* 17 times.

260. *Why a planet at perihelion does not fall to the sun.*—Since the sun's force of attraction is greatest when the distance is least, it might seem that when a planet has reached its perihelion it must inevitably fall to the sun. The planet, however, recedes from the sun, partly on account of the increased velocity near perihelion, and partly on account of the gradual change in its direction. The curvature of any part of a planetary orbit depends not solely upon the force of the sun's attraction, but also on the velocity with which the planet is moving. The greater the velocity of the planet, the less will be the curvature of the orbit.

Suppose a planet to have passed the aphelion A with so small a velocity that the sun's attraction bends the path very much, and causes it immediately to begin to approach toward the sun; the sun's attraction will increase its velocity as it moves through B, C, and D; for when the planet is at B, the sun's attractive

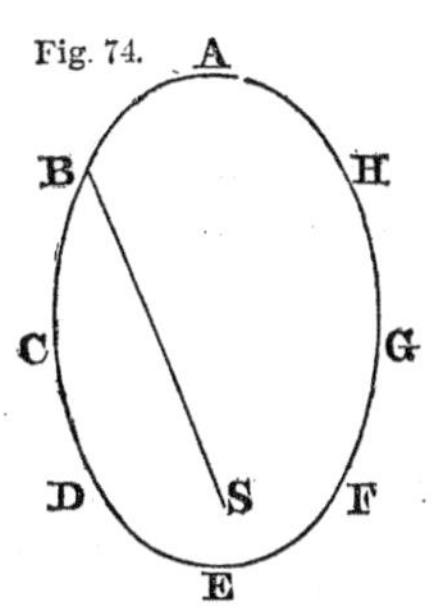

Fig. 74.

force acts in the direction BS; and, on account of the small inclination of BC to BS, the force acting in the direction of BS increases the planet's velocity. Thus the planet's velocity is continually increasing as the planet moves through B, C, and D; and although, on account of the planet's nearness, the sun's attractive force is very much increased, and tends therefore to make the orbit more curved, yet the velocity is so much increased that the orbit is no more curved at E than it was at A; and at perihelion the velocity is so great that the planet begins immediately to recede from the sun.

A similar course of reasoning will explain why, when the planet reaches its greatest distance from the sun, where the sun's attraction is least, it does not altogether fly off from the sun. As the planet passes through F, G, H, the sun's attraction, which is always directed toward S, retards the planet in its orbit, and when it has reached A its velocity is extremely small; and therefore, although the sun's attraction at A is small, yet the deflection which it produces in the planet's motion is such as to give its path the same curvature as at E. Then the planet again approaches the sun, and goes over the same orbit as before.

261. *Could the rotary and orbital motions of the earth have been caused by a single force?*—It is possible that the rotary motion of the earth, and its motion in its orbit about the sun, are both the result of a single primitive impulse. If a sphere were to receive an impulse in the direction of its centre of gravity, it would have a progressive motion without any rotation upon an axis. But if the impulse were given in any other direction, it would produce also a rotary motion. It is possible to compute at what distance from the centre of gravity an impulse must be given to produce the actual progressive and rotary motions observed in a body. In order to explain the motion of the earth in its orbit, and that of its rotation upon an axis in 24 hours, the impulse must have been given in a line passing 24 miles from the centre of the earth.

262. PROBLEMS.

Prob. 1. The mean distance of the planet Hygeia from the sun is 3.14937 (the distance of the earth being taken as unity); required its periodic time?

By Art. 250, $a^3 : a'^3 :: T^2 : T'^2$;

that is, $1^3 : 3.14937^3 :: 365.25^2 : T'^2$.

Ans. 2041.4 days.

Prob. 2. The periodic time of the planet Flora is 1193 days; required its mean distance from the sun? *Ans.* 2.2013.

Prob. 3. What would be the periodic time of a planet revolving about the sun at a mean distance of ten million miles?

Prob. 4. What would be the periodic time of a planet revolving about the sun at a mean distance of one million miles?

Prob. 5. Suppose there exists a planet revolving about the sun at a mean distance of 5000 millions of miles, what must be its periodic time?

Prob. 6. What would be the periodic time of a satellite revolving about the earth at a mean distance of 10,000 miles from the earth's centre?

Prob. 7. Suppose the earth had a satellite making one revolution in a year, what would be its mean distance from the earth?

263. *The problem of the three bodies.*—When there are only two bodies that gravitate to one another with forces inversely as the squares of their distances, they move in conic sections, and describe about their common centre of gravity equal areas in equal times. But if there are three bodies, the action of any one on the other two, changes the form of their orbits, so that the determination of their motions becomes a problem of great difficulty, distinguished by the name of *the problem of the three bodies.*

The solution of this problem, in its utmost generality, has never been effected. Under certain limitations, however, and such as are quite consistent with the condition of the heavenly bodies, it admits of being resolved. The most important of these limitations is that the force which one of the bodies exerts upon the other two is, either from the smallness of that body or its great distance, very inconsiderable, in respect of the forces which these two exert on one another.

The force of this third body is called a *disturbing force*, and its effects in changing the places of the other two bodies are called the *disturbances*, or *perturbations of the system.*

Though the small disturbing forces may be more than one, or though there be a great number of remote disturbing bodies, their combined effect may be computed, and therefore the problem of three bodies, under the conditions just stated, may be extended to any number.

264. *How the moon's elliptic motion is disturbed.*—The only body in the solar system which produces a sensible disturbing effect upon the moon is the sun; for although several of the planets sometimes come within less distances of the earth, their masses are too inconsiderable to produce any sensible disturbing effect upon the moon's motion. The mass of the sun, on the contrary, is so great, that, although the radius of the moon's orbit bears a small ratio to the sun's distance, and although lines drawn from the sun to any part of that orbit are nearly parallel, the difference between the forces exerted by the sun upon the moon and earth is quite sensible.

265. *Relative attractions of the sun and earth upon the moon.*—It was shown, Art. 252, that the earth draws the moon from a tangent 0.0534 inch in a second. If a similar calculation be made in relation to the orbit of the earth, it will be found that the sun draws the earth from a tangent 0.119 inch in a second. Also, the average force which the sun exerts upon the moon must be the same as that which it exerts upon the earth; that is, the sun exerts upon the moon a force $2\frac{1}{5}$ times as great as the earth does. The moon is therefore much more under the influence of the sun than of the earth.

266. *Mass of the sun compared with that of the earth.*—The force which the sun exerts on the earth is $2\frac{1}{5}$ times greater than that which the earth exerts on the moon. But the force of attraction varies inversely as the square of the distance, and the distance of the sun from the earth is about 400 times the distance of the moon. Hence, if the sun were at the same distance as the moon, his force of attraction would be the square of 400, or 160,000 times as great as it is now; that is, it would be $2\frac{1}{5}\times160{,}000$, or

352,000 times as great as the earth's attraction, and, consequently, must have 352,000 times as much matter.

The best determination of the sun's mass is considered to be 354,936.

267. *How the sun's attraction acts as a disturbing force.*—If the sun were at an infinite distance, the earth and moon would be attracted equally and in parallel straight lines, and, in that case, their relative motions would not be in the least disturbed. But although the distance of the sun compared with that of the moon is very great, it can not be considered infinite. The moon is alternately nearer to the sun and farther from him than the earth, and the straight line which joins her centre and that of the sun forms with the terrestrial radius vector an angle which is continually varying. Thus the sun acts unequally and in different directions on the earth and moon, and hence result inequalities in her motion, which depend on her position in respect of the sun.

268. *General effect of the sun's disturbing action.*—Let us suppose that the projectile motions of the earth and moon are destroyed, and that they are allowed to fall freely toward the sun. If the moon was in conjunction with the sun, it would be more attracted than the earth, and fall with greater velocity toward the sun, so that the distance of the moon from the earth would be increased in the fall. If the moon was in opposition, she would be less attracted than the earth by the sun, and would fall with a less velocity toward the sun than the earth, and the moon would be left behind by the earth, so that the distance of the moon from the earth would be increased in this case also. If the moon was in one of the quarters, then the earth and moon, being both attracted toward the centre of the sun, would approach the sun, and at the same time would necessarily approach each other, so that their distance from each other would in this case be diminished. Now whenever the action of the sun would increase their distance if they were allowed to fall toward the sun, it produces the same effect as if their gravity to each other was diminished; and whenever the action of the sun would diminish their distance, their gravity to each other is increased. Hence we conclude that *the sun's action increases the gravity of the moon to the earth at the quadratures, and diminishes it at the syzygies.*

269. *How to estimate the amount of the sun's disturbing force.*—We may estimate the amount of this disturbing force in the following manner:

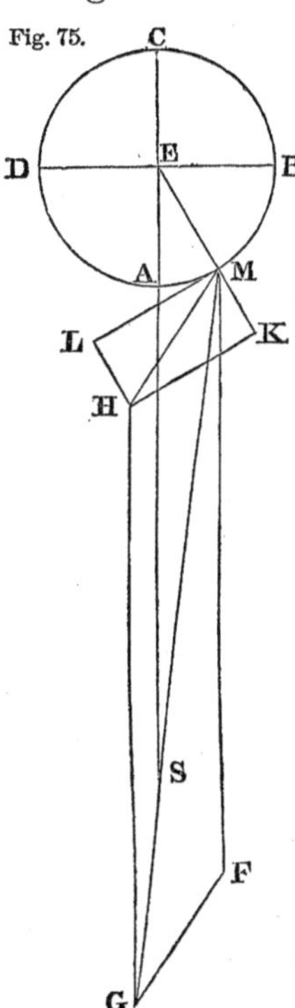

Fig. 75.

Let ABCD represent the orbit of the moon, with the earth at E, and let the sun be at S and the moon at M. Let the line SE be taken to represent the force with which the sun attracts the earth; then we may determine the magnitude of the force with which the sun acts on the moon at M by the proportion $SM^2 : SE^2 :: SE : \frac{SE^3}{SM^2}$. In the line MS, produced if necessary, take $MG=\frac{SE^3}{SM^2}$, and it will represent the force with which the sun attracts the moon. We may suppose the force MG to result from the combined action of two forces, MF and MH (MG being the diagonal of the parallelogram MFGH), of which one, MF, is equal and parallel to ES. Now if the earth and moon were only acted upon by the equal and parallel forces ES and MF, their relative motions would not be affected. Therefore it is only MH which disturbs this relative motion; that is, MH represents the quantity and direction of this disturbing force. This force, MH, may be resolved into two forces, MK, ML, the first being in the direction of the radius vector ME, and the other having the direction of a tangent to the orbit. The force MK augments or diminishes the moon's gravitation to the earth; while the force ML affects the moon's angular motion round the earth, sometimes accelerating and sometimes retarding it.

It is evident that the tangential force LM *retards* the moon's motion when going from A to B. If we construct a similar figure for each of the other quadrants, we shall find that the tangential force *accelerates* the moon's motion from D to A, and also from B to C, but retards the moon's motion when going from C to D. This force becomes zero at each of the points A, B, C, and D, and has its maximum value near the octants.

When the moon is in conjunction, the disturbing force of the sun is wholly employed in drawing the moon away from the earth; that is, in diminishing the moon's gravitation to the earth. When the moon is in opposition, the force with which the sun draws the earth is greater than that with which it draws the moon, so that the effect of the sun's attraction is to increase the distance of the moon from the earth; that is, it is the same as if the sun's force drew the moon away from the earth, or diminished the moon's gravitation to the earth.

When the moon is in quadrature, the tangential force disappears, and the disturbing force is wholly employed in augmenting the moon's gravitation to the earth. The sun attracts the earth and moon equally, but not in parallel lines. If we suppose the projectile motions of the earth and moon to be destroyed, and that they are allowed to fall freely toward the sun, the earth and moon, both moving toward the centre of the sun, would approach each other, and in one second (their distance from the sun being 400 times the radius of the moon's orbit) their distance from each other would be diminished by $\frac{1}{400}$th part of the space fallen through. Hence, if ES represents the force of the moon's gravitation to the sun, then BE will represent the augmentation of the moon's gravitation to the earth in quadratures.

270. *Numerical estimate of the sun's disturbing force.*—The ratio of the line MH to ES may be computed by Trigonometry when we know the distance of the sun and moon from the earth, and also the angular distance of the moon from the sun. Also the disturbing force of the sun upon the moon may be compared with the earth's attraction upon the moon by the following proportions:

1st. Disturbing force : sun's attraction on earth :: MH : ES;

2d. Sun's attraction on earth : earth's attr'n on sun :: 354,936 : 1;

3d. Earth's attraction on sun : earth's attraction on moon :: $EM^2 : ES^2$.

Compounding these three proportions, we have

Disturbing force : earth's attraction on moon :: $354{,}936 \times MH \times EM^2 : ES^3$.

Since the values of MH, EM, and ES are known, we can compute the ratio of the disturbing force to the earth's attraction.

Ex. 1. Compare the disturbing force of the sun upon the moon with the earth's attraction upon the moon at the time of conjunction, assuming the distance of the sun to be 399.32 times the distance of the moon, and the sun's mass 354,936 times that of the earth.

Fig. 76.

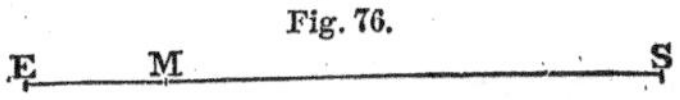

E M S

Sun's att. on moon : sun's att. on earth :: $SE^2 : SM^2$ :: 1.00502 : 1.

Hence, Disturbing force : sun's attraction on earth :: 0.00502 : 1.

And, Disturbing force : earth's attraction on moon :: 354,936 × 0.00502 : 399.32^2 :: 1 : 89;

that is, *by the disturbing action of the sun at conjunction, the moon's gravity to the earth is diminished by* $\frac{1}{89}$*th part.*

Ex. 2. Compare the disturbing force of the sun upon the moon with the earth's attraction upon the moon at the time of opposition.

Fig. 77.

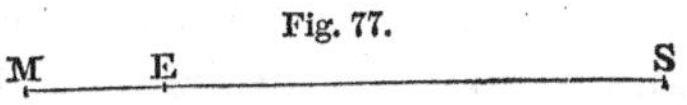

M E S

Sun's att. on moon : sun's att. on earth :: $SE^2 : SM^2$ :: .99501 : 1.

Hence, Disturbing force : sun's attraction on earth :: 0.00499 : 1.

And, Disturbing force : earth's attraction on moon :: 354,936 × 0.00499 : 399.32^2 :: 1 : 90;

that is, *by the disturbing action of the sun at opposition, the moon's gravity to the earth is diminished by* $\frac{1}{90}$*th part.*

Ex. 3. Compare the disturbing force of the sun upon the moon with the earth's attraction upon the moon at the time of quadrature.

Disturbing force : sun's attraction on earth :: 1 : 399.32. Art. 269.

Hence, Disturbing force : earth's attraction on moon :: 354,936 : 399.32^3 :: 1 : 179;

that is, *by the disturbing action of the sun at quadrature, the moon's gravity to the earth is increased by* $\frac{1}{179}$*th part.*

Thus we see that at the quadratures, the gravity of the moon to the earth is increased by about the 179th part, while at the opposition and conjunction it is diminished by about twice this quantity; and, by a computation extending to every part of the orbit, it is found that *the average effect is to diminish the moon's gravity by* $\frac{1}{358}$*th part.*

In consequence of this diminution of her gravity, the moon describes her orbit at a greater distance from the earth, with a less angular velocity, and in a longer time, than if she were urged to the earth by her gravity alone.

271. *The equation of the centre* depends upon the eccentricity of the orbit. The eccentricity of the moon's orbit was stated in Art. 208 to be $\frac{1}{18}$th, and the greatest value of the equation of the centre is 6° 18′ 17″, being more than three times that of the sun.

272. *Evection.*—After the equation of the centre, the most important inequality affecting the motion of the moon is that termed the *Evection*, the discovery of which we owe to the famous astronomer Hipparchus, in the second century before the Christian era. The evection is an inequality in the equation of the centre depending on the position of the major axis of the moon's orbit, in respect of the line drawn from the earth to the sun.

273. *Cause of evection.*—Any cause which at the perigee should have the effect to increase the moon's gravitation toward the earth beyond its mean, and at the apogee to diminish the moon's gravitation toward the earth, would *augment* the difference between the gravitation at the perigee and apogee, and, consequently, increase the eccentricity of the orbit. But any cause which at the perigee should have the effect to diminish the moon's gravitation toward the earth beyond its mean, and at the apogee to increase it, would *diminish* the difference between the two, and, consequently, diminish the eccentricity.

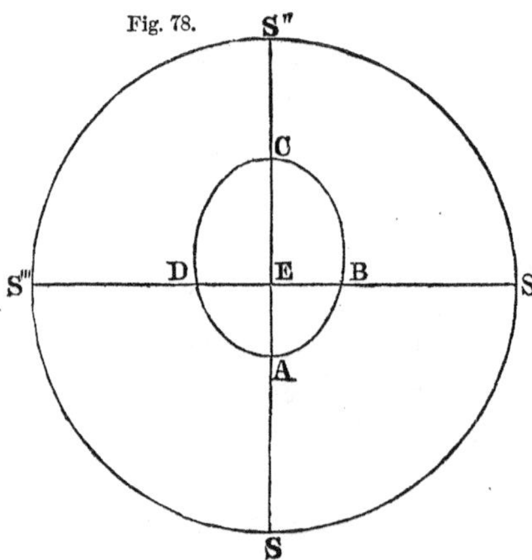

Fig. 78.

Let E represent the earth, ABCD the moon's orbit, of which A is the perigee and C the apogee, and let SS′ S″S‴ be the apparent orbit of the sun. If the sun be at S, so that the major axis of the moon's orbit is directed to the sun, the distance of the moon at A from the earth is less than if it moved in a circle, and the sun's disturbing force, computed as in *Ex.* 1, Art. 270, will be found to be *less* than $\frac{1}{89}$th of the moon's gravity. So, also, the distance of the moon from the earth at C is greater than if it moved in a circle, and the disturbing force computed, as in *Ex.* 2,

Art. 270, will be found to be *greater* than $\frac{1}{90}$th part of the moon's gravity; that is, when the transverse axis of the moon's orbit is directed to the sun, the moon's gravity to the earth when at perigee is diminished *less* than the mean, and at apogee is diminished *more* than the mean. Hence the moon, when at perigee, is drawn away from the earth by less than the mean quantity, and when at apogee, is drawn away from the earth by more than the mean quantity. Thus the inequality between the two distances of the moon from the earth is increased; that is, *the eccentricity of the moon's orbit is increased.*

But if the sun be at S′ and the moon at A, the sun's disturbing force, computed as in *Ex.* 3, Art. 270, will be found to be *less* than $\frac{1}{179}$th part of the moon's gravity; but if the moon be at C, and the sun at S′, the disturbing force of the sun will be found to be *greater* than $\frac{1}{179}$th part of the moon's gravity; that is, when the line of the apsides is in quadrature, the gravitation at the apogee is *most* augmented, and that at perigee is *least* augmented. Hence the effect of the sun's action is to diminish the inequality between the two distances of the moon from the earth at these two points; that is, *to diminish the eccentricity of the orbit.* Thus we find, in general, that *the moon's orbit is most eccentric when the line of the apsides is in syzygy, and is least eccentric when the line of the apsides is in quadrature.* The greatest value of evection is 1° 16′ 27″.

274. *Variation.*—Another large inequality in the moon's motions is called the *Variation.* By comparing the moon's observed place with the place computed from the mean motion, the equation of the centre, and the evection, Tycho Brahe, in the sixteenth century, discovered that the two places did not generally agree. They agreed only at the syzygies and quadratures, and varied most in the octants, where the inequality amounted to 39′ 30″.

275. *Cause of variation.*—This inequality is occasioned by that part of the sun's disturbing force which acts in the direction of a tangent to the moon's orbit, Art. 269. This force is nothing at the syzygies and quadratures, and is greatest near the octants. It accelerates the moon's motion in going from quadrature to conjunction; and when the moon is past conjunction, the tangential force changes its direction and retards the moon's motion.

276. The *annual equation* is an inequality in the moon's motion arising from the variation of the sun's distance from the earth. When the earth is at perihelion, the sun's disturbing force is greater than its average value; the moon's gravity to the earth is diminished more than usual; and its velocity is therefore slower than the mean. For the same reason, at aphelion the moon's velocity is greater than the mean. The period of this inequality is one year, and its maximum effect upon the moon's longitude amounts to 11′ 9″.

277. *Other inequalities in the moon's motion.*—These three inequalities, evection, variation, and annual equation, are the largest of the inequalities in the moon's motion. The other inequalities are more minute; but, in order to represent the moon's place with the greatest possible accuracy, it is necessary to take into account a large number of corrections.

The moon's place for every hour of the year is computed several years beforehand, and published in the Nautical Almanac. These places are now computed from Tables published by Professor Hansen in 1858. The average difference between the observed places of the moon and the places computed from these Tables does not exceed 3″, and only once or twice in a year does the difference amount to so large a quantity as 10″.

278. *Cause of the retrograde motion of the moon's nodes.*—The plane of the moon's orbit is inclined to the ecliptic about 5°; that is, in half of her revolution she is on the north side of the ecliptic, and in half is on the south side of the ecliptic. The sun is seldom in the plane of the moon's orbit, and his action generally has a tendency to draw the moon out of the plane in which she is moving. This oblique force may be resolved into two other forces—one lying in the plane of the ecliptic, and the other perpendicular to it. Let ENN″ represent the ecliptic, and AN a portion of the moon's orbit. Let the moon be at A, and approaching the descending node N. The sun being situated in the plane EN, his attrac-

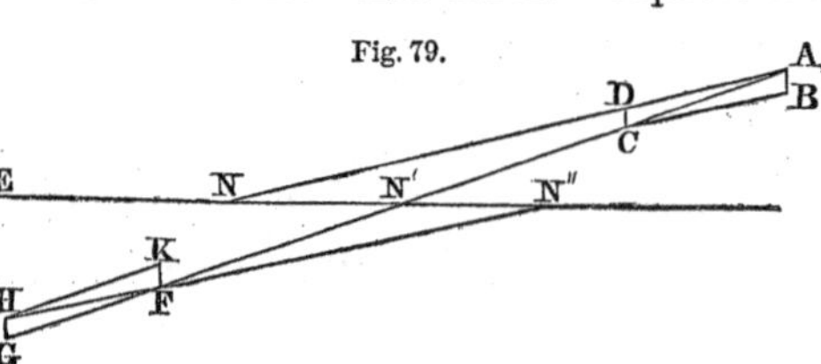

Fig. 79.

tion tends to draw the moon toward that plane. Let that part of the sun's disturbing force which is perpendicular to the plane EN be represented by AB, and suppose that in the time that the perpendicular force would cause it to describe AB, the moon, if undisturbed, would have advanced from A to D. By the combined action of these two forces, the moon will describe the diagonal AC, and cross the ecliptic in the point N′. Thus the node has shifted from N to N′ in a direction contrary to that of the moon's motion, and the inclination of the orbit to the ecliptic has increased. After the moon has crossed the ecliptic, the sun's disturbing action tends to draw the moon northward toward the ecliptic. Suppose the moon to be at F, and let that part of the sun's disturbing force which is perpendicular to the ecliptic be represented by FK, while FG represents the moon's velocity in her orbit. The resultant of these two forces will be a motion in the diagonal FH, as if the moon had come, not from N′, but from N″, a point still farther to the westward. Thus the node has traveled farther westward, but the inclination of the orbit to the plane of the ecliptic has diminished. Thus it appears that both in approaching the node, and in receding from it, *the node shifts its place in a direction contrary to that of the moon's motion;* but the inclination of the moon's orbit *increases* while the moon approaches the node, and *diminishes* while the moon is receding from it.

When the line of the nodes of the moon's orbit passes through the sun, there is no disturbing force tending to draw the moon out of the plane of its orbit; but in every other position the line of the nodes is constantly regressing, making a complete revolution in about 19 years. See Art. 240. The inclination of the plane of the orbit to the ecliptic increases and diminishes alternately. This variation is, however, confined within very narrow limits, so that there is no permanent change in the inclination of the orbit.

279. *Cause of the progression of the line of the apsides.*—The apsides of the moon's orbit are distant from each other more than 180°. This is caused by the disturbing action of the sun, which tends to diminish the moon's gravity to the earth. If the moon was only acted upon by the earth's attraction, she would describe an ellipse, and her angular motion from perigee to apogee would be just 180°; but when the effect of the sun's action is to dimin-

ish the moon's gravity, she will continually recede from the ellipse that would otherwise be described; her path will be less curved, and she must move through a greater distance before the radius vector intersects the path at right angles. She must therefore move through a greater angular distance than 180° in going from perigee to apogee, and, consequently, the apsides must advance. On the contrary, when the moon's gravity is increased by the sun's action, her path will fall *within* the ellipse which she would otherwise describe; its curvature will be increased, and the distance through which she must move before the radius vector intersects her path at right angles will be less than 180°. The apsides will therefore move backward. Now it has been shown that the sun's action alternately increases and diminishes the moon's gravity to the earth. The motion of the apsides will therefore be alternately direct and retrograde. But as the diminution of gravity has place during a much longer part of the moon's revolution, and is also greater than the increase, the direct motion will exceed the retrograde; and in one revolution of the moon, the apsides have a progressive motion of about 3°, making a complete revolution in about nine years. See Art. 239.

280. *Periodical and secular inequalities.*—The perturbations in the elliptic movements of the planets and their satellites may be divided into two distinct classes. Those of the first class depend simply on the configurations of the planets, and complete the cycle of their values upon each successive return of the same configuration. These are called *periodic inequalities.* Their periods, generally speaking, are not long; and their general effect is slightly to accelerate or retard a planet in its orbit. The perturbations of the second class depend on the configuration of the nodes and perihelia. They vary with extreme slowness, requiring centuries to complete the cycle of their values, and they are hence denominated *secular inequalities.* Laplace has indeed demonstrated that the last-mentioned inequalities are also periodic, but the periods are much longer than those of the other inequalities, and are independent of the mutual configurations of the planets.

281. *Secular acceleration of the moon's mean motion.*—The mean motion of the moon exhibits a secular inequality which has become very celebrated. By comparing the results of recent ob-

servations with the Chaldean observations of eclipses at Babylon in the years 719 and 720 before the Christian era, Dr. Halley discovered that the periodic time of the moon is now sensibly shorter than at the time of the Chaldean eclipses. The mean motion of the moon increases at the rate of more than 10″ in one hundred years. If this acceleration of her motion, and the consequent diminution of her distance, were perpetually to continue, it would follow that she would eventually be precipitated to the earth. But Laplace has shown that this acceleration of the moon is occasioned by the change in the eccentricity of the earth's orbit. It has been stated, Art. 113, that the eccentricity of the earth's orbit has been diminishing from the time of the earliest observations.

The mean action of the sun upon the moon tends to diminish the moon's gravity to the earth, and thereby causes a diminution of her angular velocity. This diminution being once supposed to occur, the angular velocity would afterward remain constant, provided the mean solar action always retained the same value. This, however, is not the case, for it depends, to a certain extent, on the eccentricity of the earth's orbit. Now the eccentricity of the earth's orbit has been continually diminishing from the date of the earliest recorded observations down to the present time; hence the sun's mean action must also have been diminishing, and, consequently, the moon's mean motion must have been continually increasing. This acceleration will continue as long as the earth's orbit is approaching toward a circular form; but as soon as the eccentricity begins to increase, the sun's mean action will increase, and the acceleration of the moon's mean motion will be converted into a retardation.

CHAPTER X.

ECLIPSES OF THE MOON.

282. *Cause of eclipses.*—An eclipse of the sun is caused by the moon passing between the sun and the earth. It can therefore only occur when the moon is in conjunction with the sun; that is, at the time of new moon. An eclipse of the moon is caused by the earth passing between the sun and moon. It can therefore only occur when the moon is in opposition; that is, at the time of full moon.

283. *Why eclipses do not occur every month.*—If the moon's orbit coincided with the plane of the ecliptic, there would be an eclipse of the sun at every new moon, since the moon would pass directly between the sun and earth; and there would be an eclipse of the moon at every full moon, since the earth would be directly between the sun and moon. But since the moon's orbit is inclined to the ecliptic about 5°, an eclipse can only occur when the moon, at the time of new or full, is at or near one of its nodes. At other times, the moon is too far north or south of the ecliptic to cause an eclipse of the sun, or to be itself eclipsed.

284. *Form of the earth's shadow.*—Since the magnitude of the sun is far greater than that of the earth, and both bodies are of a globular form, the earth must cast a conical shadow in a direction opposite to that of the sun. Let AB represent the sun, and CD

Fig. 80.

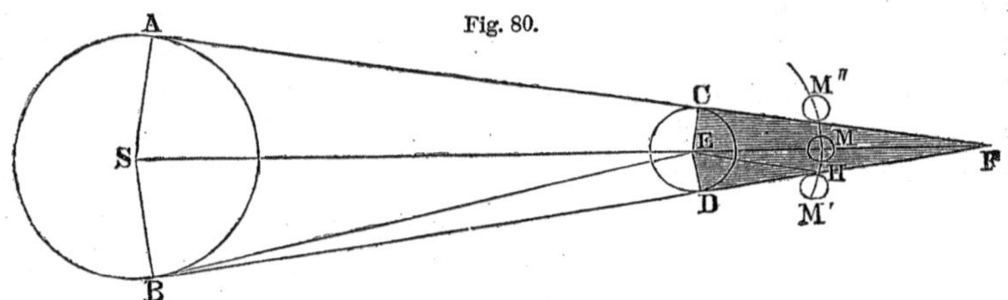

the earth, and let the tangent lines AC, BD be drawn, and produced to meet in F. Then CFD will represent a section of the earth's shadow, and EF will be its axis. If the triangle AFS be supposed to revolve round the axis SF, the tangent CF will describe the convex surface of a cone, within the whole of which the light of the sun must be intercepted by the earth.

285. *The semi-angle of the cone of the earth's shadow is equal to the sun's apparent semi-diameter, minus his horizontal parallax.*

In Fig. 80 the semi-angle of the cone of the earth's shadow is EFC or EFD. Now SEB=EFB+EBF; that is, EFB=SEB−EBD; or half the angle of the cone of the earth's shadow is equal to the sun's apparent semi-diameter, minus his horizontal parallax. Putting s for the sun's semi-diameter, and p for his horizontal parallax, we have the semi-angle of the earth's shadow, EFC $=s-p$.

286. *The length of the earth's shadow varies according to the distance of the sun from the earth; its mean length being* 856,200 *miles, or more than three times the distance of the moon from the earth.*

In the right-angled triangle EFC, right-angled at C,

$$\text{sin. EFC} : \text{CE} :: \text{R} : \text{EF} = \frac{\text{CE}}{\text{sin. EFC}} = \frac{\text{CE}}{\text{sin.}\,(s-p)}.$$

The mean value of the sun's apparent semi-diameter is 16′ 1″.8, and the sun's horizontal parallax is 8″.6; hence $s-p=15' \; 53''.2$. Also, the mean diameter of the earth $=3956.6$ miles. Hence the average length of the earth's shadow $=\frac{3956.6}{\text{sin. } 15' \; 53''.2}=856{,}200$ miles.

Since the mean distance of the moon from the earth is only 238,880 miles, the shadow extends to a distance more than three times that of the moon.

287. *The average breadth of the earth's shadow, at the distance of the moon, is almost three times the moon's diameter.*

Let M′M″ represent a portion of the moon's orbit. The apparent semi-diameter of the earth's shadow at the distance of the moon is the angle MEH. But EHD=FEH+HFE. Hence MEH =EHD−HFE. But EHD=the moon's horizontal parallax; and HFE = the sun's semi-diameter minus his horizontal parallax $(=s-p)$; therefore half the angle subtended by the section of the shadow is equal to the sum of the parallaxes of the sun and moon, minus the sun's semi-diameter. If we represent the moon's horizontal parallax by p', we shall have

$$\text{MEH}=p+p'-s.$$

The mean value of p' is 57′ 2″.3, and $s-p=15' \; 53''.2$; hence $p+p'-s=41' \; 9''.1$. The mean value of the moon's apparent semi-diameter is 15′ 39″.9. Hence the diameter of the shadow is almost three times the moon's diameter, and therefore the moon may be totally eclipsed for as long a time as she takes to describe about twice her own diameter. The eclipse will begin when the moon's disc at M′ touches the earth's shadow, and the eclipse will end when the moon's disc touches the earth's shadow at M″.

288. *Lunar ecliptic limits.*—There is a certain distance of the moon's node from the centre of the earth's shadow beyond which a lunar eclipse is *impossible*, and a certain less distance within

which an eclipse is *inevitable*. These distances are called the lunar *ecliptic limits*. The first is called the major limit, and the second the minor limit.

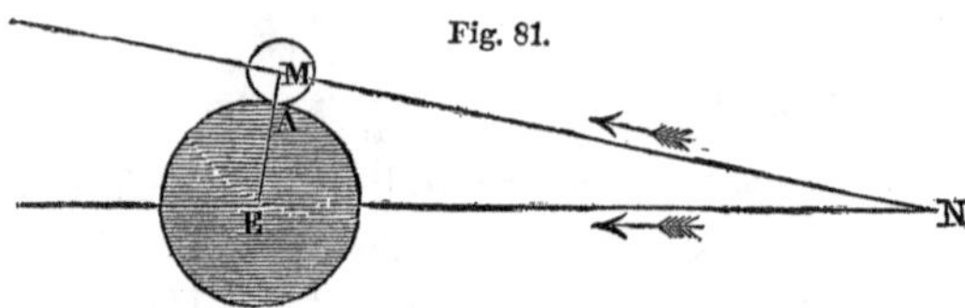

Fig. 81.

Let NE represent the ecliptic, NM the moon's orbit, and N the moon's ascending node. Let EA be the semi-diameter of the earth's shadow, and MA the semi-diameter of the moon. When the line ME, joining the centres of the moon and shadow, becomes equal to the sum of the semi-diameters, the moon will touch the earth's shadow; and if ME be less than this limit, the moon will enter the shadow, and be partially or totally eclipsed. The line NE represents that distance of the moon's node from the centre of the earth's shadow beyond which there can be no eclipse.

289. *To compute the values of the ecliptic limits.*—We may regard EMN as a spherical triangle, right-angled at M, in which EM represents the sum of the radii of the moon and of the earth's shadow, and N is the inclination of the moon's orbit to the ecliptic. Now, by Napier's rule,

$$\text{R sin. EM} = \text{sin. EN sin. N}; \text{ or sin. EN} = \frac{\text{sin. EM}}{\text{sin. N}}.$$

Since EM and N are both variable, the ecliptic limit is variable. To obtain the distance beyond which a lunar eclipse is impossible, we must employ the greatest possible value of EM, and the least possible value of N. To obtain the distance within which an eclipse is inevitable, we must employ the least possible value of EM, and the greatest possible value of N. The greatest possible value of EM is 62′ 38″, and the least inclination of the moon's orbit to the ecliptic is 5°, from which we obtain the major limit of lunar eclipses, 12° 4′. The least possible value of EM is 52′ 20″, and the greatest possible inclination of the moon's orbit to the ecliptic is 5° 17′, from which we obtain the minor limit of lunar eclipses, 9° 30′.

If, then, at the time of opposition, the moon's node is distant from the centre of the earth's shadow less than 9° 30′, or if the sun be distant from the opposite node of the moon less than 9° 30′, there will certainly be an eclipse of the moon; but if the sun be distant from the node of the moon's orbit more than 12° 4′, there can not be an eclipse. When the distance falls between these limits, it will be necessary to make a more minute calculation in order to determine whether there will or will not be an eclipse.

290. *Different kinds of lunar eclipses.*—When the moon just touches the earth's shadow, but passes by it without entering it, the circumstance is called an *appulse.* When a part, but not the whole of the moon enters the shadow, the eclipse is called a *partial* eclipse; when the moon enters entirely into the shadow, it is called a *total* eclipse; and if the moon's centre should pass through the centre of the shadow, it would be called a *central* eclipse. It is probable, however, that a strictly central eclipse of the moon has never occurred.

291. *The earth's penumbra.*—Long before the moon enters the cone of the earth's shadow, the earth begins to intercept from it a portion of the sun's light, so as to render the illumination of its surface sensibly more faint. This partial shadow is called the earth's *penumbra.* Its limits are determined by the tangent lines AD, BC produced. Throughout the space included between the

Fig. 82.

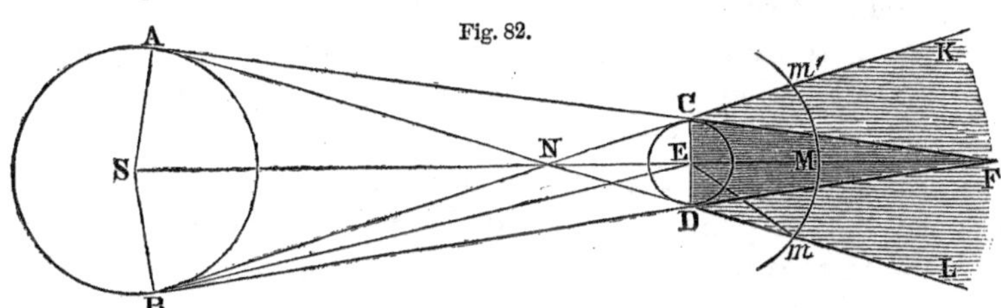

lines CK and DL, the light of the sun is more or less obstructed by the earth. If a spectator were situated at L, he would see the entire disc of the sun; but between L and the line DF, he would see only a portion of the sun's surface, and the portion of the sun which was hidden would increase until he reached the line DF, beyond which the sun would be entirely hidden from view.

292. *The semi-angle of the earth's penumbra is equal to the sun's apparent semi-diameter, plus his horizontal parallax.*

The angle KNF=BNS=BEN+NBE. But BEN is the sun's

apparent semi-diameter, and NBE is the sun's horizontal parallax. Hence the semi-angle of the penumbra is represented by $s+p$.

293. *The apparent semi-diameter of a section of the penumbra at the moon's distance is equal to the sum of the parallaxes of the sun and moon, plus the sun's semi-diameter.*

The angle $\text{ME}m=\text{EN}m+\text{E}m\text{N}$.

But $\text{EN}m=s+p$, Art. 292.

And $\text{E}m\text{N}=$the moon's horizonal parallax$=p'$.

Hence $\text{ME}m=p+p'+s$, which equals the apparent semi-diameter of the shadow, plus the sun's diameter.

294. *Effect of the earth's atmosphere.*—In obtaining the above expression for the dimensions of the earth's shadow, the shadow is assumed to be limited by those rays of the sun which are tangents to the sun and earth. It is, however, found that the observed duration of an eclipse always exceeds the duration computed on this hypothesis. This fact is accounted for in part by supposing that most of those rays which pass near the surface of the earth are absorbed by the lower strata of the atmosphere; but we must also admit that those rays of the sun which enter the atmosphere, and are so far from the surface as not to be absorbed, are refracted toward the axis of the shadow, and are thus spread over the entire extent of the geometrical shadow, thereby diminishing the darkness, but increasing the diameter of the shadow, and, consequently, the duration of the eclipse.

In consequence of the gradual diminution of the moon's light as it enters the penumbra, it is difficult to determine with accuracy the instant when the moon enters the dark shadow; and astronomers have differed as to the amount of correction that should be made for the effect of the earth's atmosphere. It is generally found necessary, however, to increase the computed diameter of the shadow by about $\frac{1}{60}$th part.

295. *Moon visible when entirely immersed in the earth's shadow.*—When the moon is totally immersed in the earth's shadow, she does not, except on some rare occasions, become invisible, but assumes a dull reddish hue, somewhat of the color of tarnished copper. This arises from the refraction of the sun's rays in passing through the earth's atmosphere, as explained in the preceding

Article. Those rays from the sun which enter the atmosphere, and are so far from the surface as not to be absorbed, are bent toward the axis of the shadow, and fall upon the moon, causing sufficient illumination to render the disc distinctly visible.

296. *Computation of lunar eclipses.*—By the solar tables we may ascertain the apparent position of the centre of the sun from hour to hour, and hence we may learn the position of the centre of the earth's shadow. From the lunar tables we ascertain, in the same manner, the position of the moon's centre from hour to hour. The eclipse will begin when the distance between the centre of the moon and that of the shadow is equal to the *sum* of the apparent semi-diameters of the moon's disc and the shadow; the middle of the eclipse will occur when this distance is least; and the eclipse will end when the distance between the centres is again equal to the sum of the apparent semi-diameters. The Nautical Almanac for each year furnishes the places of the sun for every day of the year, as computed from the solar tables, and the places of the moon are given for every hour of the year. With this assistance, it is easy to compute the times of beginning and end of an eclipse.

297. *Construction of the diagram.*—First find the time of opposition, or the time of full moon. For this time compute the declination, horizontal parallax, and semi-diameter both of the sun and moon; also the hourly motion of the moon from the sun both in right ascension and declination.

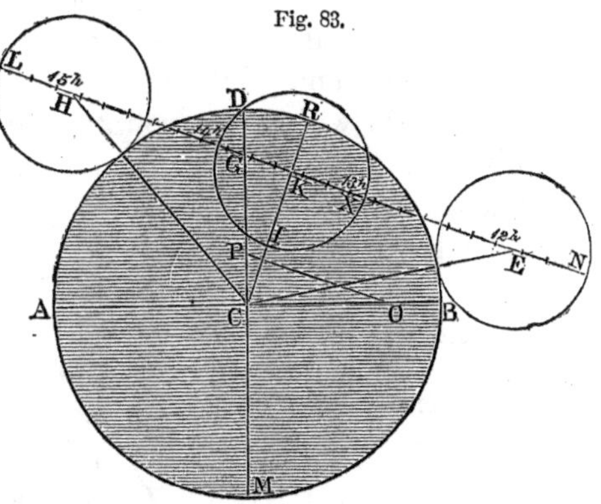

Fig. 83.

Let C represent the centre of the earth's shadow. Draw the line ACB parallel to the equator, and DCM perpendicular to it. Select a convenient scale of equal parts, and from it take CG, equal to the moon's declination, minus the declination of the centre of the shadow, and set it on CD from C to G, above the line AB, if the centre of the moon is north of the centre of the shadow, but below if south. Take CO, equal to the hourly motion of the moon from the sun in right ascension, reduced to the arc of a great circle, and set it on the line CB, to the right of C. Take CP, equal to the moon's hourly motion from the sun in declination, and set it on the line CD from C to P, above the line AB, if the moon is moving northward with respect to the shadow, but below if moving southward. Join the points O and P. The line OP will represent the hourly motion of the moon from the sun; and parallel to it, through G, draw NGL, which will represent the relative orbit of the moon, the earth's shadow being supposed stationary. On this line are to be marked the places of the moon before and after opposition, by means of the hourly motion OP, in such a manner that the moment of opposition may fall exactly upon the point G.

298. *To determine the beginning and end of the eclipse.*—The semi-diameter of the earth's shadow is equal to the horizontal parallax of the moon, plus that of the sun, minus the sun's semi-diameter, which result must be increased by $\frac{1}{60}$th part, on account of the earth's atmosphere. With this radius, describe the circle ADB about the centre C. Add the moon's semi-diameter to the radius CB, and, with this sum for a radius, describe about the centre C a circle, which, if there be an eclipse, will cut NL in two points, E and H, representing respectively the places of the moon's centre at the beginning and end of the eclipse. Draw the line CKR perpendicular to LN, and cutting it in K. The hours and minutes marked on the line LN, at the points E, K, and H, will represent respectively the times of the beginning of the eclipse, middle of the eclipse, and end of the eclipse. If the circle does not intersect NL, there will be no eclipse. With a radius equal to the moon's semi-diameter, describe a circle about each of the centres E, H, and K. If the eclipse is total, the whole of the circle about K will fall within ARB; but if part of the circle falls without ARB, the eclipse will be partial. In either case, the magnitude of the eclipse will be represented by the ratio of the obscured part RI to the moon's diameter. When the eclipse is total, the beginning and end of total darkness may be found by taking a radius equal to CB, diminished by the moon's semi-diameter, and describing with it round the centre C a circle cutting LN in two points,

representing respectively the places of the moon's centre at the beginning and end of total darkness.

Example 1.

299. Required the times of beginning, end, etc., of the eclipse of the moon, March 30, 1866, at Greenwich.

By the Nautical Almanac, the Greenwich mean time of opposition in right ascension is, March 30, 16h. 39m. 18.9s. Corresponding to this time, the Nautical Almanac furnishes the following elements:

Declination of the moon	S. 4° 12′ 55″.5
Declination of the earth's shadow	S. 4 3 42 .3
Moon's equatorial horizontal parallax	54 28 .1
Sun's horizontal parallax	8 .6
Moon's semi-diameter	14 52 .0
Sun's semi-diameter	16 2 .2
Moon's hourly motion in right ascension	28 48 .0
Sun's hourly motion in right ascension	2 16 .4
Hourly motion of moon in declination	S. 9 14 .1
Hourly motion of shadow in declination	S. 58 .1

The figure of the earth being spheroidal, that of the shadow will deviate a little from a circle, so that instead of the equatorial horizontal parallax, we should employ the horizontal parallax belonging to the mean latitude of 45°. The reduction for latitude,

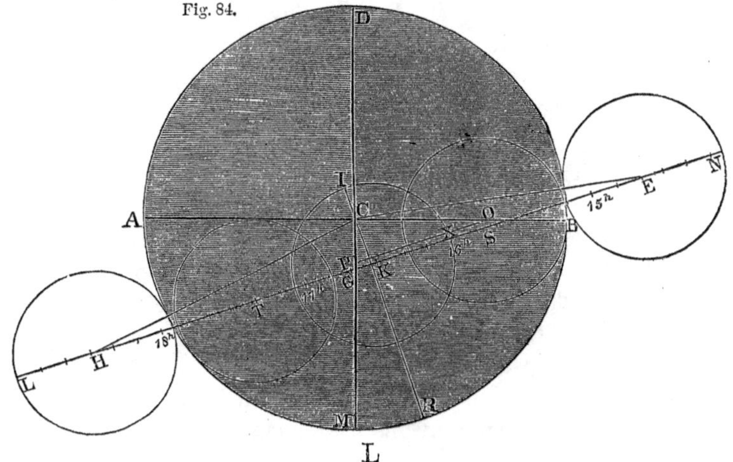

Fig. 84.

by Table VIII., is 5″.4, so that the moon's reduced parallax is 54′ 22″.7. Then, to obtain CB, the semi-diameter of the earth's shadow, we have 54′ 22″.7+8″.6−16′ 2″.2, which is equal to 38′ 29″.1. Increasing this by $\frac{1}{60}$th part of itself, or 38″.5, we have 39′ 7″.6=CB; to which adding the moon's semi-diameter, we obtain CE=53′ 59″.6. From the centre C, with a radius CB, taken from a convenient scale of equal parts, describe the circle ARB, representing the earth's shadow. Draw the line ACB to represent a parallel to the equator, and make CG perpendicular to it, equal to 9′ 13″.2, which is the moon's declination, minus the declination of the centre of the shadow; the point G being taken below C, because the centre of the moon is south of the centre of the shadow.

The hourly motion of the moon from the sun in right ascension is 26′ 31″.6, which must be reduced to the arc of a great circle by multiplying it by the cosine of the moon's declination, 4° 12′ 55″, Art. 152, thus:

$$26' \, 31''.6 = 1591''.6 = 3.201834$$
$$\text{cos. dec. } 4° \, 12' \, 55'' = 9.998824$$
$$\text{Reduced hourly motion} = 1587''.3 = \overline{3.200658}$$

Make CO = 1587″.3, and CP, perpendicular to it, equal to 8′ 16″.0, which is the hourly motion of the moon from the shadow in declination, the point P being placed below C, because the moon was moving southward with respect to the shadow. Join OP; and parallel to it, through G, draw the line NGL, which represents the path of the moon with respect to the shadow. On NL let fall the perpendicular CK. Now at 16h. 39m. 18.9s. the moon's centre was at G. To find X, the place of the moon's centre at 16h., we must institute the proportion

60m. : 39m. 18.9s. :: OP : GX;

which distance, set on the line GN, to the right of G, reaches to the point X, where the hour, 16h. preceding the full moon, is to be marked. Take the line OP, and lay it from 16h., toward the right hand, to 15h., and successively toward the left to 17h., 18h., etc. Subdivide these lines into 60 equal parts, representing minutes, if the scale will permit; and the times corresponding to the points E, K, and H will represent respectively the beginning of the eclipse, 14h. 38m.; the middle of the eclipse, 16h. 33m.; and the end of the eclipse, 18h. 28m.

If the results obtained by this method are not thought to be sufficiently accurate, we may institute a rigorous computation.

Computation of the Eclipse.

300. The phases of the eclipse may be accurately computed in the following manner:

In the right-angled triangle OCP, we have given CO=1587″.3, and CP=496″.0, to find OP and the angle CPO, thus:

$$\text{CP} : \text{R} :: \text{CO} : \text{tang. CPO}.$$

$$\begin{aligned} \text{CO}&=1587.3=3.200658 \\ \text{CP}&=\ \ 496.0=\underline{2.695482} \\ \text{CPO}=72°\ 38'\ 49''\ \text{tang.}&=0.505176 \end{aligned}$$

Also,

$$\text{sin. CPO} : \text{R} :: \text{CO} : \text{OP}.$$

$$\begin{aligned} \text{CO}&=3.200658 \\ \text{sin. CPO}&=\underline{9.979769} \\ \text{OP}=1663''.0&=3.220889 \end{aligned}$$

301. *Beginning, middle, and end of the eclipse.*—The middle of the eclipse is found by means of the triangle CGK, which is similar to CPO, because EG and OP are parallel, and CK is perpendicular to PO. Hence the angle CGK=72° 38′ 49″; and CG, the difference of declination between the moon and the centre of the shadow=9′ 13″.2=553″.2. To find CK and KG, we have the proportions

$$\text{R} : \text{CG} :: \text{sin. CGK} : \text{CK} :: \text{cos. CGK} : \text{GK}.$$

$$\begin{aligned} \text{sin. CGK}&=9.979769 & \text{cos. CGK}&=9.474593 \\ \text{CG}&=\underline{2.742882} & \text{CG}&=\underline{2.742882} \\ \text{CK}=528''.0&=2.722651 & \text{GK}=165''.0&=2.217475 \end{aligned}$$

Then, to find the time of describing GK, we say,

As OP (1663″.0) is to GK (165″.0), so is 1 hour to the time (357.2s.), 5m. 57.2s., between the middle of the eclipse and the time of opposition in right ascension, 16h. 39m. 18.9s., which gives the time of *middle* of the eclipse 16h. 33m. 21.7s.

Now, in the triangle CKE, we have the hypothenuse CE=53′ 59″.6=3239″.6, and CK=528″.0, to find KE, thus:

$$\text{KE}=\sqrt{\text{CE}^2-\text{CK}^2}=\sqrt{\overline{\text{CE}+\text{CK}}\times\overline{\text{CE}-\text{CK}}}=3196''.3.$$

To find the time of describing KE, we form the proportion

1663″.0 : 3196″.3 :: 3600s. : 6919.3s. = 1h. 55m. 19.3s.;

which, subtracted from 16h. 33m. 21.7s., the time of middle, gives 14h. 38m. 2.4s. for the *beginning* of the eclipse; and, added to the time of middle, gives for the *end* of the eclipse 18h. 28m. 41.0s.

302. *Magnitude of the eclipse.*—Subtracting CK, 8′ 48″.0, from CR, 39′ 7″.6, we have KR, 30′ 19″.6; to which adding KI, 14′ 52″.0, we obtain RI, 45′ 11″.6. Dividing this by the moon's diameter, 29′ 44″.0, we obtain the magnitude of the eclipse, 1.520 (the moon's diameter being unity); and the eclipse takes place on the moon's north limb.

The magnitude of an eclipse is sometimes expressed in *digits*, or twelfths of the moon's diameter. In the present instance, the eclipse amounts to 18 digits.

303. *Beginning and end of total darkness.*—The beginning and end of total darkness may be found in the same manner. With a radius equal to CB, diminished by the moon's semi-diameter (that is, 39′ 7″.6−14′ 52″.0, which equals 24′ 15″.6, or 1455″.6), describe about the centre C a circle cutting LN in the points S and T, which will represent the places of the moon's centre at the beginning and end of total darkness.

In the triangle CKS, CS=1455″.6, and CK=528″.0.

Hence $\quad KS=\sqrt{1455.6^2-528.0^2}=1356''.5.$

Then, to find the time of describing KS, we say,

1663″.0 : 1356″.5 :: 3600s. : 2936.4s.=48m. 56.4s.;

which, being subtracted from 16h. 33m. 21.7s., gives the beginning of total darkness 15h. 44m. 25.3s.; and, being added to the time of middle, gives for the end of total darkness 17h. 22m. 18.1s.

304. *Contacts with the penumbra.*—The contacts with the penumbra may be found in a similar manner. The semi-diameter of the penumbra is equal to the semi-diameter of the shadow, plus the sun's diameter, or 39′ 7″.6+32′ 4″.4=71′ 12″.0. If we take the circle ARB, Fig. 84, to represent the limits of the penumbra, CE will be equal to 71′ 12″.0+14′ 52″.0=86′ 4″.0.

Then, in the triangle CKE, we have given CE=5164″.0, and CK=528″.0.

Hence $\quad KE=\sqrt{5164^2-528^2}=5136''.9.$

To find the time of describing KE, we say,

1663″.0 : 5136″.9 :: 3600s. : 11120.3s.=3h. 5m. 20.3s.;

which, being subtracted from 16h. 33m. 21.7s., gives the first contact with the penumbra at 13h. 28m. 1.4s.; and, being added to the time of middle, gives for the last contact with the penumbra 19h. 38m. 42.0s.

305. *Results.*—The results thus obtained are as follows:

	h.	m.	s.	
First contact with the penumbra at	13h.	28m.	1.4s.	Mean time at Greenwich.
First contact with the umbra - -	14	38	2.5	
Beginning of total eclipse - - -	15	44	25.3	
Middle of the eclipse - - - - - -	16	33	21.7	
End of total eclipse - - - - - -	17	22	18.1	
Last contact with the umbra - -	18	28	40.9	
Last contact with the penumbra -	19	38	42.0	

Magnitude of the eclipse, 1.520 on the northern limb.

306. *Times for any other meridian.*—To obtain the times for any other place, we have only to add or subtract the longitude. For New Haven, whose longitude is 4h. 51m. 41.6s. west of Greenwich, the times will accordingly be

	h.	m.	s.	
First contact with the penumbra at	8h.	36m.	20s.	Mean time at New Haven.
First contact with the umbra - -	9	46	21	
Beginning of total eclipse - - -	10	52	44	
Middle of the eclipse - - - - - -	11	41	40	
End of total eclipse - - - - - -	12	30	37	
Last contact with the umbra - -	13	36	59	
Last contact with the penumbra -	14	47	0	

Ex. 2. Compute the phases of the eclipse of March 19, 1867, for New York city, longitude 4h. 56m. 0.2s. west of Greenwich, from the following elements:

Greenwich mean time of opposition in right ascension - - - - - - - - - - - - -	20h.	28m.	30.9s.
Declination of the moon - - - - - - - -	N. 0°	50′	56″.0
Declination of the earth's shadow - - -	N. 0	17	4 .5
Moon's equatorial horizontal parallax - -		57	2 .7
Sun's horizontal parallax - - - - - - -			8 .6
Moon's semi-diameter - - - - - - - -		15	34 .2
Sun's semi-diameter - - - - - - - - -		16	5 .3
Moon's hourly motion in right ascension -		31	25 .4
Sun's hourly motion in right ascension - -		2	16 .5
Hourly motion of moon in declination - -	S. 10		19 .7
Hourly motion of shadow in declination -	S. 0		59 .2

	h.	m.	
Ans. First contact with the penumbra at	13h.	9.3m.	Mean time at New York.
First contact with the umbra - - - - -	14	20.0	
Middle of the eclipse - - - - - - - -	15	52.8	
Last contact with the umbra - - - - -	17	25.6	
Last contact with the penumbra - - - -	18	36.3	

CHAPTER XI.

ECLIPSES OF THE SUN.

307. *Length of the moon's shadow.*—The length of the moon's shadow is about equal to the distance of the moon from the earth, being alternately a little greater and a little less.

Suppose the moon at conjunction to be at one of her nodes. Her centre will then be in the plane of the ecliptic, and in the straight line passing through the centres of the sun and earth.

Fig. 85.

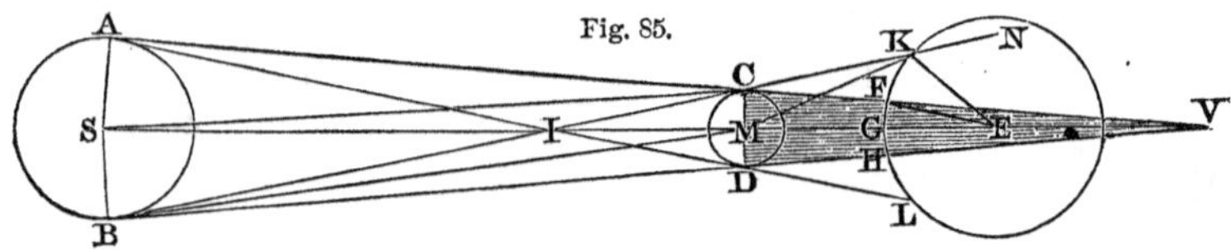

Let ASB be a section of the sun, KEL that of the earth, and CMD that of the moon interposed directly between them. Draw AC, BD, tangents to the sun and moon, and produce these lines to meet in V. Then V is the vertex of the moon's shadow; and these lines represent the outlines of a cone, whose base is AB, and whose vertex is V.

The angle $\quad SMB=MVB+MBV;$

hence $\quad MVB=SMB-MBV.$

But $\quad SMB : SEB :: SE : SM$ (Art. 111) $:: 400 : 399;$

therefore $$SMB=\frac{400}{399}\,SEB.$$

Now SEB, the sun's mean semi-diameter as seen from the earth $=16'\ 1''.8$; hence $SMB=16'\ 4''.2$, which is the sun's semi-diameter as seen from the moon.

Put $\quad p=$the sun's horizontal parallax.

$p'=$the moon's horizontal parallax.

$s'=$the moon's semi-diameter.

Since the parallaxes of bodies at different distances are inversely as the distances, Art. 136, we shall have

$$p' : p :: SE : ME,$$

or $$p'-p : p :: SM : ME.$$

But since the apparent diameters of the same body at different distances are inversely as the distances, Art. 111, we shall have

$$\text{SM} : \text{ME} :: s' : \text{MBC};$$

hence $$\text{MBC} = \frac{ps'}{p'-p}.$$

Now p, p', and s' are known quantities; hence MBC$=2''.3$, which is the sun's horizontal parallax as observed from the moon.

Hence MVB, the semi-angle of the cone of the moon's shadow, equals $16'\ 4''.2-2''.3=16'\ 1''.9$.

Then sin. $16'\ 1''.9$: DM (1080 miles) :: rad. : MV $=$ 231,590 miles.

But the mean distance of the moon from the earth's centre is 238,883 miles. Hence, when the moon is at the mean distance from the earth, her shadow will not quite reach to the earth's surface.

When the earth is at its greatest distance from the sun, the sun's apparent semi-diameter is $15'\ 45''.5$; and the angle MVB$=$ $15'\ 45''.6$.

In this case MV$=$235,582 miles. Now when the moon is nearest the earth, her distance from the centre of the earth is only 221,436 miles. Hence, when the moon is nearest to us, and her shadow is the longest, the shadow extends 14,000 miles beyond the earth's centre, or about three and a half times the earth's radius; and there must be a total eclipse of the sun at all places within this shadow.

308. *Breadth of the moon's shadow at the earth.*—The greatest breadth of the moon's shadow at the earth, when it falls perpendicularly on the surface, is about 166 miles.

In the triangle FEV, FE : EV :: sin. FVE : sin. VFE.

But when the moon is nearest, and the shadow is the longest, EV$=$14,146 miles; and the angle FVE$=15'\ 45''.6$. Also, FE $=$3956.6 miles.

In this case VFE$=56'\ 20''.9$.

But the angle FEG $=$ VFE$+$FVE $= 56'\ 20''.9 + 15'\ 45''.6 =$ $1°\ 12'\ 6''.5 =$ arc FG. Hence the arc FH$=2°\ 24'\ 13''$; and if we allow 69 miles to a degree, the breadth of the moon's shadow is 166 miles, nearly.

When the moon is at some distance from the node, the shadow falls obliquely on the earth, and its greatest breadth will evidently be increased.

309. *Breadth of the moon's penumbra at the earth.*—The greatest breadth of the moon's penumbra at the earth's surface, when it falls perpendicularly on the surface, is about 4800 miles.

If we draw the tangent lines AD, BC, and produce them to meet the earth, the sun's rays will be partially excluded from the space included between DV and DL, as also between CV and CK. Any point on the line CK will receive light from all points of the sun's disc. As the point advances toward CV, it will receive less and less of the sun's light, since a larger portion of the moon, M, will be interposed between it and the sun. At the boundary CV, all the rays of the sun are intercepted. This space, KCV, from which the sun's light is partially intercepted, is called the moon's *penumbra.*

The semi-angle of the penumbra CIM=SCB+CSM, of which SCB is the sun's apparent semi-diameter at the moon, and CSM is the sun's horizontal parallax at the moon. The breadth of the penumbra will be greatest when the moon's distance from the earth is greatest, and the sun's distance is least. The sun's greatest apparent semi-diameter at the moon is 16′ 20″.2. Hence CIM 16′ 22″.5.

In the triangle IKM, the angle CKM, when least, is 14′ 41″.0; and KM, when greatest, is 249,307 miles.

Then sin. CIM : sin. CKM :: KM : IM=223,552 miles.

Hence IE=476,815 miles.

Then, in the triangle IEK,

EK : IE :: sin. EIK : sin. EKI=144° 58′ 10″.

Hence EKN=35° 1′ 50″.

The angle KEI=EKN−EIK=34° 45′ 28″=the arc KG.

Hence the entire arc KL=69° 30′ 56″; and if we allow $69\frac{1}{6}$ miles to a degree, the breadth of the penumbra is 4808 miles, nearly.

310. *Velocity of the moon's shadow over the earth.*—The moon advances eastward among the stars about 30′ per hour more than the sun; and 30′ of the moon's orbit is about 2070 miles, which therefore we may consider as the hourly velocity with which the moon's shadow passes over the earth, or at least over that part of it on which the shadow falls perpendicularly; in every other place the velocity will be increased in the ratio of the sine of the angle which MV makes with the surface, in the direction of its

motion, to radius. But the earth's rotation upon its axis will also affect the apparent velocity of the shadow, and, consequently, the duration of the eclipse at any point of the earth. If the point be moving in the direction of the shadow, its velocity in respect to that point will be diminished, and, consequently, the time in which the shadow passes over that point will be increased; but if the point be moving in a direction contrary to that of the shadow, as may happen at places within the polar circle, the relative velocity of the shadow will be increased, and the time diminished.

311. *Different kinds of eclipses of the sun.*—A *partial* eclipse of the sun is one in which a part, but not the whole, of the sun is obscured. A *total* eclipse is one in which the sun is entirely obscured. It must occur at all those places on which the moon's shadow falls. A *central* eclipse is one in which the axis of the moon's shadow, or the axis produced, passes through a given place. An *annular* eclipse is one in which a part of the sun's disc is seen as a ring surrounding the moon.

The apparent discs of the sun and moon, though nearly equal, are subject to small variations, corresponding to their variations of distance, in consequence of which the disc of the moon is sometimes a little greater, and sometimes a little less than that of the sun. If the centres of the sun and moon coincide, and the disc of the moon be less than that of the sun, the moon will cover the central portion of the sun, but will leave uncovered around it a regular ring or annulus, as shown in Fig. 86. This is called an annular eclipse.

Fig. 86.

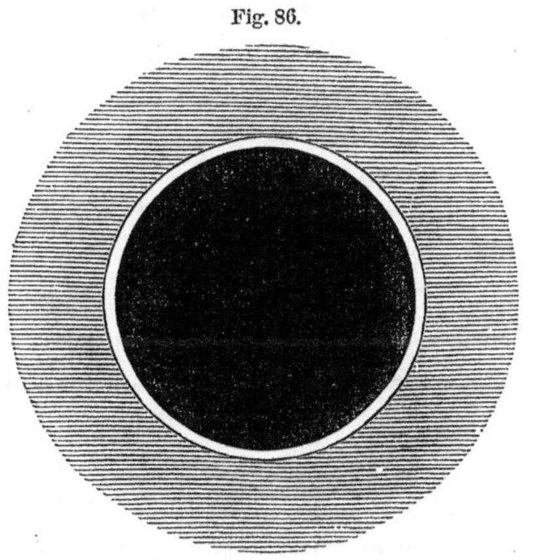

312. *Duration of total and annular eclipses.*—The greatest value of the apparent radius of the moon, as seen from the earth's centre, is 1006″, which may be increased by the moon's elevation above the horizon, Art. 215, to 1024″; and the least value of the radius of the sun is 945″. Their difference is 79″. The greatest

possible duration of a total solar eclipse will be the time required for the centre of the moon to gain upon that of the sun twice 79″, or 158″, which would be about 5m. if the earth did not rotate upon an axis; but, allowing for the earth's rotation, the greatest possible time during which the sun can be totally obscured is 7m. 58s. This will be the duration at the equator. In the latitude of Paris, the greatest possible duration of a total eclipse is 6m. 10s.

The greatest apparent radius of the sun being 978″, and the least apparent radius of the moon being 881″, the greatest possible breadth of the annulus, when the eclipse is central, is 97″. The greatest interval during which the eclipse can continue annular is the time required for the centre of the moon to gain upon that of the sun twice 97″, or 194″, which would be about 7m. if the earth did not rotate; but, by the earth's rotation, this quantity may be increased to 12m. 24s. at the equator. In the latitude of Paris, the greatest possible duration of an annular eclipse is 9m. 56s.

Since the visual directions of the centres of the sun and moon vary with the position of the observer on the earth's surface, an eclipse which is total at one place may be partial at another, while at other places no eclipse whatever may occur.

Since the moon's apparent diameter increases as her elevation above the horizon increases, it sometimes happens, when the apparent diameters of the sun and moon are very nearly equal, that the apparent diameter of the moon, when near the horizon, is a little less than that of the sun, but becomes a little greater than that of the sun as it approaches the meridian; that is, an eclipse which is *annular* at places where it occurs near sunrise, may be *total* at places where it occurs near midday.

313. *To compute the values of the solar ecliptic limits.*—No eclipse of the sun can take place unless some part of the globe of the moon pass within the lines AC and BD, which touch externally

Fig. 87.

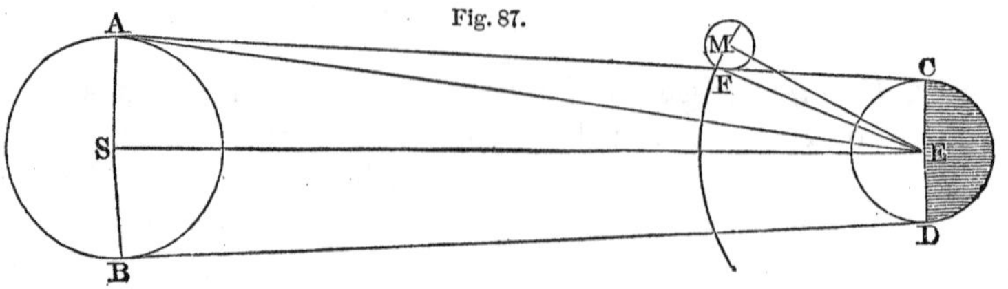

the globes of the sun and earth. The apparent distance, MES, of the moon's centre from the ecliptic at this limit is equal to AEF+AES+FEM. But AEF=EFC−EAC. Hence MES= EFC−EAC+AES+FEM$=p'-p+s'+s$; that is, the sum of the apparent semi-diameters of the sun and moon, plus the difference of their horizontal parallaxes. Taking the greatest and least values of these quantities, we obtain

the major limit of MES$=1° 34' 14''$,
and the minor limit$=1° 24' 19''$.

Computing the corresponding distances from the moon's node, as in Art. 289, we find that if, at the time of conjunction, the sun's distance from the moon's node is more than 18° 20′, an eclipse is *impossible;* and if its distance from the node is less than 15° 25′, an eclipse is *inevitable.* Between these limits an eclipse may or may not occur, according to the magnitude of the parallaxes and apparent diameters.

Since, then, an eclipse can only take place within a few degrees of the moon's node, and the sun passes the two nodes of the moon at opposite seasons of the year, it is evident that if an eclipse occurs in January, one or more eclipses may be expected in July; but no eclipse, either of the sun or moon, could possibly happen in April or October of the same year.

314. *Number of eclipses in a year.*—There may be seven eclipses in a year, and can not be less than two. When there are seven, five of them are of the sun and two of the moon; when there are but two, they are both of the sun.

A solar eclipse is inevitable if conjunction takes place within 15° 25′ on either side of the moon's node, comprehending an arc of longitude of 30° 50′. Now, during a synodic revolution of the moon, the sun's mean motion in longitude is 29° 6′, and in this time the moon's nodes move backward 1° 31′. Hence the sun's motion with reference to the moon's node, in one lunation, is 30° 37′, which is less than 30° 50′. Hence at least one solar eclipse must occur near each node of the moon's orbit, and therefore there must be at least two solar eclipses annually. But it *may* happen that two solar eclipses shall occur near each node, and also one lunar eclipse; and this will happen if opposition takes place very near the moon's node. In this case the moon will be almost centrally eclipsed; and since the sun's motion in reference to the node during half a lunation is only 15° 18′, it is evident that, both at the previous and following new moons, the sun may be within the ecliptic limits from the node, and may therefore be eclipsed at each of these new moons. At the full moon, which occurs in a little less than six months after the former, the sun will be near the other node of the moon's orbit. Consequently, there must be a large eclipse of the moon, and there may be an eclipse of the sun both at the previous and following new moons. At the new moon which occurs five and a half lunations after this latter full moon, and therefore a little before the close of the year, the sun will be near the node again, and must therefore be eclipsed. Thus there may be two eclipses of the moon and five of the sun within a period of twelve months, and these may all be embraced in one calendar year.

315. In the space of eighteen years there are usually about 70 eclipses, 29 of the moon and 41 of the sun. These numbers are nearly in the ratio of two to three. Nevertheless, more lunar than solar eclipses are visible in any particular place, because a lunar eclipse is visible to an entire hemisphere, while a solar is only visible to a part.

The next eclipse of the sun, which will be total in any part of the United States, will occur August 7, 1869, and will be total in Virginia. The next annular eclipse will occur in 1875, and will be annular in Massachusetts. See the list of eclipses, page 325.

316. *Period of eclipses.*—At the expiration of a period of 223 lunations, or about 18 years and 10 days, eclipses, both of the sun and moon, return again in nearly the same order as during that period.

The time from one new moon to another is 29.53 days, and, consequently, 223 lunations include 6585.32 days.

The mean period in which the sun moves from one of the moon's nodes to the same node again is 346.62 days, because the node shifts its place to the westward 19° 35′ per annum. This period is called the synodical revolution of the moon's node. Now 19 synodical revolutions of the node embrace a period of 6585.78 days. Hence, whatever may be the distance of the sun from one of the moon's nodes at any new or full moon, he must, at the end of 223 lunations, be nearly at the same distance from the same node. Hence, after a period of 6585.32 days (which is 18 years $11\frac{1}{3}$ days when there are four bissextile years in the period, or 18 years $10\frac{1}{3}$ days when there are five), eclipses must occur again in nearly the same order as during that period. This period was known to the Chaldæan astronomers. It was by them called the *Saros*, and was used in predicting eclipses.

On page 327 is given a list of eclipses, which will illustrate the period of the Saros, and also show that seven eclipses may occur within a period of twelve months.

317. *Occultations.*—When the moon passes between the earth and a star or planet, she must, during the passage, render the body invisible to some parts of the earth. This phenomenon is called an occultation of the star or planet. The moon, in her monthly course, occults every star which is included in a zone extending to a quarter of a degree on each side of the apparent path of her centre. From new moon to full, the moon moves with the dark edge foremost; and from full moon to new, it moves with the bright edge foremost. During the former period, stars disappear at the dark edge, and reappear at the bright edge; while during the latter period they disappear at the bright edge, and reappear at the dark edge. The disappearance of a star at the dark limb is very sudden and startling, the star appearing to be instantly annihilated at a point of the sky where nothing is seen to interfere with it.

318. *Darkness attending a total eclipse of the sun.*—During a total eclipse of the sun, the darkness is generally so great as to render the brighter stars and planets visible. Each of the five brighter planets has been repeatedly seen during the total obscuration of the sun; all the stars of the first magnitude have in turn been seen, and, on some occasions, a few stars of the second magnitude have been detected. During a total eclipse, the degree of darkness is therefore somewhat less than that which prevails at night in presence of a full moon; but the darkness appears much greater than this, on account of the sudden transition from day to night.

This darkness, however, has little resemblance to the usual darkness of the night, but is attended by an unnatural gloom, which is sometimes tinged with green, sometimes red, and sometimes a yellowish-crimson. The color of the sky changes from its usual azure blue to a livid purple or violet tint. The color of surrounding objects becomes yellowish, or of a light olive or greenish tinge; and the figures of persons assume an unearthly, cadaverous aspect.

319. *Moon sometimes visible in an eclipse of the sun.*—During a total eclipse of the sun, the moon's surface is sometimes faintly illumined by a purplish-gray light, spreading over every part of the disc, so that the light of the disc is quite noticeable to the naked eye. In the eclipse of May 3, 1733, lunar spots were distinctly observed by Vassenius at Gottenberg. This effect is produced by the sun's light reflected from the earth to the moon; for the side of the earth which at such times is presented to the moon is wholly illumined by the sun, and the light of the earth is about 14 times that of the full moon.

320. *Bright points on the moon's disc.*—During the total eclipse of June 24, 1778, about a minute and a quarter before the sun began to emerge from behind the moon's disc, Ulloa discovered, near the northwest part of the moon's limb, a small point of light, estimated as equal to a star of the fourth magnitude. This point gradually increased, and became equal to a star of the second magnitude, when it united with the edge of the sun, which at that instant emerged from behind the moon. This phenomenon was doubtless due to the sun's rays shining through a deep valley on the moon's limb, and the long continuance of this light was due to the moon's motion being nearly parallel to that portion of the sun's circumference.

A similar phenomenon was seen by M. Valz, of Marseilles, during the eclipse of July 8, 1842.

Again the same phenomenon was seen during the eclipse of July 18, 1860, in Algeria, by two French observers, one with the naked eye, and the other with a telescope. The bright point gradually increased, until it blended with the light of the sun's disc as it emerged from behind the moon.

During the eclipse of May 15, 1836, about 25 seconds before the middle of the eclipse, Professor Bessel, with the Königsberg heliometer, observed a faint point of light near the edge of the moon's limb. The point became brighter, and other similar points appeared beside it, which soon united, and in this manner rendered visible the whole of the moon's border between the extremities of the sun's cusps.

Analogous phenomena have been observed in the occultation of stars by the moon. When a star just grazes the northern or southern limb of the moon, it sometimes disappears behind a lunar mountain, and reappears through an adjacent valley, to disappear again behind the next mountain. Several such disappearances and reappearances have been observed within an interval of a few minutes.

321. *The corona.*—During the total obscuration of the sun, the dark body of the moon appears surrounded by a ring of light called the *corona.* This ring is of variable extent, and resembles the "glory" with which painters encircle the heads of saints. It is brightest next to the moon's limb, and gradually fades to a distance equal to one third of her diameter, when it becomes confounded with the general tint of the heavens. Sometimes its breadth is nearly equal to that of the moon's diameter. The corona generally begins 5 or 6 seconds before the total obscuration of the sun, and continues a few seconds after the sun's reappearance. Sometimes the corona is distinctly seen at places where the eclipse is not quite total.

The color of the corona has been variously described. Sometimes it is compared to the color of tarnished silver. Sometimes it is described as of a pearl white; sometimes of a pale yellow; sometimes of a golden hue; sometimes peach-colored, and sometimes reddish.

The intensity of the light of the corona is sometimes such that the eye is scarcely able to support it; but generally it is described as precisely similar to that of the moon.

The corona generally presents somewhat of a radiated appearance. Sometimes these rays are very strongly marked; and long beams have occasionally been traced to a distance of 3° or 4° from the moon's limb.

322. *Cause of the corona.*—Some have maintained that this corona is caused by the diffraction of the sun's light in its passage near the edge of the moon. But the diffracted light, surrounding an opaque circular disc, consists of concentric rings exhibiting a

regular succession of colors—pale blue, yellow, and red. If the corona seen in solar eclipses were due to diffraction, it ought to exhibit a series of concentric colored rings, like those seen surrounding the moon when obscured by a thin haze. Such is not the appearance actually observed. It is more probable that this corona is due to an atmosphere surrounding the sun, extending to a height of several thousand miles above its disc, and reflecting a portion of the sun's light.

The radiated appearance of the corona is probably analogous to the rays which are frequently seen in the western sky after sunset, and which are caused by the shadows of clouds situated near, or perhaps below our visible horizon. In like manner, the clouds which float in the solar atmosphere intercept a portion of the light of the sun's disc, and the space behind them is less bright than that portion of space which is illumined by the unobstructed rays of the sun.

323. *Baily's beads.*—When, in the progress of the eclipse, the sun's disc has been reduced to a thin crescent, this crescent often appears as a band of brilliant points, separated by dark spaces, giving it the appearance of a string of brilliant beads. The same peculiarity is noticed in annular eclipses a few seconds previous to the formation, and again a few seconds previous to the rupture, of the annulus. This phenomenon was first clearly described by Sir Francis Baily on occasion of the annular eclipse of May 15, 1836, and it has hence acquired the name of *Baily's beads.* This appearance is generally ascribed to the inequalities of the moon's surface. The outline of the moon's disc is, not a perfect circle, but is full of notches; and these inequalities are easily seen when the moon's disc is projected upon that of the sun. Just before the commencement of the total eclipse, the tops of the lunar mountains extend to the edge of the sun's disc, but still permit the sun's light to glimmer through the hollows between the mountain ridges.

Fig. 88.

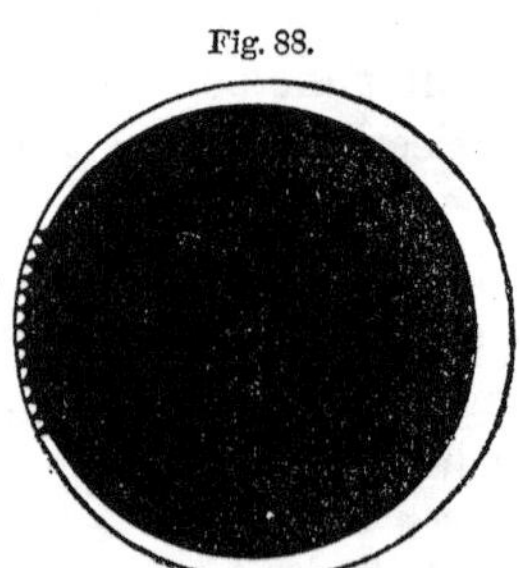

These appearances are materially modified by the color of the glass through which the observations are made. They are most conspicuous through a red glass, and through certain colored glasses are scarcely noticed at all. This peculiarity is probably due to the unequal penetrating power of the differently colored rays of the sun. The red rays of the sun are less readily absorbed than any other rays of the spectrum; and a glass which transmits only the red rays will allow the sun's light to appear through minute crevices in the edge of the moon, when rays of any other color would be entirely absorbed by the colored glass through which the observation is made.

324. *Flame-like protuberances.* — Immediately after the commencement of the total obscuration, red protuberances, resembling flames, may be seen to issue from behind the moon's disc. These appearances were noticed in the eclipse of May 3, 1733, and they have been re-observed during every total solar eclipse which has taken place since that time. They did not, however, attract much attention before the eclipse of July 8, 1842, when they were carefully observed and delineated in accurate diagrams. They were again made the subject of special study in the eclipse of July 28, 1851, and also in that of July 18, 1860.

The *forms* of these protuberances are very various, and some of them quite peculiar. Many of them are nearly conical, the height being frequently greater than the breadth of the base. Others resemble the tops of a very irregular range of hills stretching continuously along one sixth of the moon's circumference. Some of these protuberances reach to a vast height, and show remarkable curvature. One has been compared to a sickle; a second to a Turkish cimeter; a third to a boomerang, with one extremity extending off horizontally far beyond the support of the base; while a fourth was of a circular form, entirely detached from the moon's limb by a space nearly equal to its own breadth.

The *size* of these protuberances is very various. Some have been estimated to have an apparent height of 3′, which would imply an absolute height of 80,000 miles; while others have every intermediate elevation down to the smallest visible object.

The colors of these protuberances have been variously described. Some have been called simply reddish, while others have been characterized as rose-red, purple, or scarlet; and a few have been represented as nearly white.

During the solar eclipses of 1842, 1851, and 1860, the largest of these protuberances were seen by the unassisted eye. In 1860, some of them were observed several seconds before the total obscuration; and in 1842, as well as in 1851, some of them remained visible from 5s. to 7s. after the sun's emersion.

325. *These protuberances emanate from the sun.*—These protuberances emanate from the disc of the sun, and not from that of the moon. This is proved by the following observations made in 1851. The protuberances seen near the eastern limb decrease in dimensions from the commencement of the total eclipse to its close, while those near the western limb increase from the commencement to the close; indicating that the moon covers more and more the protuberances on the eastern side of the sun's disc, and gradually exposes a larger and larger portion of the protuberances on the western side. Again, during the eclipse of 1860, the astronomers who went to Spain to observe the eclipse obtained two excellent photographs, in which these flame-like protuberances were faithfully copied; and it was found that the protuberances retained a fixed position with reference to the sun as the moon glided before it; and they did not change their form, except as the moon, by passing over them, shut them off on the eastern side, while fresh ones became visible on the western side. See Plate III.

326. *Nature of these protuberances.*—That these protuberances are not *solid* bodies like mountains is proved by their peculiar forms, the tops frequently extending horizontally far beyond the support of the base; and they sometimes appear entirely detached from the sun's disc without any visible support.

The same argument proves that they are not liquid bodies; and hence we must conclude that they are gaseous, or are sustained in a gaseous medium.

These flame-like emanations seem to be analogous to the clouds which float at great elevations in our own atmosphere; and we are naturally led to infer that the sun is surrounded by a transparent atmosphere, rising to a height exceeding one tenth of his diameter; and in this atmosphere there are frequently found cloudy masses of extreme tenuity floating at various elevations, and sometimes rising to the height of 80,000 miles above the luminous surface of the sun.

CHAPTER XII.

DIFFERENT METHODS OF FINDING THE LONGITUDE OF A PLACE.

327. *Difference of time under different meridians.*—The sun, in his apparent diurnal motion from east to west, passes successively over the meridians of different places; and noon occurs later and later as we travel westward from any given meridian. If we start from the meridian of Greenwich, then the sun will cross the meridian of a place 15° west of Greenwich one hour later than it crosses the Greenwich meridian—that is, at one o'clock, Greenwich time. A difference of longitude of 15° corresponds to a difference of one hour in local times. In order, then, to determine the longitude of any place from Greenwich, we must accurately determine the local time, and compare this with the corresponding Greenwich time.

328. *Method of chronometers.*—Let a chronometer which keeps accurate time be carefully adjusted to the time of some place whose longitude is known—for example, Greenwich Observatory. Then let the chronometer be carried to a place whose longitude is required, and compared with the correct time reckoned at that place. The difference between this time and that shown by the chronometer will be the difference of longitude between the given place and Greenwich.

It is not necessary that the chronometer should be so regulated as neither to gain nor lose time. This would be difficult, if not impracticable. It is only necessary that its rate should be well ascertained, since an allowance can then be made for its gain or loss during the time of its transportation from one place to the other.

The manufacture of chronometers has attained to such a degree of perfection that this method of determining difference of longitude, especially of stations not very remote from each other, is one of the best methods known. The most serious difficulty in the application of the method consists in determining the rate of the chronometer during the journey; for chronometers generally have a different rate, when transported from place to place, from that which they maintain in an observatory. For this reason, when great accuracy is required, it is customary to employ a large number of chronometers as checks upon each other; and the chronometers are transported back and forth a considerable number of times.

This is the method by which the mariner commonly determines his position at sea. Every day, when practicable, he measures the sun's altitude at noon, and hence determines his latitude. About three hours before or after noon he measures the sun's altitude again, and from this he computes his local time by Art. 145. The chronometer which he carries with him shows him the true time at Greenwich, and the difference between the two times is his longitude from Greenwich.

329. *By eclipses of the moon.*—An eclipse of the moon is seen at the same instant of absolute time in all parts of the earth where the eclipse is visible. Therefore, if at two distant places the times of the beginning of the eclipse are carefully observed, the difference of these times will give the difference of longitude between the places of observation; but, on account of the gradually increasing darkness of the penumbra, it is impossible to determine the precise instant when the eclipse begins, and therefore this method is of no value except under circumstances which preclude the use of better methods.

330. *By the eclipses of Jupiter's satellites.*—The moons of Jupiter are eclipsed by passing into the shadow of Jupiter in the same manner as our moon is eclipsed by passing into the shadow of the earth. These eclipses begin at the same instant of absolute time for all places at which they are visible. If, then, the times of the beginning of an eclipse be accurately observed at two different places, the difference of these times will be the difference of longitude of the places. Since, however, the light of a satellite diminishes gradually while entering the shadow, and increases gradually on leaving it, the *observed* time of beginning or ending of the eclipse must depend on the power of the telescope used, and also upon the eye of the observer. This method, therefore, is of no

value at fixed observatories, where better methods are always available.

331. *By an eclipse of the sun or the occultation of a star.*—The times of the beginning and end of an eclipse of the sun, or of the occultation of a star or planet at any place, depend on the position of the place. We can not, therefore, use a solar eclipse as an instantaneous signal for comparing directly the local times at two stations; but we may deduce by computation from the observed beginning and end of an eclipse, the time of true conjunction of the sun and moon—that is, the time of conjunction as seen from the centre of the earth; and this is a phenomenon which happens at the same absolute instant for every observer on the earth's surface. If the eclipse has been observed under two different meridians, we may determine the instant of true conjunction from the observations at each station; and since the absolute instant of this phenomenon is the same for both places, the difference of the results thus obtained is the difference of longitude of the two stations. This is one of the most accurate methods known to astronomers for determining the difference of longitude of two stations remote from each other. This is especially true when the moon crosses a cluster containing a large number of stars, as the Pleiades.

332. *By moon culminating stars.*—Certain stars situated near the moon's path, and passing the meridian at short intervals before or after the moon, are called *moon culminating stars.* The moon's motion in right ascension is very rapid, amounting to about half a degree, or two minutes in time, during a sidereal hour—that is, during the interval that elapses from the time a star is on the meridian of any place, till it is on the meridian of a place whose longitude is 15° west of the former. Hence the intervals between the passages of the moon and a star over the meridians of two places differing an hour in longitude must differ about two minutes; and for other differences of longitude there must be a proportional difference in the intervals. Hence, if the intervals between the passages of the moon and a star over the meridians of two places be accurately observed, the difference of their longitude may be found by means of the moon's hourly variation in right ascension.

The chief disadvantage of this method consists in this circumstance, that an error in the observed increase of right ascension will produce an error nearly 30 times as great in the computed longitude. Hence, to obtain a satisfactory result by this method requires a series of observations made with the utmost care, and continued through a long period of time.

333. *By lunar distances.*—The Nautical Almanac furnishes for each day the distance of the moon from the sun, the larger planets, and several stars situated near the moon's path. These distances are given for Greenwich time, and are such as they would appear to a spectator placed at the centre of the earth. A mariner on the ocean measures with a sextant the distance from the moon to one of the stars mentioned in the Almanac. He corrects this distance for refraction and parallax, and thus obtains the true lunar distance as it would be seen at the centre of the earth. By other observations, he knows the local time at which this distance was measured, and, by referring to the Nautical Almanac, he finds the Greenwich time at which the lunar distance was the same. The difference between the local time and the Greenwich time represents the longitude of the place of observation from Greenwich. This method of finding the longitude may be practiced at sea, and in long voyages should always be resorted to as a check upon the method by chronometers.

334. *By the electric telegraph.*—The difference of the local times of two places may be determined by means of any signal which can be seen or heard at both places at the same instant. When the places are not very distant, the explosion of a rocket, or the flash of gunpowder, or the flight of a shooting star may serve this purpose.

The electric telegraph affords the means of transmitting signals to a distance of a thousand miles or more with scarcely any appreciable loss of time. Suppose that there are two observatories at a considerable distance from each other, and that each is provided with a good clock, and with a transit instrument for determining its error; then, if they are connected by a telegraph wire, they have the means of transmitting signals at pleasure from either observatory to the other for the purpose of comparing their local times. For convenience, we will call the most eastern station E, and the western W. The following is one mode of comparing their local times.

335. *Mode of comparing the local times.*—The plan of operations having been previously agreed upon, the astronomer at E strikes the key of his register, and makes a record of the time according to his observatory clock. Simultaneously with this signal at E, the armature of the magnet at W is moved, producing a click like the ticking of a watch. The astronomer at W hears the sound, and notes the instant by his clock. The difference between the time recorded at E and that at W is the difference between the two clocks. A single comparison of this kind will furnish the difference of longitude to the nearest second; but to obtain the fraction of a second with the greatest precision requires many repetitions, and this is accomplished as follows:

At the commencement of the minute by his clock, the astronomer at E strikes his signal key, and the time of the signal is recorded both at E and W. At the close of 10 seconds the signal is repeated, and the observation is recorded at both stations. The same thing is done at the end of 20 seconds, of 30 seconds, and so on to 20 repetitions. The astronomer at W then transmits a series of signals in the same manner, and the times are recorded at both stations.

336. *The velocity of the electric fluid.*—This double set of signals not only furnishes an accurate comparison of the two clocks, but also enables us to measure the velocity of the electric fluid. If the fluid requires no time for its transmission, then the apparent difference between the two clocks will be the same, whether we determine it by signals transmitted from E to W, or from W to E. But if the fluid requires time for its transmission, these results will differ. Suppose the true difference of longitude between the places is one hour, and that it requires one second for a signal to be transmitted from E to W. Then, if at 10 o'clock a signal be made and recorded at E, it will be a second before the signal is heard and recorded at W—that is, the time recorded at W will be 9 hours and 1 second; and the apparent difference between the two clocks will be 59 minutes and 59 seconds. But if a signal be made at W at nine o'clock, it will be heard at E at 10 hours and 1 second; and the apparent difference between the two

clocks will be 1 hour and 1 second. Thus the differences between the two clocks, as derived from the two methods of comparison, differ by *twice* the time required for the transmission of a signal from E to W. Numerous observations, made on the longest lines and with the greatest care, have shown that the velocity of the electric fluid upon the telegraph wires is about 16,000 miles per second. The mean of the results obtained by signals transmitted in both directions, gives the true difference between the two clocks, independent of the time required in the transmission of signals.

337. *How the clock may break the electric circuit.*—The most accurate method of determining difference of longitude consists in employing one of the clocks to break the electric circuit each second. This may be accomplished in the following manner: Near the lower extremity of the pendulum place a small metallic cup containing a globule of mercury, so that once in every vibration the pointer at the end of the pendulum may pass through the mercury. A wire from one pole of the battery is connected with the supports of the pendulum, and another wire from the other pole of the battery connects with the cup of mercury. When the pointer is in the mercury, the electric circuit will be complete through the pendulum; but as soon as it passes out of the mercury, the circuit will be broken.

When the connections are properly made, there will be heard a click of the magnet at each station, simultaneously with the beats of the electric clock. If each station be furnished with an ordinary Morse register, there will be traced upon the paper a series of lines, of equal length, separated by short intervals, thus:

——— ——— ——— ——— ———

The mode of using the register for marking the date of any event is to strike the key of the register at the required instant, when an interruption will be made in one of the lines of the graduated scale; and its position will indicate not only the second, but the fraction of a second at which the signal was made.

We now employ the same electric circuit for telegraphing transits of stars. A list of stars having been selected beforehand, and furnished to each observer, the astronomer at E points his transit telescope upon one of the stars as it is passing his meridian, and strikes the key of his register at the instant the star passes successively each wire of his transit, and the dates are recorded, not only upon his own register, but also upon that at W. When the same star passes over the meridian of W, the observer there goes through the same operations, and his observations are printed upon both registers. These observations furnish the difference of longitude of the two stations, independently of the tabular place of the star employed, and also independently of the absolute error of the clock.

CHAPTER XIII.

THE TIDES.

338. *Definitions.*—The alternate rise and fall of the surface of the sea twice in the course of a lunar day, or of 24h. 51m. of mean solar time, is the phenomenon known by the name of the *tides.* When the water is rising it is said to be *flood* tide, and when it is falling, *ebb* tide. When the water is at its greatest height it is said to be *high* water, and when at its least height, *low* water.

339. *Spring and neap tides.*—The time from one high water to the next is, at a mean, 12h. 25m. 24s. Near the time of new and full moon the tide is the highest, and the interval between the consecutive tides is the least, viz., 12h. 19m. Near the quadratures, or when the moon is 90° distant from the sun, the tides are the least, and the interval between them is the greatest, viz., 12h. 30m. The former are called the *spring tides*, and the latter the *neap tides.* At New York the average height of the spring tides is 5.4 feet, and of the neap tides 3.4 feet.

340. *The establishment of a port.*—The time of high water is mostly regulated by the moon; and for any given place, the hour of high water is always nearly at the same distance from that of the moon's passage over the meridian. The mean interval between the moon's passage over the meridian, and high water at any port on the days of new and full moon, is called the *establishment* of the port. The mean interval at New York is 8h. 13m., and the difference between the greatest and the least interval occurring in different parts of the month is 43 minutes.

341. *Tides at perigee and apogee.*—The height of the tide is affected by the distance of the moon from the earth, being highest near the time when the moon is in perigee, and lowest near the time when she is in apogee. When the moon is in perigee, at or near the time of a new or full moon, unusually high tides occur.

342. *Cause of the tides.*—The facts just stated indicate that the moon has some agency in producing the tides. The tides, however, are not due to the simple attraction of the moon upon the earth, but to the *difference* of its attraction on the opposite sides

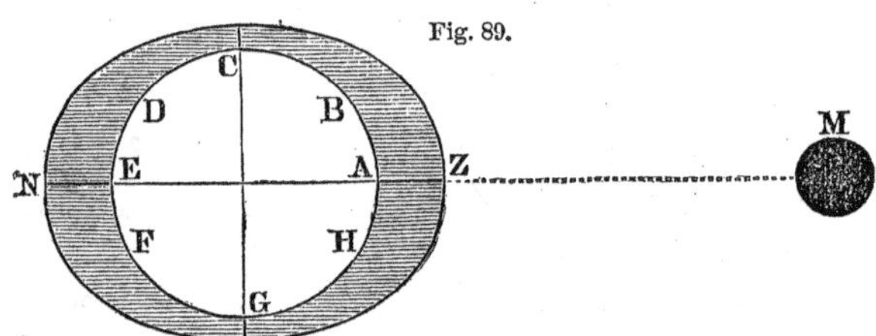

Fig. 89.

of the earth. Let ACEG represent the earth, and let us suppose its entire surface to be covered with water; also, let M be the place of the moon. The different parts of the earth's surface are at unequal distances from the moon. Hence the attraction which the moon exerts at A is greater than that which it exerts at B and H, and still greater than that which it exerts at C and G; while the attraction which it exerts at E is least of all. The attraction which the moon exerts upon the mass of water immediately under it, near the point Z, is greater than that which it exerts upon the solid mass of the globe. The water will therefore heap itself up over A, forming a convex protuberance—that is, high water will take place immediately under the moon. The water which thus collects at A will flow from the regions C and G, where the quantity of water must therefore be diminished—that is, there will be low water at C and G.

The water at N is *less* attracted than the solid mass of the earth. The solid mass of the earth will therefore recede from the waters at N, leaving the water behind, which will thus be heaped up at N, forming a convex protuberance, or high water, similar to that at Z. The sea is therefore drawn out into an ellipsoidal form, having its major axis directed toward the moon.

343. *Effect of the sun's attraction.*—The attraction of the sun produces effects similar to those of the moon, but less powerful in raising a tide, because the *inequality* of the sun's attraction on different parts of the earth is very small. It has been computed that the tidal wave due to the action of the moon is about double that which is due to the sun.

There is, therefore, a solar as well as a lunar tide wave, the latter greater than the former, and each following the luminary from which it takes its name. When the sun and moon are both on the same side of the earth, or on opposite sides, that is, when it is new or full moon, their effects in producing tides are combined, and the result is an unusually high tide, called *spring tide.*

When the moon is in quadrature, the action of the sun tends to produce low water where that of the moon produces high water, and the result is an unusually small tide, called *neap tide.*

344. *Effect of the moon's declination on the tides.*—The height of the tide at a given place is influenced by the declination of the moon. When the moon has no declination, the highest tides should occur along the equator; and the heights should diminish from thence toward the north and south; but the two daily tides at any place should have the same height. When the moon has north declination, as shown in Fig. 90, the highest tides on the side of the earth next the moon will be at places having a cor-

Fig. 90.

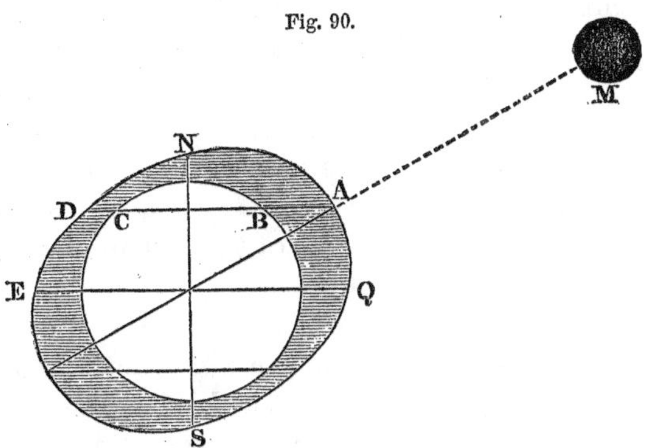

responding north latitude, as at B, and on the opposite side at those which have an equal south latitude. And of the two daily tides at any place, that which occurs when the moon is nearest the zenith should be the greatest. Hence, when the moon's declination is north, the height of the tide at a place in north latitude should be greater when the moon is above the horizon than when she is below it. At the same time, places south of the equator have the highest tides when the moon is below the horizon, and the least when she is above it. This is called the *diurnal inequality*, because its cycle is one day; but it varies greatly in amount at different places.

The great wave which constitutes the tide is to be considered as an undulation of the waters of the ocean, in which (except when it passes over shallows or approaches the shores) there is little or no *progressive* motion of the water.

345. *Why the phenomena of the tides are so complicated.*—The actual phenomena of the tides are far more complicated than they would be if the earth were entirely covered with an ocean of great depth. The water covers less than three quarters of the earth's surface, and a considerable part of this water is less than a mile in depth. Two great continents extend from near the north pole to a great distance south of the equator, thus interrupting the regular progress of the tidal wave around the globe. In the northern hemisphere, the waters of the Atlantic can communicate with those of the Pacific only by Behring's Strait, a channel 36 miles in breadth, which effectually prevents the transmission of any considerable wave from the Atlantic to the Pacific through the northern hemisphere. In the southern hemisphere, the American continent extends to 56° of S. latitude, and in about latitude 60° commences a range of islands, near which are indications of an extensive antarctic continent, leaving a passage only about 500 miles in breadth. Through this passage the motion of the tidal wave (as we shall presently see) is eastward, and not westward; whence we conclude that the tides of the Atlantic are not propagated into the Pacific.

346. *Cotidal lines.*—The phenomena of the tides, being thus exceedingly complicated, must be learned chiefly from observations; and in order to present the results of observations most conveniently upon a map, we draw a line connecting all those places which have high water at the same instant of absolute time. Such lines are called *cotidal* lines. The accompanying map, Plate I., shows the cotidal lines for nearly every ocean, drawn at intervals of 3 hours, and expressed in Greenwich time.

347. *Origin of the tidal wave.*—By inspecting this map, we perceive that the great tidal wave originates in the Pacific Ocean, not far from the western coast of South America, in which region high water occurs about two hours after the moon has passed the meridian. The wave thus formed, if left undisturbed, would travel, like ordinary waves, with a velocity depending upon the depth of water. When the breadth of a wave is very great in comparison with the depth of water, the velocity of its progress is equal to that which a heavy body would acquire in falling by gravity through half the depth of the liquid. The velocity of such a wave for different depths of the ocean is as follows:

When the depth of the water is	25 feet,	the velocity of the wave is	19 miles per hour.
	100 "		39 " "
	250 "		61 " "
	1,000 "		122 " "
	5,000 "		273 " "
	20,000 "		547 " "
	50,000 "		865 " "

348. *Progress and velocity of the tidal wave.*—Since the moon travels westward at the rate of 1000 miles per hour over the equator, it tends to carry high water along with it at the same rate. But the shallow water of most parts of the ocean does not allow the tidal wave to travel with this velocity. The wave of high water, first raised near the western coast of South America, travels toward the northwest through the deep water of the Pacific at the rate of 850 miles per hour, and in about ten hours reaches the coast of Kamtschatka. On account of more shallow water, the same wave travels westward and southwestward with less velocity, and it is about 12 hours old when it reaches New Zealand, having advanced at the rate of about 400 miles per hour. Passing south of Australia, the tidal wave travels westward and northward into the Indian Ocean, and is 29 hours old when it reaches the Cape of Good Hope. Hence it is propagated through the Atlantic Ocean, traveling northward at the rate of about 700 miles per hour, and in 40 hours from its first formation it reaches the shallow waters of the coast of the United States, whence it is propagated into all the bays and inlets of the coast. The wave which enters at the eastern end of Long Island Sound is about 4 hours in reaching the western end, so that the wave is 44 hours old when it arrives at New Haven.

349. *Tides of the North Atlantic.*—A portion of the great Atlantic wave advances up Baffin's Bay, and at the end of 56 hours reaches the latitude of 78°. The principal part of the Atlantic wave, however, turns eastward toward the Northern Ocean, and in 44 hours brings high water to the western coast of Ireland. After passing Scotland, a portion of this wave turns southward with diminished velocity into the North Sea, and thence follows up the Thames, bringing high water to London at the end of 66 hours from the first formation of this wave in the Pacific Ocean.

350. *Velocity of the tidal wave in shallow water.*—As the tidal wave approaches the shallow water of the coast, its velocity is speedily reduced from 500, or perhaps 900 miles per hour, to 100 miles, and soon to 30 miles per hour; and in ascending bays and rivers its velocity becomes still less. From the entrance of Chesapeake Bay to Baltimore the tide travels at the average rate of 16 miles per hour, and it advances up Delaware Bay with about the same velocity. From Sandy Hook to New York city the tide advances at the rate of 20 miles per hour, and it travels from New York to Albany in 9h. 9m., being at the average rate of nearly 16 miles per hour.

From New York Bay the tidal wave is propagated through East River until it meets the wave which has come in from the Atlantic through the eastern end of the Sound. This place of meeting is only 21 miles from New York, showing that the velocity of the tidal wave through East River is only 7½ miles per hour—a result which must be ascribed to the narrowness and intricacy of the channel.

351. *Tidal wave on the western coast of South America.*—The tidal wave which we have thus traced through oceans, bays, and rivers, has every variety of direction; in some places advancing westward, and in others eastward; in some places northward, and in others southward; but in each case it may be regarded as a continuous forward movement, and the change in its direction results from a change in the direction of the channel. But there is one exception to this general rule. We have traced the origin of the tidal wave to a region about 1000 miles west of the coast of South America. From this point high water is not only propagated westward around the globe, but also eastward toward Cape

Horn. In this region the motion of the tidal wave appears to be similar to that of the wave produced by throwing a stone upon the surface of a tranquil lake, the wave traveling off in all directions from the first point of disturbance.

352. *Is the tidal wave a free or a forced oscillation?*—If the moon should suddenly cease its disturbing action upon the waters of the ocean, the tidal wave already formed would travel forward with a velocity depending solely upon the depth of water, and this would be called a *free* wave. Now the moon continually tends to form high water directly beneath it—that is, it tends to carry high water westward at the rate of 1000 miles per hour over the equator. Such a wave, if it could actually be formed, would be called a *forced* oscillation, because its velocity would be independent of the depth of water. Is, then, the great tidal wave a free or a forced oscillation? We may answer this question by observing the velocity of the tidal wave in the Atlantic Ocean, whose depth has been approximately determined. The velocity of the tidal wave in the North Atlantic, from the equator to latitude 50°, is about 640 miles per hour, corresponding to a depth of 27,500 feet, which is somewhat greater than the average depth of the Atlantic. The velocity of the tidal wave in the Atlantic appears to be about one third greater than that of a free wave, and this excess of velocity is probably due to the immediate action of the sun and moon; in other words, the tidal wave is, to some extent, a *forced* oscillation, but its rate of progress appears to be determined mainly by the depth of water.

353. *Height of the tides.*—At small islands in mid-ocean the tides never rise to a great height, sometimes even less than one foot; and the average height of the tides for the islands of the Atlantic and Pacific Oceans is only 3½ feet. Upon approaching an extensive coast where the water is shallow, the velocity of this tidal wave is diminished, the cotidal lines are crowded more closely together, and the height of the tide is thereby increased; so that while in mid-ocean the average height of the tides does not exceed 3½ feet, the average in the neighborhood of continents is not less than 4 or 5 feet. According to theory, the height of the wave should vary inversely as the fourth root of the depth; that is, in water 100 feet deep, the wave should be twice as high as in

water 1600 feet deep. Fig. 91 shows the change in the form of waves in approaching shallow water.

Fig. 91.

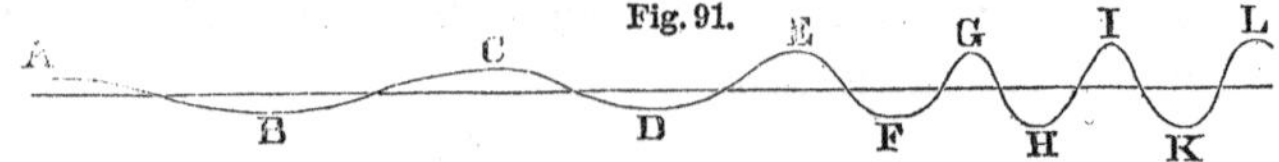

354. *Height of the tides modified by the conformation of the coast.*—Along the coast of an extensive continent the height of the tides is greatly modified by the conformation of the shore line. When the coast is indented by broad bays which are open in the direction of the tidal wave, this wave, being contracted in breadth, must necessarily increase in height, so that at the head of a bay the height of the tide may be several times as great as at the entrance. The operation of this principle is exhibited at numerous places upon the Atlantic coast. Thus, if we draw a straight line from Cape Hatteras to the southern part of Florida, it will cut off a bay about 200 miles in depth. At Cape Hatteras and Cape Florida the tide rises only 2 feet; at Cape Fear and St. Augustine it rises 4 feet; while at Savannah it rises 7 feet.

Fig. 92.

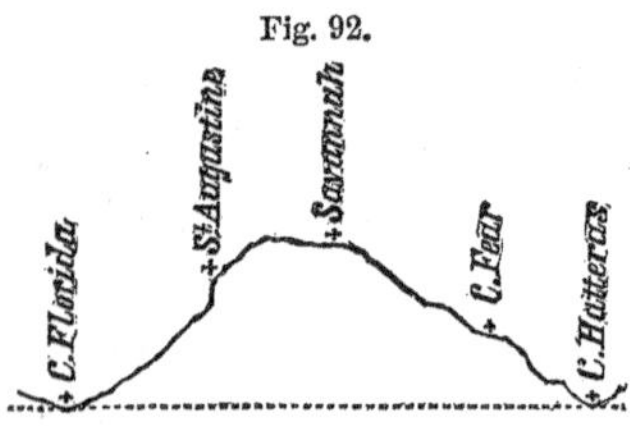

355. *Tides in the Bay of Fundy.*—If we draw a straight line from Nantucket to Cape Sable, it will cut off a bay in which the phenomena of the tides are still more remarkable. At Nantucket the tide rises only 2 feet; at Boston it rises 10 feet; near the entrance to the Bay of Fundy, 18 feet; while at the head of the bay it sometimes rises to the height of 70 feet. This result is due mainly to the contraction of the channel through which the tidal wave advances.

Fig. 93.

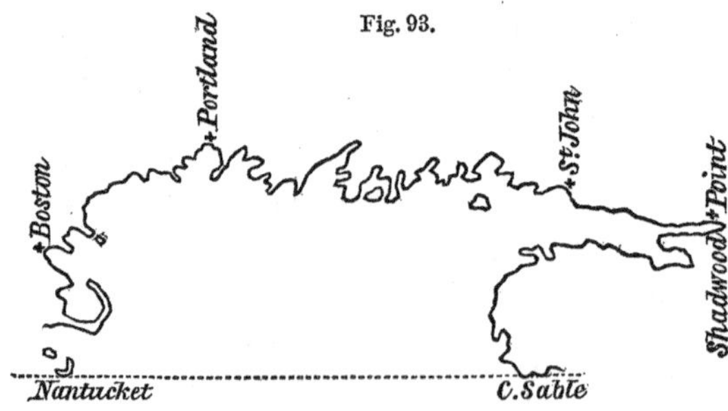

356. *Tides of Long Island Sound, etc.*—So, also, at the east end of Long Island Sound, the tide rises only 2 feet; but in its progress westward through the Sound the height continually increases, until at the west end the height is more than 7 feet.

At the entrance to Delaware Bay the tide rises only 3½ feet, while at New Castle it rises 6½ feet.

The tide from the North Atlantic is propagated through the Gulf of St. Lawrence, and thence through the River St. Lawrence, at the average rate of about 70 miles per hour, being 12 hours from the ocean to Quebec. This tide increases in height as it advances, being only 9 feet at the mouth of the St. Lawrence, while it is 20 feet at Quebec.

357. *Tides modified by a projecting promontory.*—A promontory, as A, projecting into the ocean, so as to divide the tidal wave and throw it off upon either side, not only causes the tide at B and C to rise above the mean height, but sometimes reduces the tide at A below the mean height. Thus, at Cape Hatteras, the tide rises less than 2 feet in height, while along the coast on either side the tide rises to the height of 5 or 6 feet. So, also, on the south side of Nantucket, the tides are less than 2 feet in height, while along the coast north of Cape Cod the tide rises 10 feet in height.

Fig. 94.

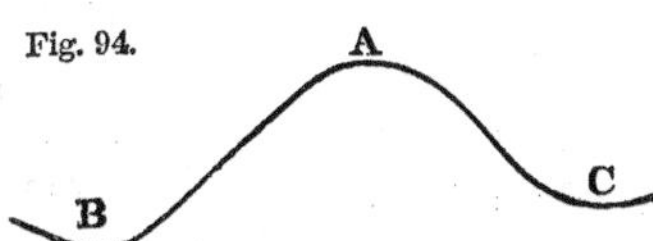

358. *Tides on the coast of Ireland.*—So, also, on the southwest

Fig. 95.

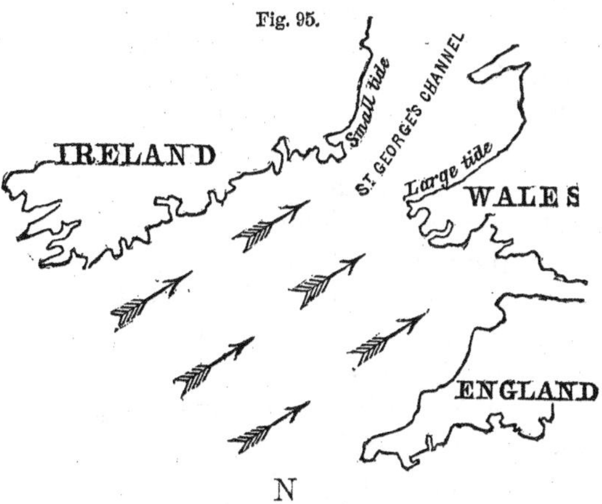

coast of Ireland, where the tidal wave from the Atlantic first strikes the coast, the tide is less than it is at a short distance along the coast either eastward or northward.

In some cases the form and position of a promontory are such as to divert the tidal wave from some part of the coast, and leave it almost destitute of a tide. Such a case occurs on the east coast of Ireland. The wave from the Atlantic, being forced up St. George's Channel, is driven upon the coast of Wales, where the tide rises to the height of 36 feet, while it is almost wholly diverted from the opposite coast of Ireland, where the range of the tide is only 2 feet.

359. *Tides of rivers.*—The tides of rivers exhibit the operation of similar principles. In a channel of uniform breadth and depth, the height of the tide should gradually *diminish*, in consequence of the effect of friction. But if the channel contracts or shoals rapidly, the height of the tide will *increase*. There is, then, a certain rate of contraction, with which the range of the tides will remain stationary. If the river contracts more rapidly, the height of the tides will increase; if the channel expands, the height of the tides will diminish. Hence, in ascending a long river, it may happen that the height of the tides will increase and decrease alternately.

Thus, at New York, the mean height of the tide is 4.3 feet; at West Point, 55 miles up the Hudson River, the tide rises only 2.7 feet; at Tivoli, 98 miles from New York, the tide amounts to 4 feet; while at Albany it rises only 2.3 feet.

360. *The diurnal inequality in the height of the tides.*—If the sun and moon moved always in the plane of the equator, and the earth were entirely covered with water to a great depth, the two daily tides should have nearly the same height; but when they are out of the equator, the two daily tides should generally be unequal. The moon sometimes reaches 28° north declination, in which case it tends to raise the highest tide at a station in latitude 28° north, while the highest tide on the opposite side of the earth should be in latitude 28° south. Hence the two tides which are formed in the northern hemisphere under opposite meridians must be of unequal heights—that is, the morning and evening tides at a given place should be unequal. The same would be true for the southern hemisphere, but on the equator there would be no such diurnal inequality.

361. *Diurnal inequality in the North Atlantic Ocean.*—Along the Atlantic coast of the United States, when the moon has its greatest declination, the difference between high water in the forenoon and afternoon averages about 18 inches; but this difference almost entirely disappears when the moon is on the equator.

On the coast of Ireland, the diurnal inequality, at its maximum, is only one foot, while the average height of the tides is nine feet. On some parts of the European coast the diurnal inequality is still smaller, and can with difficulty be detected in a long series of observations.

362. *Diurnal inequality on the Pacific coast.*—On the Pacific coast of the United States, when the moon is far from the equator, there is one large and one small tide during each day. In the Bay of San Francisco, the difference between high and low water in the forenoon is sometimes only *two inches*, while in the afternoon of the same day the difference is $5\frac{1}{2}$ feet. When the moon is on the equator this inequality disappears, and the two daily tides are nearly equal.

Fig. 96.

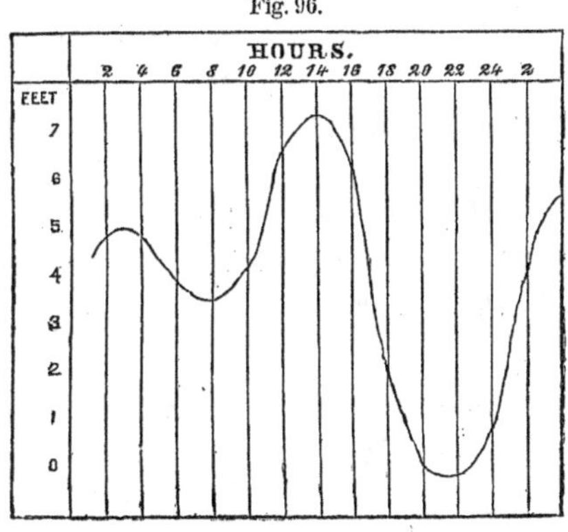

At other places on the Pacific coast this inequality in the two daily tides is still more remarkable. At Port Townsend, near Vancouver's Island, when the moon has its greatest declination, there is *no* descent corresponding to morning low water, but merely a temporary check in the rise of the tide. Thus one of the two daily tides becomes obliterated; that is, we find but one tide in the 24 hours. Similar phenomena occur at

Fig. 97.

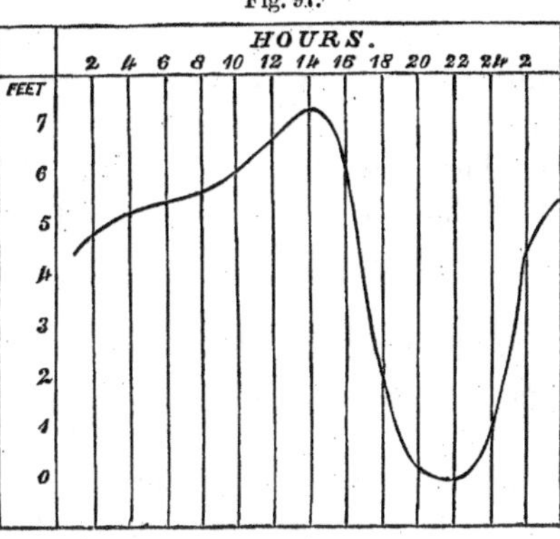

other places upon the Pacific coast, and also on the coast of Kamtschatka.

363. *Cause of these variations in the diurnal inequality.*—The tide actually observed at any port is the effect, not simply of the *immediate* action of the sun and moon upon the waters of the ocean, but is rather the resultant of their continued action upon the waters of the different seas through which the wave has advanced from its first origin in the Pacific until it reaches the given port, embracing an interval sometimes of one or two days, and perhaps even longer. During this period the moon's action tends sometimes to produce a large tide, and sometimes a small one; and in a tide whose age is more than 12 hours, these different effects are combined so as sometimes partly to obliterate the diurnal inequality, and sometimes to exaggerate it. This is probably the reason why the diurnal inequality is less noticeable in the North Atlantic than in the North Pacific.

364. *Four tides in 24 hours.*—In some places the tide rises and falls four times in 24 hours. This happens on the east coast of Scotland, where the form of the tidal wave is such as is represented by the annexed figure. This anomaly is ascribed to the superposition of two tidal waves, one traveling round the north of Scotland, and advancing southward through the North Sea, while the other passes through the English Channel, and thence advances northward into the same sea. At some places these two waves arrive nearly at the same hour, and are so superposed as not to be distinguished from each other; but at other places one arrives 2 or 3 hours behind the other, thus presenting the appearance of high water 4 times in 24 hours.

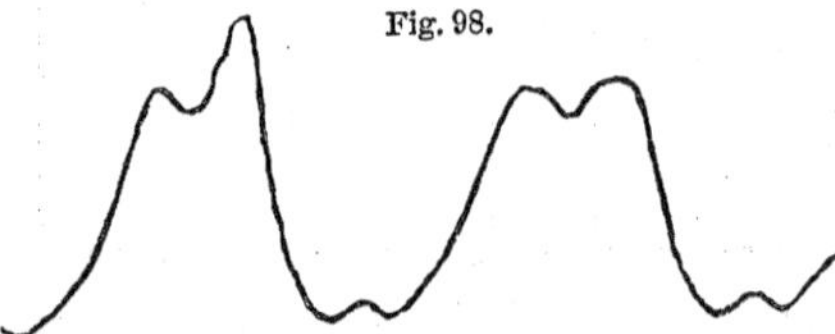
Fig. 98.

365. *Small tides of the Pacific Ocean.*—Near the middle of the Pacific Ocean, in the neighborhood of the Society Islands, from latitude 13° to 18° S., and from longitude 140° to 176° W., the tides are smaller than have been found in any other portion of the open sea, averaging less than one foot in height. At Tahiti

(latitude 17° 29′ S., longitude 149° 29′ W.), the tides at full moon rise to the height of about 15 inches, and at the quadratures only about 3 inches. There are two high waters daily occurring near noon and midnight, being seldom earlier than 10 A.M., or later than 2½ PM.

366. *Cause of these peculiarities.*—It is uncertain what is the cause of this small height of the tides, but it is believed that the following consideration will explain it, at least in part. The original tide wave, starting from the eastern part of the Pacific Ocean, reaches Tahiti about six hours after the moon's transit over that meridian. Hence, when the main tidal wave of the Pacific reaches that port, the *immediate* effect of the moon is to produce *low* water at the same hour; and the superposition of these two waves produces a nearly uniform level of the water.

The occurrence of high water within about two hours of noon every day seems to indicate that the power of the sun to raise a tide is here nearly equal to that of the moon. In the Atlantic Ocean, the influence of the moon upon the tides is generally about double that of the sun; but this ratio appears to be a variable one.

367. *Tides of the Gulf of Mexico.*—The Gulf of Mexico is a shallow sea, about 800 miles in diameter, almost entirely surrounded by land, and communicating with the Atlantic by two channels, each about 100 miles in breadth. It is by the Florida channel that the tidal wave from the Atlantic is chiefly propagated into the Gulf, but its progress is so much obstructed by the West India Islands that its height is very much reduced. Between Florida and Cuba the tidal wave advances slowly westward; but after passing the channel it moves more rapidly, and reaches the western side of the Gulf in seven hours, showing an average progress of 125 miles per hour.

The tides in the Gulf are every where quite small. At Mobile and Pensacola the average height is only one foot. The diurnal inequality is also quite large, so that at most places (except when the moon is near the equator) one of the daily tides is well-nigh inappreciable, and the tide is said to ebb and flow but once in 24 hours.

368. *Tides of the Mediterranean.*—The tides of the Mediterranean are generally so small as not to be regarded by navigators. Their average height does not exceed 18 inches. In the neighborhood of the Strait of Gibraltar the tide rises from 2 to 4 feet; at Venice it rises from 18 inches to 4 feet; and at Tunis it sometimes rises to the height of 3 feet.

The length of the Mediterranean is 2400 miles, or nearly one third the diameter of the earth; and the average height of the tides is here at least one third what it is in the open sea.

369. *Tides of inland seas.*—In small lakes and seas which do not communicate with the ocean there is a daily tide, but so small that it requires the most accurate observations to detect it. The existence of a tide in Lake Michigan has been proved by a series of observations made at Chicago in 1859. The average height of this tide is $1\frac{3}{4}$ inches; and the average time of high water is 30 minutes after the time of the moon's transit.

The length of Lake Michigan is 350 miles, or $\frac{1}{23}$d of the earth's diameter, and its tide is about $\frac{1}{23}$d of that which prevails in mid-ocean.

CHAPTER XIV.

THE PLANETS — THEIR APPARENT MOTIONS. — ELEMENTS OF THEIR ORBITS.

370. *Number, etc., of the planets.*—The planets are bodies of a globular form, which revolve around the sun as a common centre, in orbits which do not differ much from circles. The name *planet* is derived from the Greek word πλανήτης, signifying a wanderer, and was applied by the ancients to these bodies because their apparent movements were complicated and irregular. Five of the planets—Mercury, Venus, Mars, Jupiter, and Saturn—are very conspicuous, and have been known from time immemorial. Uranus was discovered in 1781, and Neptune in 1846, making eight planets including the earth. Besides these there is a large group of small planets, called asteroids, situated between the orbits of Mars and Jupiter. The first of these was discovered in 1801, and the number known in 1867 amounts to 94.

The orbits of Mercury and Venus are included within the orbit of the earth, and they are hence called *inferior planets*, while the others are called *superior planets*.

371. *The satellites.*—Some of the planets are the centres of secondary systems, consisting of smaller globes, revolving round them in the same manner as they revolve around the sun. These are called *satellites* or *moons*. The primary planets which are thus attended by satellites carry the satellites with them in their orbits around the sun. Of the satellites known at the present time, four revolve around Jupiter, eight around Saturn, four around Uranus, and one around Neptune. The moon is also a satellite to the earth.

372. *The orbits of the planets.*—The orbit of each of the planets is an ellipse, of which the sun occupies one of the foci. That point of its orbit at which a planet is nearest the sun is called the *perihelion*, and that point at which it is most remote is called the *aphelion*.

The *eccentricity* of a planetary orbit is the distance of the sun from the centre of the ellipse which the planet describes, expressed in terms of the semi-major axis regarded as a unit; or, in other words, it is the quotient of the distance between the centre and focus, divided by the semi-major axis. The eccentricities of most of the planetary orbits are so minute that, if the form of the orbit were exactly delineated on paper, it could not be distinguished from a circle except by careful measurement.

373. *Geocentric and heliocentric places.*—The motion of a planet as it appears to an observer on the earth is called the *geocentric* motion, while its motion as it would appear if the observer were transferred to the sun is called its *heliocentric* motion. The motions of the planets can not be observed from the sun as a centre, but from the geocentric motions, combined with the relative distances of the earth and planet from the sun, we may deduce the heliocentric motions by the principles of Geometry.

The geocentric place of a body is its place as seen from the centre of the earth, and the heliocentric place is its place as seen from the centre of the sun.

374. *Elongation, conjunction, and opposition of a planet.*—The

angle formed by lines drawn from the earth to the sun and a planet is called the *elongation* of the planet from the sun; and it is east or west, according as the planet is on the east or west side of the sun.

A planet is said to be in *conjunction* with the sun when it has the same longitude, being then in nearly the same part of the heavens with the sun. It is said to be in *opposition* with the sun when its longitude differs from that of the sun 180°, being then in the quarter of the heavens opposite to the sun. A planet is said to be in *quadrature* when it is distant from the sun 90° in longitude.

A planet which is in conjunction with the sun passes the meridian about noon, and is therefore above the horizon only during the day. A planet which is in opposition with the sun passes the meridian about midnight, and is therefore above the horizon during the night. A planet which is in quadrature passes the meridian about 6 o'clock either morning or evening.

An inferior planet is in conjunction with the sun when it is between the earth and the sun, as well as when it is on the side of the sun opposite to the earth. The former is called the *inferior* conjunction, the latter the *superior* conjunction.

375. *Why the apparent motions of the planets differ from the real motions.*—If the planets could be viewed from the sun as a centre, they would all be seen to advance invariably in the same direction, viz., from west to east, in planes only slightly inclined to each other, but with very unequal velocities. Mercury would advance eastward with a velocity about one third as great as our moon; Venus would advance in the same direction with a velocity less than half that of Mercury; the more distant planets would advance still more slowly; while the motions of Uranus and Neptune would be scarcely appreciable except by comparing observations made at long intervals of time. None of the planets would ever appear to move from east to west.

The motions of the planets, as they actually appear to us, are very unlike those just described, first, because we view them from a point remote from the centre of their orbits, in consequence of which the distances of the planets from the earth are subject to great variations; and, second, because the earth itself is in motion, and the planets have an apparent motion, resulting from the real motion of the earth.

376. *The apparent motion of an inferior planet.*—In order to deduce the apparent motion of an inferior planet from its real motion, let CKZ represent a portion of the heavens lying in the plane of the ecliptic; let *a*, *b*, *c*, *d*, etc., be the orbit of the earth; and

Fig. 99.

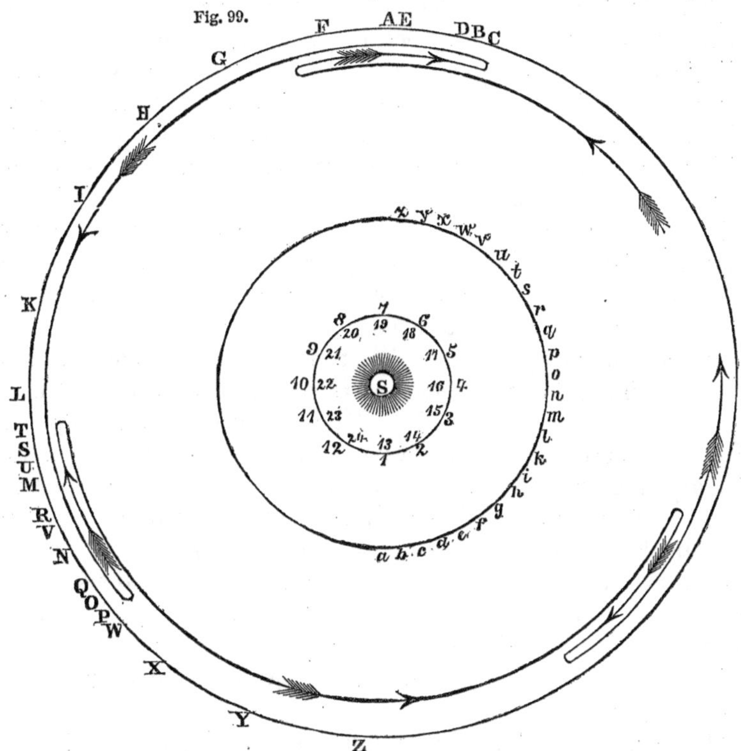

1, 2, 3, 4, etc., the orbit of Mercury. Let the orbit of Mercury be divided into 12 equal parts, each of which is described in $7\frac{1}{3}$ days; and let *ab*, *bc*, *cd*, etc., be the spaces described by the earth in the same time. Suppose Mercury to be at the point 1 in his orbit when the earth is at the point *a*; Mercury will then appear in the heavens at A, in the direction of the line *a* 1. In $7\frac{1}{3}$ days Mercury will have arrived at 2, while the earth has arrived at *b*, and therefore Mercury will appear at B. When the earth is at *c*, Mercury will appear at C, and so on. By laying the edge of a ruler on the points *c* and 3, *d* and 4, *e* and 5, and so on, the successive apparent places of Mercury in the heavens will be obtained. We thus find that from A to C, his apparent motion is from east to west; from C to P, his apparent motion is from west to east; from P to T it is from east to west; and from T to Z the apparent motion is from west to east.

377. *Direct and retrograde motion.*—When a planet appears to move in the direction in which the sun appears to move in the ecliptic, its apparent motion is said to be *direct;* and when it appears to move in the contrary direction, it is said to be *retrograde.* The apparent motion of an inferior planet is always direct, except within a certain elongation east and west of the inferior conjunction, when it is retrograde.

If we follow the movements of Mercury during several successive revolutions, we shall find its apparent motion to be such as is indicated by the arrows in the preceding diagram, viz., while passing from its greatest western to its greatest eastern elongation, it appears to move in the same direction as the sun toward P. As it approaches P its apparent motion eastward becomes gradually slower, until it stops altogether at P, and becomes stationary. It then moves westward, returning to T, where it again becomes stationary, after which it again moves eastward, and continues to move in that direction through an arc about equal to CP, when it again becomes stationary. It again moves westward through an arc about equal to PT, when it again becomes stationary, and so on. The middle point of the arc of retrogression, PT, is that at which the planet is in inferior conjunction; and the middle point of the arc of progression, CP, is that at which the planet is in superior conjunction.

These apparently irregular movements suggested to the ancients the name of *planet*, or wanderer.

378. *Apparent motion of a superior planet.*—In order to deduce the apparent motion of a superior planet from the real motions of the earth and planet, let S be the place of the sun; 1, 2, 3, etc., be the orbit of the earth; *a*, *b*, *c*, etc., the orbit of Mars; and CGL a part of the starry firmament. Let the orbit of the earth be divided into 12 equal parts, each of which is described in one month; and let *ab*, *bc*, *cd*, etc., be the spaces described by Mars in the same time. Suppose the earth to be at the point 1 when Mars is at the point *a*, Mars will then appear in the heavens in the direction of the line 1 *a*. When the earth is at 3 and Mars at *c*, he will ap-

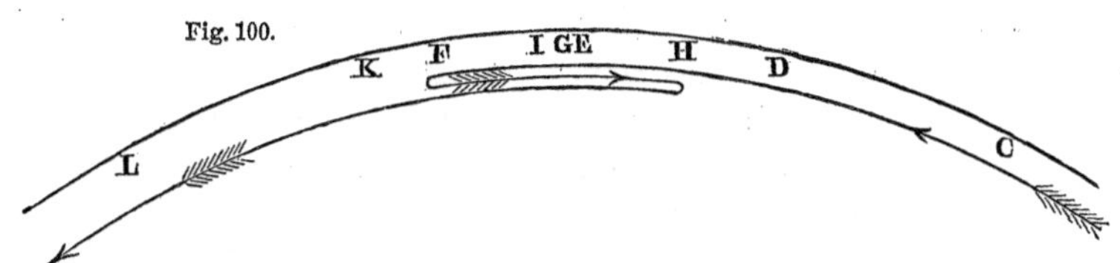

pear in the heavens at C. When the earth arrives at 4, Mars will arrive at *d*, and will appear in the heavens at D. While the earth moves from 4 to 5 and from 5 to 6, Mars will appear to have advanced among the stars from D to E and from E to F, in the direction from west to east. During the motion of the earth from 6 to 7 and from 7 to 8, Mars will appear to go backward from F to G and from G to H, in the direction from east to west. During the motion of the earth from 8 to 9 and from 9 to 10, Mars will appear to advance from H to I and from I to K, in the direction from west to east, and the motion will continue in the same direction until near the succeeding opposition.

The apparent motion of a superior planet projected on the heavens is thus seen to be similar to that of an inferior planet, except that, in the latter case, the retrogression takes place near inferior conjunction, and in the former it takes place near opposition.

379. *Conditions under which a planet is visible.*—One or two of the planets are sometimes seen when the sun is above the horizon; but generally, in order to be visible without a telescope, a planet must have an elongation from the sun greater than 30°, so as to be above the horizon before the commencement of the morning twilight, or after the close of the evening twilight.

The greatest elongation of the inferior planets never exceeds 47°. If they have eastern elongation, they pass the meridian in the afternoon, and, being visible above the horizon after sunset, are called *evening stars.* If they have western elongation, they pass the meridian in the forenoon, and, being visible above the eastern horizon before sunrise, are called *morning stars.*

A superior planet, having every degree of elongation from 0 to 180°, may pass the meridian at any hour of the day or night. At opposition the planet passes the meridian at midnight, and is therefore visible from sunset to sunrise.

380. *Phases of a planet.*—That hemisphere of a planet which is presented to the sun is illumined, and the other is dark. But if the same hemisphere which is turned toward the sun is not also presented to the earth, the hemisphere of the planet which is presented to the earth will not be wholly illumined, and the planet will exhibit *phases.*

The inferior planets exhibit the same variety of phases as the moon. At the inferior conjunction, the dark side of the planet is turned directly toward the earth. Soon afterward the planet appears a thin crescent, which increases in breadth until at quadrature it becomes a half moon. From quadrature the planet becomes gibbous, and at superior conjunction it beomes a full moon.

The distances of the superior planets from the sun are, with but one exception, so much greater than that of the earth, that the hemisphere which is turned toward the earth is sensibly the same as that turned toward the sun, and these planets always appear full.

381. *Elements of the orbit of a planet.*—There are seven different quantities necessary to be known in order to compute the place of a planet for a given time. These are called the *Elements of the orbit.* They are,

1. The periodic time.

2. The mean distance from the sun, or the semi-major axis of the orbit.

3. The longitude of the ascending node.

4. The inclination of the plane of the orbit to that of the ecliptic.

5. The eccentricity of the orbit.

6. The longitude of the perihelion.

7. The place of the planet in its orbit at a particular epoch.

If the mass of a planet is either known or neglected, the mean distance can be computed from the periodic time by means of Kepler's third law, so that the number of independent elements is reduced to six.

The orbits of the planets can not be determined in the same manner as the orbit of the moon, Art. 207, because the centre of the earth may be regarded as a fixed point relative to the moon's orbit, but it is *not* fixed relative to the planetary orbits. The methods therefore employed for determining the orbits of the planets are in many respects quite different from those which are applicable to determining the orbit of the moon, and also that of the earth.

382. *To find the periodic time. First method.*—Each of the planets, during about half its revolution around the sun, is found to be on one side of the ecliptic, and during the other half on the other side. The period which elapses from the time that a planet is at one of its nodes, till it returns to the same node (allowance being made for the motion of the nodes), is the sidereal period of the planet. When a planet is at either of its nodes, it is in the plane of the ecliptic, and its latitude is then nothing. Let the right ascension and declination of a planet be observed on several successive days, near the period when it is passing a node, and let its corresponding longitudes and latitudes be computed. From these we may obtain, by a proportion, the time when the planet's latitude is nothing. If similar observations are made when the planet passes the same node again, we shall have the time of a revolution.

Example.—The planet Mars was observed to pass its ascending node as follows:

1862, December, 5d. 22h. 17m.
1864, October, 22d. 21h. 58m.

The interval is 686.986 days, which differs but a few minutes from the most accurate determination of its period.

When the orbit of a planet is but slightly inclined to the ecliptic, a small error in the observations has a great influence on the computed time of crossing the ecliptic. A more accurate result will be obtained by employing observations separated by a long interval, and dividing this interval by the number of revolutions of the planet.

383. *Second method.*—The synodical period of a planet is the interval between two consecutive oppositions, or two conjunctions of the same kind. The sidereal period may be deduced from the synodical by a method similar to that of Art. 205. Let p be the sidereal period of a planet, p' the sidereal period of the earth, and s the time of a synodic revolution, all expressed in mean solar days. The daily motion of the planet, as seen from the sun, is $\frac{360°}{p}$, while that of the earth is $\frac{360°}{p'}$; and if p be a superior planet, the earth will gain upon the planet daily $\frac{360°}{p'}-\frac{360°}{p}$. But in a synodic revolution the earth gains upon the planet 360°; that is, its daily gain is $\frac{360°}{s}$. Hence we have the equation

$$\frac{360}{p'}-\frac{360}{p}=\frac{360}{s}.$$

Hence

$$sp-sp'=pp',$$

or

$$p=\frac{sp'}{s-p'}.$$

For an inferior planet, we shall find in like manner

$$p=\frac{sp'}{s+p'}.$$

384. *How to obtain the mean synodic period.*—Since the angular motion of the planets is not uniform, the interval between two successive oppositions will not generally give the *mean* synodical period. But if we take two oppositions, separated by a long interval, when the planet was found in the same position relatively to some fixed star, and divide the interval by the number of revolutions, we may obtain the mean synodical period very accurately.

Example.—The planet Mars was observed in opposition as follows:

1864, November, 30d. 17h. 58m.
1817, December, 8d. 9h. 15m.

The interval is 17159.37 days, which divided by 22, the number of synodic revolutions, gives for the mean time of one synodic revolution 779.97 days. By comparing the observations of Ptolemy, A.D. 130, with recent observations, the time of one synodical revolution is found to be 779.936 days; from which, according to the formula given above, the mean sidereal period of Mars is found to be 686.980 days. And in the same manner the periods of the other planets may be found.

The following table shows the time of a synodical, as well as of a sidereal revolution of the planets:

	Synodical Revolution.	Sidereal Revolution.	Mean daily Motion.
	Days.	Days.	
Mercury . .	115.877	87.969 or 3 months.	4° 5′ 32″.6
Venus . . .	583.921	224.701 " $7\frac{1}{2}$ "	1 36 7 .8
Earth		365.256 " 1 year.	0 59 8 .3
Mars	779.936	686.980 " 2 years.	0 31 26 .7
Jupiter . . .	398.884	4332.585 " 12 "	0 4 59 .3
Saturn . . .	378.092	10759.220 " 29 "	0 2 0 .6
Uranus . . .	369.656	30686.821 " 84 "	0 0 42 .4
Neptune . .	367.489	60126.722 " 164 "	0 0 21 .6

385. *To find the distance of a planet from the sun.*—The mean distance of a planet, whose periodic time is known, can be computed by Kepler's third law. It can, however, be determined independently by methods like the following:

The distance of an inferior planet from the sun may be determined by observing the angle of greatest elongation.

Fig. 101.

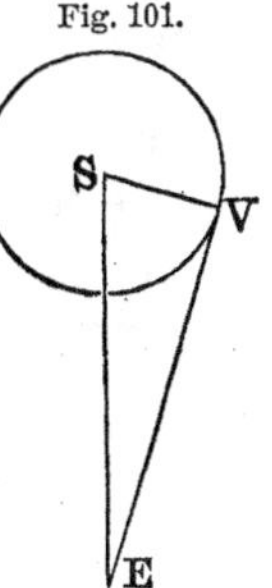

In the triangle SEV, let S be the place of the sun, E the earth, and V an inferior planet at the time of its greatest elongation. Then, since the angle SVE is a right angle, we have

SV : SE :: sin. SEV : radius;

or SV=SE sin. SEV.

If the orbits of the planets were exact circles, this method would give the mean distance of the planet from the sun; but since this is not the case, we must observe the greatest elongation in different parts of the orbit, and thus obtain its average value. The average value of the greatest elongation of Venus is 46° 20′; whence the mean distance of Venus is found to be .7233, the distance of the earth from the sun being called unity.

386. *Distance of a superior planet.*—The distance of a superior planet, whose periodic time is known, may be found by measuring the retrograde motion of the planet in one day at the time of opposition. Let S be the place of the sun, E the earth, and M the planet on the day of opposition, when the three bodies are situated in the same straight line. Let EE′ represent the earth's motion in one day from opposition, and MM′ that of the planet in the same time. The angles ESE′ and MSM′ are known from the periodic times. Draw E′B parallel to SM; join E′M′, and produce the line to meet SM in A. The angle SAE′, which equals AE′B, is the retrogradation of the planet in one day, and is supposed to be known from observations. In the triangle E′SM′, the side E′S and the angle E′SM′ are known, and E′M′S=M′SA+M′AS; from these we can compute SM′.

Fig. 102.

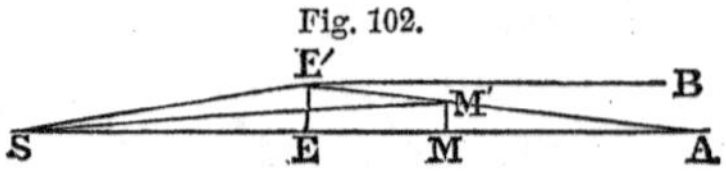

If we only know the periodic time of the planet, we are obliged, in the first approximation, to assume the orbit to be a circle in order to compute the angle MSM′; but if we observe the retrograde motion at a large number of oppositions in different parts of the orbit, we may obtain the average value of the arc of retrogradation, and hence we may compute the mean distance.

Example. The average arc of retrogradation of Mars on the day of opposition is 21′ 25″.7. If we take the mean daily motions of the earth and Mars, as given on page 207, we shall find the mean distance of Mars to be $\frac{\sin. 80' \, 34''}{\sin. 52' \, 52''.4}=1.52369$, the distance of the earth from the sun being called unity.

The following table shows the mean distances of the planets from the sun, expressed in miles, and also their relative distances, the distance of the earth being called unity:

	Mean Distance from the Sun.	Relative Distance.
Mercury . .	37,000,000 miles.	0.387
Venus . . .	69,000,000 "	0.723
Earth. . . .	95,000,000 "	1.000
Mars	145,000,000 "	1.524
Jupiter . . .	496,000,000 "	5.203
Saturn . . .	909,000,000 "	9.539
Uranus. . .	1,828,000,000 "	19.183
Neptune . .	2,862,000,000 "	30.037

387. *Diameters of the planets.*—Having determined the distances of the planets, it is only necessary to measure their apparent diameters, and we can easily compute their absolute diameters in miles. The apparent diameters of the planets are of course variable, since they depend upon the distances which are continually varying. The following table shows the *mean* apparent diameters, and also the absolute diameters of the planets, as well as their volumes, that of the earth being called unity:

	Equatorial Diameters.		Volume.
	Apparent.	In Miles.	
Mercury . . .	7″	3,000	$\frac{1}{17}$
Venus	17	7,700	$\frac{9}{10}$
Earth.		7,926	1
Mars	7	4,500	$\frac{1}{5}$
Jupiter	38	92,000	1412
Saturn	17	75,000	770
Uranus. . . .	4	36,000	96
Neptune . . .	2	35,000	90

388. *To determine the position of the nodes of a planetary orbit.*—Let the longitude of a planet be determined when it is at one of its nodes; this longitude will be the geocentric longitude of the node. Also, by means of the solar tables, let the longitude of the sun and the radius vector of the earth be found for the time the planet is at the node. When the planet returns to the same node again, let its longitude be again determined, as also the longitude of the sun and the radius vector of the earth. From these data (the node in the interval being supposed to remain fixed) the position of the line of the nodes may be determined, and also the distance of the planet from the sun at the times of observation.

Let S be the place of the sun, E the earth, and P a superior planet at its node; and let E′ be the place of the earth after the

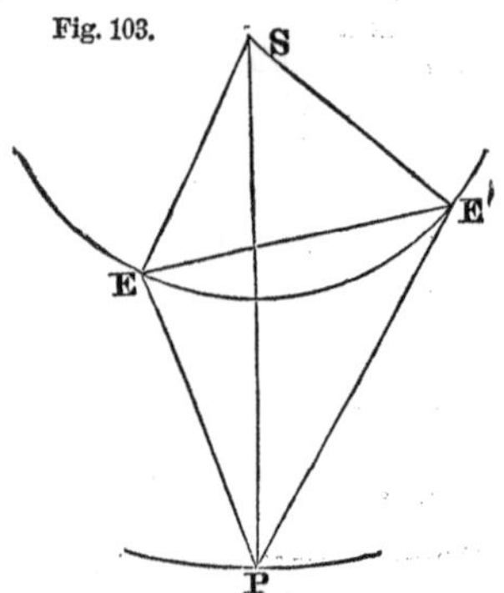

Fig. 103.

planet has made an entire revolution, and returned to the point P. Then from the solar tables we can determine SE and SE′, as also the angle ESE′. Hence EE′ can be computed, as also the angles SEE′, SE′E. Now, since the angles SEP, SE′P are determined by the observations, we can obtain the angles PEE′, PE′E. Then, in the triangle PEE′, having two angles and one side, we can compute PE. Hence, in the triangle PES, we have two sides and the included angle, from which we can compute SP, and also the angle ESP, which, added to the longitude of the earth when at E, will give the heliocentric longitude of the planet when at its node.

When observations of this kind are made at a considerable distance of time from one another, it is found that the nodes of every planet have a slow motion retrograde, or in a direction contrary to the order of the signs. The most rapid motion of the nodes is in the case of Mercury, amounting to about 70′ in a century.

389. *To determine the inclination of an orbit to the ecliptic.*—Let the time at which the sun's longitude is the same as the heliocentric longitude of the node be found by means of the solar tables, and let the longitude and latitude of the planet be determined at the same time.

Fig. 104.

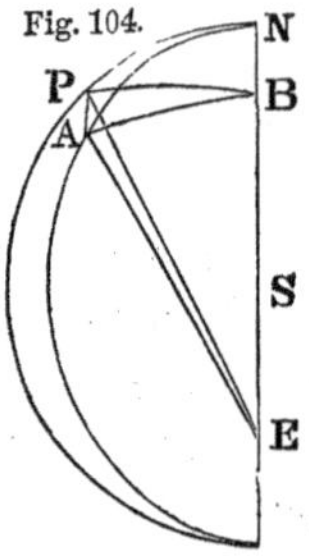

Let NSE be the line of a planet's nodes, S the sun, E the earth, and P the planet's place in its orbit. From E as a centre, with a radius PE, suppose a sphere to be described whose surface meets the line NE in B; and let PA be an arc of a great circle perpendicular to the ecliptic. Then PBA will be a spherical triangle right-angled at A; the angle PBA will measure the inclination of the plane of the planet's orbit to the ecliptic; PA will measure PEA, the geocentric latitude of the planet; and AB will measure AEB, the difference between the longitudes of the sun and planet.

Then, by Napier's rule, we have

$$R \times \sin.\ AB = \text{tang.}\ PA\ \cot.\ PBA;$$

or

$$\text{tang.}\ PBA = \frac{\text{tang.}\ PA}{\sin.\ AB};$$

that is, *the tangent of the inclination of the orbit is equal to the tangent of the planet's geocentric latitude, divided by the sine of the planet's elongation from the sun*, the earth being in the line of the planet's nodes. If, at the time of observation, the elongation of the planet from the sun was 90°, its geocentric latitude would be the inclination of its orbit to the ecliptic; and the results of this method will be the more reliable the farther the planet is from its node.

The orbits of the planets have generally small inclinations to the ecliptic. The orbit of Mercury is inclined about 7°, while all the other planets (with the exception of the asteroids) are inclined less than 4°. Four of the asteroids have inclinations exceeding 20°, and one has an inclination of 34°.

390. *To determine the heliocentric longitude and latitude of a planet.* —When the place of the ascending node and the inclination of the orbit of a planet are known, the heliocentric longitude and latitude of a planet, and also its radius vector, may be deduced from the geocentric longitude and latitude.

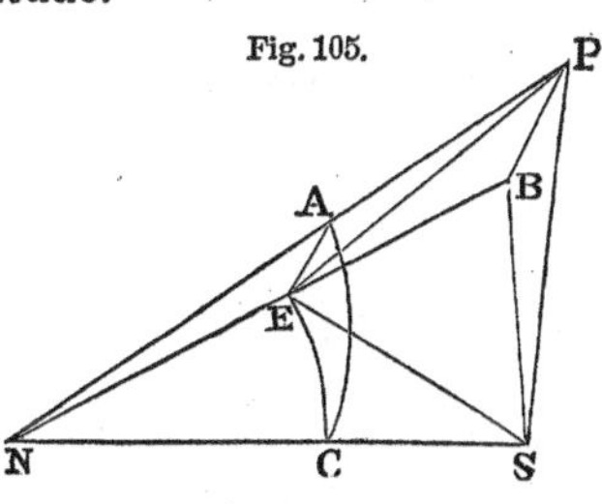

Fig. 105.

Let S be the place of the sun, E the earth, P the planet, and NS the line of the nodes of the planet's orbit. From P draw PB perpendicular to the ecliptic, and let a plane pass through E, P, and B, intersecting the line of the nodes in N. With N as a centre, and NE as a radius, let a sphere be described, cutting the planes PNS, ENS, and PNE in the right-angled spherical triangle AEC. The angle PEB will be the geocentric latitude of the planet, BES will be the difference between the longitudes of the planet and sun, and the spherical angle ACE will measure the inclination of the planet's orbit to the ecliptic.

1st. In the triangle NES, the angle NES is known, being the supplement of BES; also ESN can be derived from the solar tables when the place of the node is given, and ES is also known; hence we can compute EN, NS, and the angle ENS.

2d. In the spherical triangle AEC, right-angled at E, the angle

ACE is given, and also EC, which measures ENC; hence AE, which measures ANE, can be computed.

3d. In the triangle PNE, we know NE, ENP, and NEP, the supplement of the planet's geocentric latitude; hence PN can be computed.

4th. In the right-angled triangle NPB, we know NP and the angle PNB; hence PB and NB can be computed.

5th. In the triangle BNS, NB, NS, and the angle BNS are known; hence we can compute SB and NSB, which is the difference between the heliocentric longitude of the planet and that of its node. Hence the heliocentric longitude of the planet is determined.

6th. In the right-angled triangle PBS, we know PB and BS, from which we can compute the angle PSB, the planet's heliocentric latitude, and also PS, its distance from the sun.

391. *To determine the longitude of the perihelion, the eccentricity, etc.*—Assuming the orbit of the planet to be an ellipse, if we determine, by Art. 390 or Art. 388, the length and position of three radii vectores of the planet, we can determine the form and dimensions of the ellipse.

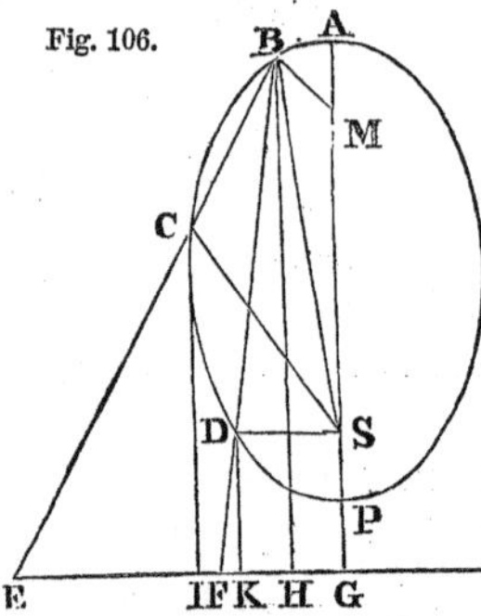

Let SB, SC, SD be three radii vectores of the planet, given in length and position. Draw the lines BC, BD, and produce them, making SB : SD :: BF : DF; and SB : SC :: BE : CE; then

$$SB-SD : SB :: BD : BF=\frac{SB.BD}{SB-SD};$$

and

$$SB-SC : SB :: BC : BE=\frac{SB\times BC}{SB-SC}.$$

Then the straight line passing through the points E and F will be the directrix of the ellipse. For BH, CI, DK being drawn perpendicular to EF, the triangles BEH, CEI are similar; therefore BH : CI :: BE : CE. Now, by construction, BE : CE :: SB : SC; hence BH : CI :: SB : SC; or BH : SB :: CI : SC; also BH : DK :: BF : DF :: SB : SD. Therefore the perpendiculars BH, CI, DK are always in the same proportion as the lines SB, SC, SD; consequently, EF is the directrix of the ellipse, passing through B, C, and D. (Geom., Ellipse, Prop.

22.) Through S draw ASG perpendicular to FE; take GA : AS :: CI : CS, and GP : SP :: CI : CS; then CI+CS : CS :: GS : SP= $\frac{SC \times SG}{CI+CS}$, and AS$=\frac{SC \times SG}{CI-CS}$; then A and P will be the vertices of the ellipse.

The lengths of SP and SA can accordingly be computed; their sum gives the major axis; and their difference, MS, divided by the major axis, is the eccentricity of the ellipse. Also, in the triangle BSM, we know BS, SM, and BM=PA−SB; whence the angle BSA is determined, which gives the position of the major axis relatively to SB.

CHAPTER XV.

THE INFERIOR PLANETS, MERCURY AND VENUS.—TRANSITS.

392. *Greatest elongations of Mercury and Venus.*—Mercury and Venus having their orbits far within that of the earth, their elongation or angular distance from the sun is never great. They appear to accompany the sun, being seen in the west soon after sunset, or in the east a little before sunrise.

Fig. 107.

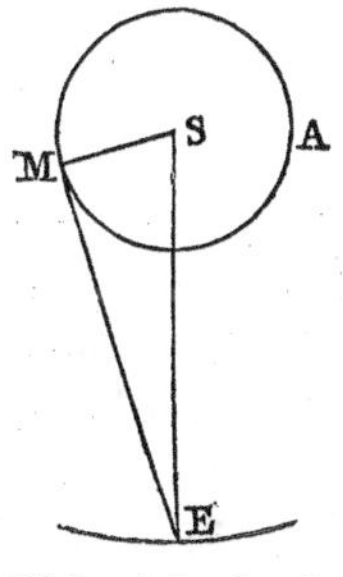

Let S be the place of the sun, MA the orbit of Mercury, E the place of the earth, and M the place of the planet when at its greatest elongation, at which time the angle EMS is a right angle. Since the distances of the planet and the earth from the sun both vary, the greatest elongation must also vary. The elongation will be the greatest possible when SM is greatest and SE is the least; that is, when Mercury is at its aphelion and the earth at perihelion. Combining the greatest value of SM with the least value of SE, we find the greatest possible value of Mercury's greatest elongation to be 28° 20′. Combining the least value of SM with the greatest value of SE, we find the least possible value of Mercury's greatest elongation to be 17° 36′.

In a similar manner, we find the greatest elongation of Venus to vary from 45° to 47° 12′.

393. *Phases of Mercury and Venus.*—The planets Mercury and

Venus exhibit to the telescope phases similar to those of the moon. At the greatest elongations eastward or westward, we see only half the disc illuminated, as in the case of our own satellite

Fig. 108.

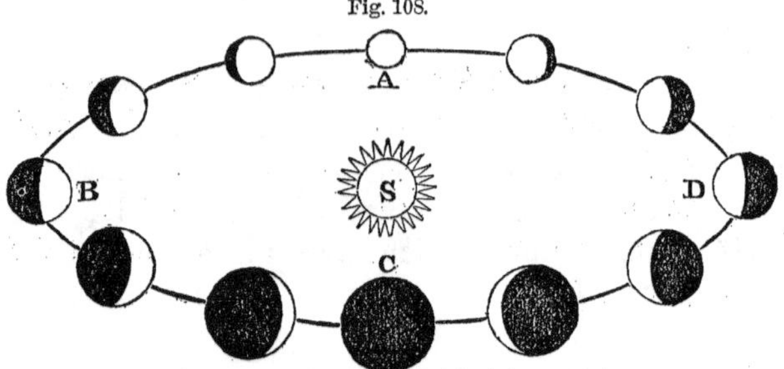

at first or last quarter. As they move toward the superior conjunction, at A, their form becomes *gibbous*, and the outline of the disc becomes more nearly circular the nearer they approach the superior conjunction. Owing to the intensity of the sun's light, we lose the planets for a little time before and after the conjunction, but on emerging from the sun's rays we find the form still *gibbous*. The illumined part diminishes as the planets approach their greatest elongation, near which time they again appear as a half moon; and as they advance toward the inferior conjunction, the form becomes more nearly that of a crescent, until it is again lost in the sun's rays at C.

MERCURY.

394. *Period, distance from sun*, etc.—Mercury performs its revolution round the sun in a little less than three months; but its synodic period, or the time from one inferior conjunction to another, is 116 days. Its mean distance from the sun is 37 millions of miles.

The eccentricity of its orbit is much greater than in the case of any other of the large planets. At perihelion Mercury is only 29 millions of miles from the sun, while in aphelion it is distant 44 millions, making a variation of 15 millions of miles, which is about one fifth of the major axis of the orbit.

When between the earth and the sun, the disc of this planet subtends an angle of about twelve seconds of arc; but as the planet approaches the opposite part of the orbit, its breadth does not exceed five seconds. The real diameter of Mercury is about 3000 miles.

395. *Visibility of Mercury.*—Since the elongation of Mercury from the sun never exceeds 28° 20′, this planet is seldom seen except in strong twilight either morning or evening; and it does not ever appear conspicuous to the naked eye, although it sometimes shines with the brilliancy of a star of the first magnitude. Supposing the atmosphere clear, the other circumstances that favor its visibility are that the greatest elongation should occur at the season when the twilight is shortest; that it should then be near the aphelion of its orbit; and that its distance from the north pole should be several degrees less than that of the sun.

396. *Greatest brightness of Mercury.*—Mercury does not appear most brilliant when its disc is circular like a full moon, because its distance from us is then too great; neither when it is nearest to us, because then it appears as a thin crescent, and almost the entire illumined part is turned away from the earth. The greatest brightness must then occur at some intermediate point. Assuming the orbits of the planets to be circular, and that the quantity of light received at the earth varies directly as the area of the visible part of the planet, and inversely as the square of the distance from the earth, it has been computed that Mercury is brightest between its greatest elongation and superior conjunction, when the elongation from the sun is 22°. When the planet is seen after sunset, the greatest brightness occurs a few days *before* the greatest elongation; when it is seen before sunrise, the greatest brightness occurs a few days *after* the greatest elongation.

397. *Rotation on its axis.*—By observing Mercury with powerful telescopes, some astronomers think they have discovered indications of mountains on its surface, and by examining them at various times it has been concluded that the planet has a rotation upon its axis in 24h. 5m. 28s. Other astronomers, with equally good means of observation, have never remarked upon the planet's surface any spots by which they could approximate to the time of rotation. There is but little difference between the polar and equatorial diameters, the compression probably not exceeding $\frac{1}{150}$.

VENUS.

398. Venus, the most brilliant of the planets, is generally called

the *evening* or the *morning* star. The evening and morning star, or the Hesperus and Phosphorus of the Greeks, were at first supposed to be different. The discovery that they are the same is ascribed to Pythagoras.

399. *Period, distance, and diameter.*—Venus revolves round the sun in about 7½ months; but its synodic period, or the time from one inferior conjunction to another, is 584 days, or about 19 months. Its mean distance from the sun is 69 millions of miles; and since the eccentricity of its orbit is very small, this distance is subject to but slight variation.

The apparent diameter of Venus varies much more sensibly than that of Mercury, owing to the greater variation of its distance from the earth. At inferior conjunction its disc subtends an angle of about 64 seconds of arc, while at superior conjunction it is less than 10 seconds. The real diameter of Venus is about 7700 miles, or nearly the same as that of the earth.

400. *Venus sometimes visible during the full light of day.*—The greatest elongation of Venus from the sun amounts to 47°, and, on account of its proximity to the earth, it is, next to the sun and moon, the most conspicuous and beautiful object in the firmament. When it rises before the sun, it is called the *morning* star; when it sets after the sun, it is called the *evening* star. When most brilliant, it can be distinctly seen at midday by the naked eye, especially if at the time it is near its greatest north latitude. Its brightness is greatest about 36 days before and after inferior conjunction, its elongation being then about 40°, and the enlightened part of the disc not over a fourth part of the whole. At these periods the light is so great that objects illumined by it at night cast perceptible shadows.

401. *Rotation on an axis.*—Astronomers have frequently seen dusky spots upon Venus, which have been watched with the view of ascertaining the time of a rotation. It is concluded that this time is about 23h. 21m.; but these observations are exceedingly difficult on account of the glaring light of the planet.

402. *Twilight on Venus.*—By observing the concave edge of the crescent, which corresponds to the boundary of the illuminated

and dark hemispheres, it is found that there is a gradual fading away of the light into the darkness, caused probably by an atmosphere illuminated by the sun and producing the phenomena of twilight.

403. *Suspected satellite.*—Several observers of the last two centuries concurred in maintaining that they had seen a satellite of Venus. But Sir W. Herschel perceived no traces of a satellite; neither did Schröter, though he was most assiduous in his observations of Venus. It is therefore probable that the supposed appearances recorded by former observers were illusive.

TRANSITS OF MERCURY AND VENUS.

404. When either Mercury or Venus, being in inferior conjunction, has a distance from the ecliptic less than the sun's semi-diameter, it will appear projected upon the sun's disc as a black round spot. The apparent motion of the planet being then retrograde, it will appear to move across the disc of the sun from east to west, in a line sensibly parallel to the ecliptic. Such a phenomenon is called a *transit* of the planet.

405. *When transits are possible.*—Transits can only take place when the planet is within a small distance of its node. Let N be the node of the planet's orbit; S the centre of the sun's disc on the ecliptic, and at such a distance from the node that the edge of the disc just touches the orbit, NP, of the planet. A transit can only take place when the sun's centre is at a less distance than NS from the node. The mean value of the sun's semi-diameter being 16′, and the inclination of Mercury's orbit to the ecliptic being 7°, and that of Venus $3\frac{1}{3}$°, we find that a transit of Mercury can only take place within 2° 11′ of the node, and a transit of Venus within 4° 30′.

Fig. 109.

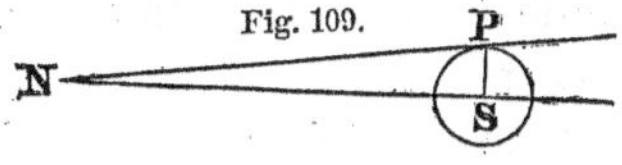

406. *Transits of Mercury.*—The longitudes of Mercury's nodes are about 46° and 226°, at which points the earth arrives about the 10th of November and the 7th of May. The transits of Mercury must therefore occur near these dates; those at the ascending node taking place in November, and those at the descending node in May.

The following are the dates of the transits of Mercury for the remainder of the present century:

1868, November 4.
1878, May 6.
1881, November 7.
1891, May 9.
1894, November 10.

407. *Intervals between the transits.*—In each of these cases the interval between two transits at the same node is 13 years. The reason is that 13 revolutions of the earth are made in nearly the same time as 54 revolutions of Mercury.

For $365.256 \times 13 = 4748.33.$

And $87.9692 \times 54 = 4750.34.$

When, therefore, a transit has occurred at one node, after an interval of 13 years, the earth and Mercury will return to nearly the same relative situation in the heavens, and another transit may occur. Transits sometimes occur at the same node at intervals of 7 years, and a transit at either node is generally preceded or followed, at an interval of $3\frac{1}{2}$ years, by one at the other node.

408. *Transits of Venus.*—The longitudes of the nodes of Venus are about 75° and 255°, at which points the earth arrives about the 5th of June and the 7th of December. The transits of Venus must therefore occur near these dates; those at the ascending node taking place in June, and those at the descending node in December.

The following list contains all the transits of Venus, from that which took place in 1639 (the first that was ever known to have been seen by any human being) to the end of the present century:

1639, December 4.
1761, June 5.
1769, June 3.
1874, December 8.
1882, December 6.

409. *Intervals between the transits.*—The interval between two transits at the same node is either 8 or 235 years. The reason of the first interval is that 8 revolutions of the earth are accomplished in nearly the same time as 13 revolutions of Venus.

For $365.256 \times 8 = 2922.05.$

And $224.701 \times 13 = 2921.11.$

Hence a transit at either node is generally preceded or followed, at an interval of 8 years, by another at the same node.

The period of 235 years is still more remarkable.

For $365.256 \times 235 = 85835.3$.

And $224.701 \times 382 = 85835.7$.

Hence, after an interval of 235 years, during which time Venus has made 382 revolutions, the earth and Venus return almost exactly to the same relative situation in the heavens.

410. *Sun's parallax and distance.*—The transits of Venus are important from their supplying data by which the sun's distance from the earth can be determined with far greater precision than by any other known method. The transits of Mercury supply similar data, but much less reliable, on account of the greater distance of that planet from the earth.

The *relative* distances of the planets from the sun may be computed by Kepler's third law, when we know their periods of revolution. In this manner we ascertain that the distances of the earth and Venus from the sun are in the ratio of 1000 to 723. Hence, when Venus is interposed between the earth and sun, the ratio of its distances from the earth and sun is that of 277 to 723.

Fig. 110.

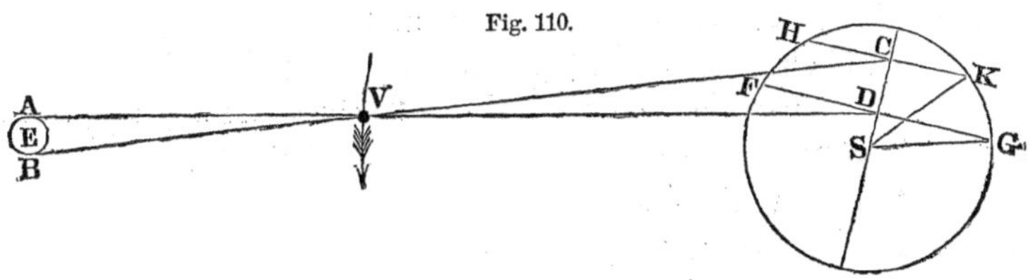

Let the circle FHKG represent the sun's disc; let E represent the earth, and A and B the places of two observers supposed to be situated at the opposite extremities of that diameter of the earth which is perpendicular to the ecliptic; also, let V be Venus moving in its orbit in the direction represented by the arrow. At present we will disregard the earth's rotation; that is, we will suppose the positions A and B to remain fixed during the transit. The planet will then appear to the observer at A to describe the chord FG, and to the observer at B the parallel chord HK. Also, when to the observer at A the centre of the planet appears to be at D, it will to the observer at B appear to be at C.

Now AB was supposed to be perpendicular to the plane of the ecliptic; and since the plane of the sun's disc is also very nearly perpendicular to the ecliptic, the line AB may be regarded as parallel to CD, and hence we have

$$CD : AB :: DV : AV :: 723 : 277 :: 2.61 : 1.$$

Therefore CD (expressed in miles) = 2.61 AB.

The *apparent* distance between the points C and D on the sun's surface may be derived from the observed times of beginning and ending of the transit at A and B. Let the observer at A note the time when the disc of the planet first appears to touch the sun's disc on the outside at L, and also the time when it first appears at M wholly within the sun's disc. L is called the *external*, and M the *internal* contact. Also, let both the internal and external contacts at N and P be observed when the planet is leaving the sun's disc. Then, since the planet's rate of motion as well as that of the sun is already accurately known from the tables, the number of seconds of a degree in the chord described by the planet can be ascertained. In the same manner, the number of seconds in the chord described by the planet as observed at B can be ascertained. Knowing the length of DG, which is the half of FG, and knowing also SG, the apparent radius of the sun, we can compute SD. In the same manner, from the length of the chord HK, we can compute SC. The difference between these lines is the value of CD, supposed to be expressed in seconds. But we have already ascertained the value of CD in miles. Hence we can determine the linear value of 1″ at the sun as seen from the earth, which is found to be 462 miles; and hence the angle which the earth's radius subtends at the sun will be $\frac{3963}{462}$, or 8″.58. This angle is called the *sun's horizontal parallax;* and from it, when we know the radius of the earth, we can compute the distance of the earth from the sun.

Fig. 111.

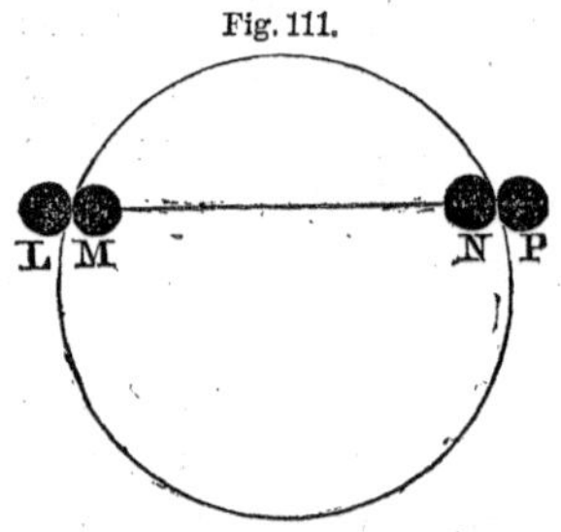

It is not necessary that the observers should be situated at the extremities of a diameter of the earth, but it is important that the two stations should differ widely in latitude; and allowance must also be made for the diurnal motion of the earth.

The transit of Venus in 1769 was observed with the greatest care at a large number of stations, extending from Lapland, latitude 70° 22′ N., to Otaheite, latitude 17° 25′ S., and the value of the sun's parallax resulting from these observations (8″.58) is that which, until recently, has generally been accepted by astronomers.

The mean distance of the earth from the sun, resulting from this value of the sun's parallax, is 95,300,000 miles. An accurate knowledge of this distance is of the greatest importance, since it serves as our base line for estimating the distances of all bodies situated beyond the limits of our solar system. See Art. 551. As there is still some uncertainty respecting the exact value of this quantity, astronomers generally call the mean distance of the earth from the sun *unity*, and estimate all distances in the planetary system by reference to this unit.

411. *Other determinations of the sun's parallax.*—When Mars is on the same side of the sun with the earth, it approaches comparatively near to the earth, and has a large horizontal parallax. Observations on the position of Mars have repeatedly been made at various observatories, both in the northern and southern hemispheres, from which the parallax of this planet has been deduced; and hence the parallax of the sun is easily computed, since the relative distances of the earth and Mars from the sun may be determined from the times of revolution. The horizontal parallax of the sun which has been deduced from these observations is 8″.95.

Considerations derived from the known velocity of light have led to nearly the same result; and it seems therefore probable that the value of the sun's horizontal parallax deduced from the last transit of Venus, and which has hitherto been generally received, will require to be somewhat increased. The effect will be to diminish slightly all the distances and magnitudes of the bodies of the solar system, except such as refer to the earth and moon.

CHAPTER XVI.

THE SUPERIOR PLANETS.—THEIR SATELLITES.

412. *How the superior planets are distinguished from the inferior.*—The superior planets, revolving in orbits without that of the earth, never come between us and the sun—that is, they have no inferior conjunction; but they are seen in superior conjunction and in opposition. Nor do they exhibit to us phases like those

of Mercury and Venus. The disc of Mars, about the period of his quadratures, appears decidedly gibbous; but the other planets are so distant that their enlightened surfaces are always turned almost entirely toward the earth, and the gibbous form is not perceptible.

MARS.

413. *Distance, period*, etc.—The mean distance of Mars from the sun is 145 millions of miles; but, on account of the eccentricity of its orbit, this distance is subject to a variation of nearly one tenth its entire amount. Its greatest distance from the sun is 158 millions of miles, and its least distance 132 millions.

The distance of this planet from the earth at opposition is sometimes reduced to 35 millions of miles, while at conjunction it is sometimes as great as 255 millions. Its apparent diameter varies in the same ratio, viz., 3½″ to 24″.

Mars makes one revolution about the sun in 687 days; but its synodic period, or the interval from opposition to opposition, is 780 days. The inclination of its orbit to the plane of the ecliptic is 1° 51′.

The real diameter of this planet is 4500 miles, and its volume about one fifth that of the earth.

414. *Phases, rotation*, etc.—At opposition and conjunction, the same hemisphere being turned to the earth and sun, the planet appears like a full moon, as shown at M1 and M5. In all other positions it appears slightly gibbous; but the deficient portion never exceeds about one ninth of a hemisphere.

Fig. 112.

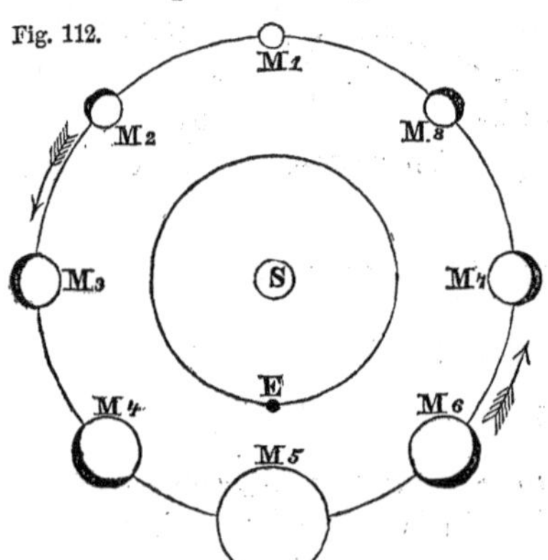

When viewed with a good telescope, the surface of Mars presents outlines of what are supposed to be continents and seas; and by observing these marks, the planet has been found to make a rotation upon its axis in 24h. 37m., and its axis is inclined to the axis of its orbit about 29°.

Hence the days and nights on Mars are nearly of the same length as on the earth; the year is diversified by seasons; and the surface of the planet by climates, not very different from those which prevail on our own globe.

415. *Spheroidal form.*—There is a sensible difference between the equatorial and polar diameters of Mars, amounting, according to some astronomers, to one fiftieth, and, according to others, to one thirty-ninth of the equatorial diameter. This is much greater than corresponds to the figure of equilibrium of a liquid planet making one rotation in 24h. 37m.

416. *Telescopic appearance.*—Many of the spots on this planet retain the same forms, with the same varieties of light and shade, even at the most distant intervals of time. But about the polar regions are sometimes seen white spots, with a well-defined outline, which undergo important changes from one season to another, and which may be explained by supposing them to proceed from polar snows, accumulated during the long winter, and which are partially dissolved during the equally protracted summer.

417. *Color of Mars.*—Mars usually shines with a red or fiery light; but this redness is much more remarkable to the naked eye than when viewed with a telescope. A very dense atmosphere has been supposed to surround this planet; but recent observations indicate that the atmosphere, though moderately dense, is not very extensive.

418. *Has Mars a satellite?*—Astronomers have not yet succeeded in discovering a satellite to this planet; and, if one exist, it is probably very small, and close to the planet.

419. *Sun's parallax.*—Mars being sometimes very near to us when in opposition, and the ratio of his distance from the sun to that of the earth being easily obtained, astronomers have sought, by means of his parallax, to determine the sun's horizontal parallax. A comparison of observations made at the Pulkova Observatory and at the Cape of Good Hope gives a solar parallax of 8″.96, which is sensibly greater than that deduced from the transit of Venus.

THE MINOR PLANETS, OR ASTEROIDS.

420. *A deficient planet between Mars and Jupiter.*—Nearly three centuries ago Kepler pointed out something like a regular progression in the distances of the planets as far as Mars, which was broken in the case of Jupiter.

In 1772, Professor Bode announced the singular relation between the distances of the planets from the sun, which has since been known as Bode's law. This law is as follows: If we set down the number 4 several times in a row, and to the second 4 add 3, to the third 4 add twice 3 or 6, to the next 4 add twice 6 or 12, and so on, the resulting numbers will represent nearly the relative distances of the planets from the sun. This law clearly indicated a deficient planet between Mars and Jupiter; and an association of astronomers was formed for the special purpose of searching for this unknown body.

On the 1st of January, 1801, Piazzi discovered the planet Ceres, and its distance was found to correspond very nearly with that required by Bode's law.

In 1802, Dr. Olbers, in searching for Ceres, discovered another planet, whose orbit was found to have nearly the same dimensions as that of Ceres. This planet was called Pallas.

On account of the close resemblance in appearance between these small planets and the fixed stars, Herschel proposed to designate them by the name *Asteroid*—a term which has been very extensively adopted. Some astronomers employ the term *Planetoïd;* but the term *minor planet* is more descriptive, and is now in common use among astronomers.

421. *Olbers's hypothesis respecting the origin of the asteroids.*—Dr. Olbers immediately advanced the hypothesis that a single planet formerly existed between Mars and Jupiter—that it was broken into fragments by volcanic action or by some internal force—that Ceres and Pallas were two of its fragments—and that probably other fragments existed, some of which might hereafter be discovered.

In 1804, Professor Harding discovered another planet, whose mean distance was found to be nearly the same as that of Ceres and Pallas. This planet was named Juno.

In 1807, Dr. Olbers discovered still another planet, whose orbit was found to be analogous to those of Ceres, Pallas, and Juno. This planet was named Vesta.

422. *Number of the asteroids.*—The search for planets was prosecuted till 1816 without farther success, when it was discontinued; but in 1845, Hencke, an observer in Prussia, resumed the search, and discovered another small planet, which has been named Astræa. Since that time the progress of discovery has been astonishingly rapid, the total number of asteroids known in 1864 amounting to 82. Of these, 24 were discovered in France, 21 in Germany, 17 in Great Britain and its colonies, 10 in America, and 10 in Italy. These bodies are all extremely minute, the largest of them probably not exceeding 300 miles in diameter. Vesta is the only one among them which is ever visible to the naked eye, and this only under the most favorable circumstances.

423. *Brightness of the asteroids.*—The asteroids closely resemble small stars, and can only be distinguished from fixed stars by their motion. One of them, when near the opposition, is of the sixth magnitude; two are of the seventh magnitude; five of the eighth; sixteen of the ninth; twenty-five of the tenth; twenty-four of the eleventh; five of the twelfth; and four of the thirteenth magnitude. Many of them can be seen only near the opposition, even by the largest telescopes. The reason that no asteroids were discovered for so long a period after 1807 was that the search was conducted with too little system, and with inadequate instruments.

424. *Distance of the asteroids.*—The average distance of the 82 asteroids from the sun is 2.667, or 254 millions of miles; but their distances differ widely from each other. The asteroid nearest to the sun is Flora, with a mean distance of 209 millions of miles; the asteroid most remote from the sun is Cybele, with a mean distance of 326 millions of miles. The orbit of Flora is therefore nearer to that of Mars than to that of Cybele.

425. *Total number of the asteroids.*—It is probable that there is a multitude of asteroids yet remaining to be discovered. From an examination of the influences exerted by the group of asteroids upon the planet Mars, Le Verrier has concluded that the entire mass of the asteroids between Mars and Jupiter may amount to

one third part of the mass of the earth. Now it would require over 500 bodies as large as the largest of the asteroids to make a body one third of the size of the earth; and, since many of the asteroids are extremely minute, their number probably amounts to many thousands.

426. *Is Olbers's hypothesis admissible?*—The hypothesis of Olbers has lost most of its plausibility since the discovery of so many asteroids. If these bodies ever composed a single planet, which burst into fragments, then, since the orbits all started from a common point, each must return to the same point in every revolution; in other words, all the orbits should have a common point of intersection. Such, however, is far from being the case. The orbits are spread over a large extent, and the smallest known orbit is every where distant from the largest by at least 50 millions of miles.

427. *What was the origin of the asteroid system?*—These bodies, however, exhibit striking resemblances, which point to some peculiar relationship. If we represent all the orbits under the form of material hoops, or rings, these rings are so interlocked as to hang together as one system, so that if we take hold of any one of the rings, we shall lift all the others with it. This feature distinguishes the asteroid orbits from all the other orbits of the solar system. It has been conjectured that all the planets once existed in the condition of gaseous matter, which gradually solidified into spherical masses. If such were the case, it is conceivable that the same causes which determined the gaseous matter, once occupying an immense space in the heavens, to collect into a single body and form a large planet, like Jupiter, should, in another part of space, have produced a division into an immense number of small masses, each of which solidified separately, thus forming the group of asteroids.

428. *Are there asteroids within the orbit of Mercury?*—The study of the motions of the planet Mercury has led Le Verrier to the conclusion that within the orbit of Mercury there exists either an undiscovered planet, whose mass is nearly equal to that of Mercury, or else a ring of minute planets with the same aggregate mass. The latter supposition is regarded as the most probable,

since a bright planet nearly equal in size to that of Mercury ought certainly to have been visible during total eclipses of the sun.

JUPITER.

429. *Distance, period*, etc.—The mean distance of Jupiter from the sun is 496 millions of miles; and, since the eccentricity of its orbit is about $\frac{1}{20}$th, this distance is augmented in aphelion, and diminished in perihelion by 24 millions of miles. On account of its distance from the sun being so much greater than that of the earth, Jupiter has no sensible phases.

Jupiter makes one revolution about the sun in $11\frac{7}{8}$ years; and the time from one opposition to another is 399 days.

430. *Diameter.*—Jupiter is the largest of the planets, its volume exceeding the sum of all the others. Its equatorial diameter is 92,000 miles, or 11 times that of the earth; and its volume is 1400 times that of the earth. Its apparent diameter varies from 30″ to 48″. When near opposition, Jupiter is a more conspicuous object in the heavens than any other planet except Venus, and is easily seen in the presence of a strong twilight.

431. *Rotation on an axis, spheroidal form.*—Permanent marks have been occasionally seen on Jupiter's disc, by means of which its rotation has been distinctly proved. The time of one rotation is 9h. $55\frac{1}{2}$m. A particle at the equator of Jupiter must therefore move with a velocity of more than 450 miles per minute, or 27 times as fast as a place on the terrestrial equator.

The Jovian day is less than half the terrestrial day; and since the period of Jupiter is 4332 terrestrial days, it consists of 10,485 Jovian days.

Jupiter's equator is but slightly inclined to the plane of its orbit, and hence the difference between the length of the days in summer and winter is very small; and the change of temperature with the seasons is also small.

The disc of Jupiter is oval, the polar diameter being to the equatorial as 16 to 17. This oblateness is found by computation to be the same as would be produced upon a liquid globe, making one rotation in about 10 hours.

432. *Belts of Jupiter.*—When viewed with a good telescope,

Jupiter's disc exhibits a light yellowish color, having a series of brownish-gray streaks, called *belts*, running nearly parallel to the equator of the planet. Two belts are generally most conspicuous, one north and the other south of the equator, separated by a bright yellow zone. These belts are commonly visible, without much change of form, during an entire rotation of the planet. Occasionally one of the belts appears broken sharply off, presenting an extremity so well defined as to afford the means of determining the time of the planet's rotation.

Near the poles the streaks are more faint, narrower, and less regular, and can only be seen with good telescopes. All the belts become less distinct toward the eastern or western limb, and disappear altogether at the limb itself. These belts, although tolerably permanent, are subject to slow but decisive variations, so that, after the lapse of some months, the appearance of the disc is totally changed.

433. *Cause of the belts.*—From long-continued observations, it is inferred that Jupiter is surrounded by an atmosphere which is continually charged with vast masses of clouds, which almost completely conceal the surface of the planet, and that these clouds have a permanence of form and position much greater than exists in terrestrial clouds.

The brightest portion of Jupiter's disc probably consists of dense clouds which reflect the light of the sun, while the darker spots and streaks are portions of the atmosphere, either free from clouds, and showing the surface of the planet more or less distinctly, or they are clouds of inferior reflecting power.

The distribution of the clouds in lines parallel to the equator is probably due to the prevalence of atmospheric currents, analogous to the trade winds, and arising from a like cause, but having a constancy and intensity far greater than prevail on the earth, on account of the more rapid rotation and greater diameter of Jupiter.

434. *Jupiter's satellites; their distances, periods,* etc.—Jupiter is attended by four moons, or satellites, revolving around the primary as our moon revolves around the earth, but with a much more rapid motion. They are numbered 1, 2, 3, 4, in the order of their distances from the primary.

The nearest moon completes a revolution in 42 hours, in which time, as seen from Jupiter, it goes through all the phases of thin crescent, half moon, gibbous, and full moon. Its distance from Jupiter is 280,000 miles. The distance of the second satellite is 440,000 miles, and it completes a revolution in 85 hours. The distance of the third satellite is 700,000 miles, and its time of revolution 172 hours. The distance of the fourth satellite is 1,200,000 miles, and its time of revolution is 400 hours, or 16 days and 16 hours.

These satellites were discovered by Galileo, at Padua, on the 8th of January, 1610. When viewed with a telescope of moderate power, they present the appearance of minute stars, ranged nearly in the direction of a line coinciding with the planet's equator. Their distances from the primary are so small that they are all included in the field of a telescope of moderate magnifying power, the distance of the most remote one being only 13 times the diameter of the planet.

The real diameter of the smallest satellite is 2200 miles, being the same as the diameter of our moon; and the diameter of the largest satellite is 3500 miles.

The satellites shine with the brilliancy of stars of between the sixth and seventh magnitude; but, owing to their proximity to the planet, which overpowers their light, they are in general invisible without the aid of the telescope. On high mountains, where the air is extremely rare, they have, however, been detected by the naked eye.

The orbits of the satellites are nearly circular, and are but slightly inclined to the plane of Jupiter's orbit. Hence their apparent motion is oscillatory, going alternately from their greatest elongation on one side to the greatest elongation on the other, nearly in a straight line.

Our moon makes one rotation on its axis in the same time that it requires to revolve around the earth. It is thought that Jupiter's moons also rotate on their axes in the time of their respective revolutions round the planet. This is inferred from periodical fluctuations in the brightness of the satellites, the periods corresponding with the times of revolution of the satellites.

435. *Eclipses of the satellites.*—Jupiter's satellites frequently pass into the shadow of the primary, and become invisible. Let JJ′

Fig. 113.

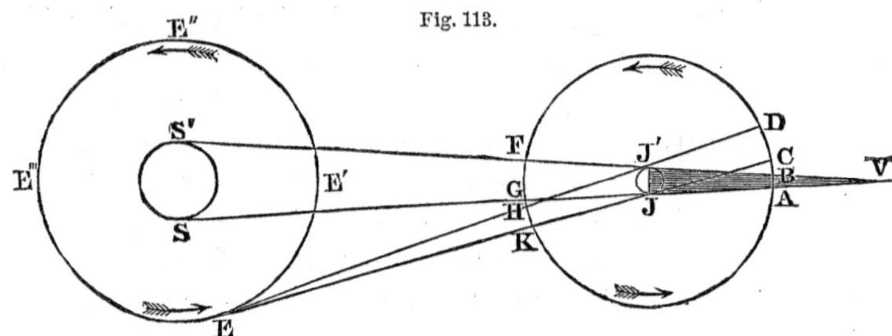

represent the planet Jupiter; JVJ′ its conical shadow; SS′ the sun; E and E″ the positions of the earth when the planet is in quadrature. Let ADFK represent the orbit of one of the satellites, whose plane we will suppose to coincide with the ecliptic. From E draw the lines EJ, EJ′, meeting the path of the satellite at H and K, as also at C and D. Let A and B be the points where the path of the satellite crosses the limits of the shadow. By a computation similar to that employed in the case of the earth, Art. 286, we find that the length of Jupiter's shadow is more than 50 millions of miles; and, since the distance of the most remote satellite is but little over one million miles, the satellites pass through the shadow at every revolution. In extremely rare cases, the fourth satellite, on account of the inclination of its orbit to the ecliptic, passes through opposition without entering the shadow.

436. *Eclipses, occultations, transits,* etc.—In the revolution of the satellites about the planet, four different classes of phenomena are observed:

1st. When the satellites pass into the shadow of the planet they are said to be *eclipsed.* Their entrance into the shadow at A is called the *immersion;* their passage out of the shadow at B is called the *emersion.*

2d. When the satellites pass between the lines SJ and S′J′ from F to G, their shadows are projected on the surface of the planet in the same manner as the shadow of the moon is projected on the earth in a solar eclipse; and in this case the shadow may be seen moving across the disc of the planet as a small round and black spot. This is called a *transit of the shadow.*

3d. When a satellite, passing behind the planet, is between the lines EJC and EJ′D, drawn from the earth at E, it is concealed from the observer by the interposition of the body of the planet. It disappears on one side of the planet's disc, and reappears on the other. This phenomenon is called an *occultation* of the satellite by the planet.

4th. When a satellite, being between the earth and planet, passes between the lines EJ and EJ′, drawn from the earth to the planet, its disc is projected on that of the planet; and it may sometimes be seen passing across the disc, being brighter or darker than the ground on which it is viewed, according as it is projected on a dark or bright belt. This is called a transit of the satellite. The entrance of the satellite upon the disc is called its *ingress*, and its departure is called its *egress*.

When the planet is in quadrature, all these phenomena may be witnessed in the revolution of the satellites. The immersion and emersion of the third and fourth satellites at A and B may both be witnessed on the same side of the planet when the planet is near quadrature, but only the *immersion* of the first and second satellites is visible. The view of their *emersion* is intercepted by the body of the planet, and they do not reappear until after having passed behind the planet.

437. *Longitude determined by observations of the eclipses.*—The times of occurrence of all these phenomena are calculated beforehand with the greatest precision, and are recorded in the Nautical Almanac. The mean time of their occurrence at Greenwich is there given; so that, if the time at which any of them occur at any other station be observed, the difference between the local time and that registered in the Almanac will give the longitude of the place from the meridian of Greenwich.

This method of determining longitude is, however, not very accurate; for, since the light of a satellite decreases gradually while entering the shadow, and increases gradually on leaving it, the *observed* time of disappearance or reappearance of a satellite must depend on the power of the telescope employed.

438. *Configurations of the satellites.*—The configurations of the satellites of Jupiter are continually varying. Sometimes they all appear on one side of the planet; frequently not more than two or three of the satellites are visible; sometimes only one satellite is visible; and a few instances are on record when all four have been invisible for a short time.

439. *Relation of the mean motions of the first three satellites.*—If the mean angular velocity of the first satellite be added to twice that of the third, the sum will be equal to three times that of the second. From this it follows that, if from the sum of the mean longitude of the first and twice that of the third, three times that of the second be subtracted, the remainder will always be the same quantity; and from observation it is found that this quantity is 180°. Hence it also follows that the first three satellites can never all be eclipsed at once; but while two of them are eclipsed, the third may be between the earth and Jupiter, in which position a satellite is often entirely invisible unless to the best telescope.

440. *Transmission of light.*—Soon after the invention of the telescope, Roemer, a Danish astronomer, computed a table showing the time of occurrence of every eclipse of the satellites of Jupiter for a period of twelve months. He then observed the moments of their occurrence, and compared his observations with the times registered in his table. At the commencement of his observations the earth was at E′, where it is nearest to Jupiter. As the earth moved toward E″, it was found that the eclipses occurred a *little later* than the time computed. As the earth moved toward E‴, the occurrence of the eclipses was more and more retarded, until at E‴ they occurred about 16 minutes later than the computed time. While the earth moved from E‴ to E′, the observed time was always later than the computed time; but this difference became less and less, until, on arriving at E′, the observed time agreed exactly with the computed time.

Thus it appeared that the lateness of the eclipse depended entirely upon the increased distance of the earth from Jupiter. When the earth was at E‴, the eclipse was observed 16 minutes later than when the earth was at E′; and, since the diameter of the earth's orbit is 190 millions of miles, the observation of the eclipse was delayed one second for every 200,000 miles that the earth's distance from Jupiter was increased. Now, since the eclipse must commence as soon as the satellite enters Jupiter's shadow, the delay in the observed time must be due to the time required for the light, which left the satellite just before its extinction, to reach the eye.

By more exact observations, it is found that light requires 16m.

26.6s. in crossing the earth's orbit; and hence the velocity of light is 192,000 miles per second.

SATURN.

441. *Distance, period*, etc.—The mean distance of Saturn from the sun is 909 millions of miles; and, on account of the eccentricity of its orbit, this distance is augmented at aphelion, and diminished at perihelion by more than $\frac{1}{20}$th of its whole amount, varying therefore from 858 millions to 960 millions of miles.

Saturn makes one revolution about the sun in 29½ years; and the interval between two successive oppositions is 378 days.

442. *Diameter, real and apparent.*—Saturn is the largest of all the planets except Jupiter. Its equatorial diameter is 75,000 miles, being more than nine times that of the earth; and its volume is nearly 800 times that of the earth.

The mean value of its apparent diameter is about 17″; and it appears as a star of the first magnitude, with a faint reddish light. Its disc is oval, the equatorial diameter being $\frac{1}{10}$th greater than the polar. The disc is traversed by streaks of light and shade parallel to its equator; but these belts are much more faint than those of Jupiter. These belts indicate the existence of an atmosphere surrounding the planet, and attended with the same system of currents which prevail on Jupiter.

443. *Rotation.*—Saturn makes one rotation upon its axis in 10½ hours; and the inclination of the planet's equator to the plane of the ecliptic is 28°. Thus the year of Saturn is diversified by the same succession of seasons as prevail on our globe. The year in Saturn is equal to 10,700 terrestrial days, or 24,700 Saturnian days.

444. *Saturn's rings.*—Saturn is surrounded by a very thin plate of matter in the form of a ring, which is nearly concentric with the planet, and in the plane of its equator. It is therefore inclined to the ecliptic at an angle of 28°, and intersects it in two points, which are called the ascending and descending nodes of the ring. With powerful telescopes certain dark streaks are seen upon its surface, bearing some resemblance to the belts of the planet. One of these is permanent in position, and indicates that

the ring consists of two concentric rings of unequal breadth, one placed outside the other, without any mutual contact.

445. *Dimensions of the rings.*—The distance from the surface of Saturn to the inside of the nearest ring is about 19,000 miles; the breadth of this ring is 17,000 miles; the interval between the two rings is 1800 miles; and the breadth of the exterior ring is 10,300 miles. The greatest diameter of the outer ring is 172,000 miles. The thickness of the rings is extremely small, and it is believed that it can not exceed 50 or 100 miles.

446. *Varying appearance of the rings.*—While this planet moves in its orbit round the sun, the plane of the rings is carried parallel to itself, so that during a revolution it undergoes changes of position analogous to those which the earth's equator exhibits. Twice in every revolution—that is, at intervals of 15 years, the plane of the rings must pass through the sun; and the ring, if seen at all, must appear as a straight line. As the planet advances in its orbit, the ring appears as a very eccentric ellipse. This eccentricity diminishes until Saturn is distant 90° from the nodes of the ring, when the minor axis of the ellipse becomes equal to about half the major axis; from which time the minor axis decreases, until, at the end of half a revolution, the ring again appears as a straight line.

These different positions of Saturn's ring are represented in the annexed diagram, where S represents the sun, MN the orbit of the

Fig. 114.

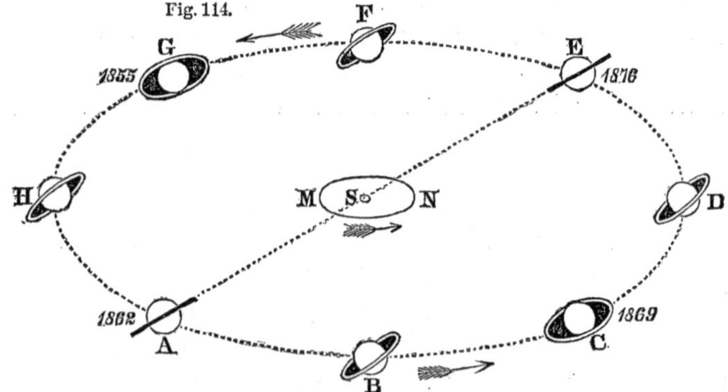

earth, and A, B, C, D, etc., different positions of Saturn. When Saturn is at A and E, the plane of the ring passes through the sun, and only the edge of the ring can be seen, as represented in the figure; when Saturn arrives at B, the ring appears as an ellipse; and when it arrives at C, the minor axis of the ellipse is equal to about half the major axis. After this the minor axis decreases, and when the planet reaches E the ring appears again as a straight line.

When the planet is in quadrature, a portion of the shadow which it projects on the ring is visible on one side of the disc; and, in certain cases, there is seen a portion of the shadow of the ring projected on the planet's disc. These phenomena prove that both the planet and the ring derive their illumination from the sun.

447. *Disappearance of the rings.*—The rings of Saturn may become invisible from the earth either because the parts turned toward the earth are not illumined by the sun, or, being illumined, subtend no sensible angle. *First*, When the plane of the rings passes through the sun, only the edge of the ring is illumined, and this is too thin to be seen by any but the most powerful telescopes. *Second*, When the plane of the rings passes through the earth, the ring, for the same reason, disappears to ordinary telescopes. *Third*, When the sun and the earth are on opposite sides of the plane of the rings—that is, when the plane of the rings, if produced, passes between the sun and the earth, the dark side of the rings is turned toward the earth, and the rings entirely disappear.

Two such disappearances usually take place during the year in which the plane of Saturn's rings crosses the earth's orbit. To

Fig. 115.

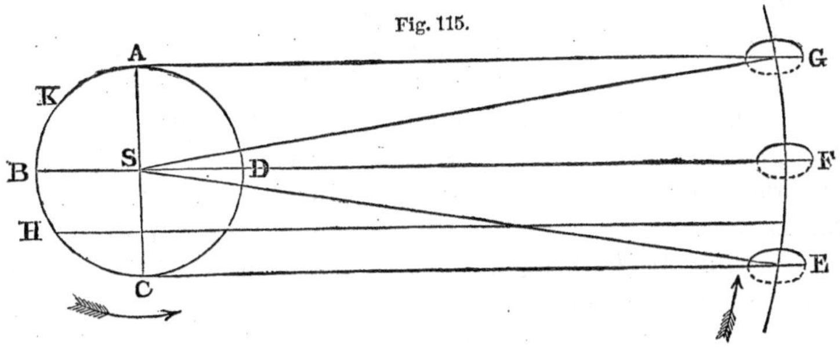

show this, let S be the sun, ABCD the earth's orbit, EFG a part of Saturn's orbit, and F the position of Saturn when the plane of the rings, if produced, would pass through the sun. Draw AG, CE, parallel to SF, touching the earth's orbit in A and C. Then, since the ring always preserves its parallelism, its plane can nowhere intersect the earth's orbit, and therefore no disappearance can take place, unless the planet be between E and G. Now, since SE, the distance of Saturn from the sun, is to SC, that of the earth, as 9.54 to 1, the angle SEC or ESF is found by computation equal to 6° 1′, and the whole angle ESG=12° 2′; and, as Saturn's periodic time in his orbit is 10,759 days, he will be 359½ days in describing 12° 2′—that is, about 6 days less than a complete year. The earth, then, describes very nearly an entire revolution within the limits of time when a disappearance of the ring is possible.

448. *Number and duration of the disappearances.*—The number of these disappearances and their duration will depend upon the position of the earth in its orbit when the planet arrives at E.

If, when Saturn arrives at E, the earth is at A, the earth will encounter the plane of the ring, advancing parallel to CE, somewhere in the quadrant BC, as at H. The ring will then *disappear*, and the disappearance will continue as the earth proceeds toward C, because the dark side of the ring is toward the earth. This disappearance will last about two months, and close when the plane of the ring at F passes the sun, for after that time the illumined side will be toward the earth. While the earth proceeds from C through D to A, the plane of the ring will move from FS to GA, and will pass A six days before the planet reaches that point. In this case there will be but one period of disappearance of the ring, lasting about two months.

If, when Saturn is at E, the earth is at K, it will meet and pass through the advancing plane of the ring somewhere in the quadrant BC, after which the dark side will be toward the earth. The plane of the ring will pass the sun when the earth is on the quadrant CD, after which the bright side will be presented to the earth. But the earth will overtake the nodal line before it reaches A, and therefore look again upon the dark side until it recrosses the line in some point of the quadrant AD. Thus there will be two periods of disappearance. These two periods may unite in a single period of about 8 months' duration, and this will happen when the earth and the nodal line pass D at the same instant; for then the plane of the ring is between the earth and sun both before and after passing the point D.

If, when the planet is at F, the earth is at B, then the illumined side of the ring must have been turned toward the earth during the whole time that the planet was moving from E to F; and the illumined side of the ring will also be turned toward the earth during the whole time that the planet moves from F to G; that is, there is only a momentary disappearance of the ring; and even this can never be observed, because the planet, being in conjunction with the sun, is lost in the splendor of the sun's light.

In general, during the year in which the line of the ring's nodes passes the earth, there are two periods of disappearance, arising from the third cause mentioned in Art. 447, each beginning and ending with a disappearance from the first or second cause.

449. *Observations near the periods of disappearance.*—The last disappearance of Saturn's ring took place in 1862. Observations near these periods of disappearance have indicated the existence of great inequalities on the rings. The rings frequently present the appearance of a broken line of light projecting from each side of the planet's disc. This broken appearance may be explained by supposing inequalities of surface, rendering some parts of the ring so thick as to be visible, while others are so thin as to be invisible when presented edgeways to the observer. It is probable, also, that the rings are not situated exactly in the same plane.

A dark line has sometimes been seen dividing the outer ring into two, which seems to indicate that the ring is really triple; and some observers have thought that they had discovered evidence of a still greater number of divisions.

450. *An inner ring discovered by Professor Bond.*—In 1850, Professor Bond, at Cambridge, discovered an inner ring, composed of matter which reflects light much more imperfectly than the planet or the other rings; and is transparent to such a degree that the body of the planet can be seen through it. This ring is situated between the planet and the bright rings, and approaches within about 8000 miles of the body of the planet. This ring has since been seen by numerous observers both in this country and in Europe, and its existence is unquestioned. In order to account for the fact that this ring has never been seen before, it has been

conjectured to be of recent formation. It appears at least probable that this ring has undergone some important change since the time of Sir William Herschel.

The discovery of this new ring, together with the apparently variable number of the divisions of the brighter rings, seem to render it probable that the rings consist of matter in the liquid condition; or, if they consist of solid matter, that this substance is divided into a very large number of small portions which have no cohesion, each revolving in its orbit as a satellite to the primary.

451. *What sustains Saturn's rings?*—Saturn's rings are sustained in precisely the same manner as our moon is sustained in its revolution about the earth. We may conceive two moons to revolve about the earth in the same orbit as the present one, and they would be sustained by the same law of attraction. In the same manner, three, four, or a hundred moons might be sustained. Indeed, we may suppose as many moons arranged around the earth as would complete a circle, so as to form a ring of moons in contact with each other. They would all be sustained in the same manner as our present moon is sustained. If we conceive these moons to be cemented together by cohesion, we shall have a continuous solid ring; and the ring would rotate about its axis in the same time as a moon situated near the middle of its breadth would revolve about the primary. Observations have actually indicated that the rings of Saturn have a revolution round their common centre, and in their own plane, in a period of 10h. 32m.

452. *Appearance of the rings from the planet itself.*—The rings of Saturn must present a magnificent spectacle in the firmament of that planet, appearing as vast arches spanning the sky from the eastern to the western horizon. Their appearance varies with the position of an observer upon the planet. To an observer stationed at Saturn's equator, the ring will pass through the zenith at right angles to the meridian, descending to the horizon at the east and west points. If the observer be stationed a few degrees from the equator, on the same side of the ring as the sun, the ring will present the appearance of an arch in the heavens, bearing some resemblance in form to a rainbow. If we suppose the observer to travel from the equator toward the pole, the elevation of the bow

will diminish, and near latitude 63° it will descend entirely below the horizon. Beyond this parallel, all view of the rings will be intercepted by the convexity of the planet. Near latitude 37° the rings are seen in their greatest splendor, forming an arch 15° in breadth.

453. *Saturn's satellites; their distance, period,* etc.—Saturn is attended by eight satellites, all of which, except the most distant one, move in orbits whose planes coincide very nearly with the plane of the rings. The satellites are numbered 1, 2, 3, etc., in the order of their distance from the primary.

The sixth satellite is the largest, and was first discovered by Huygens in 1655. Its distance from the centre of the planet is 778,000 miles, and the time of one revolution is about 16 days. Its diameter is about 3000 miles. It shines like a star of the eighth magnitude, and in powerful telescopes exhibits a decided disc.

The eighth satellite was discovered by Cassini in 1671. Its distance from the centre of the planet is 2,268,000 miles, which is nearly twice that of the farthest satellite of Jupiter, and the time of one revolution is 79 days. Its diameter is estimated at about 1800 miles. The plane of its orbit is inclined 10° to the plane of the ring.

Cassini noticed that this satellite regularly disappeared during half its revolution when to the east of Saturn. The improvement of telescopes has enabled more recent observers to follow the satellite through the entire extent of its orbit; but it is only with the greatest difficulty that it can be seen on the eastern side of the planet. It is hence inferred that this satellite rotates on its axis in the time of one revolution round the primary; and it is probable that the variations in its brightness are owing to some parts of its surface being less capable of reflecting the sun's light than others. At maximum brightness, this satellite appears like a star of the ninth magnitude.

The fifth satellite was discovered by Cassini in 1672. Its distance from the primary is 336,000 miles, and its period of revolution 4½ days. Its diameter is estimated at 1200 miles. It generally shines like a star of the tenth or eleventh magnitude.

The fourth satellite was discovered by Cassini in 1684. Its distance from the primary is 240,000 miles, and its period of revolution is 2¾ days. When brightest, it appears as a star of the eleventh magnitude.

The third satellite was discovered by Cassini in 1684. Its distance from the primary is 188,000 miles, and its period is 1 day and 21 hours. It generally resembles a star of the thirteenth magnitude. The diameters of the third and fourth satellites have been estimated at 500 miles.

The second satellite was discovered by Sir William Herschel in 1787. Its distance from the centre of the primary is 152,000 miles, and its period of revolution is 1 day 9 hours. It appears as a star of the fifteenth magnitude.

The first satellite was discovered by Sir W. Herschel in 1789. Its distance from the centre of the primary is 118,000 miles, and its period is 22½ hours. This satellite describes 360° of its orbit in 22½ hours, being at the rate of 16° per hour. Its motion, as seen from the primary, must therefore be so rapid as to resemble that of the hour-hand of an immense time-piece. In two minutes it moves over a space equal to the apparent diameter of our moon. This satellite is an extremely faint object, and can only be seen by the largest telescopes under the most favorable circumstances.

The seventh satellite was first discovered by Professor Bond, of Cambridge, September 16, 1848; and, two days later, it was seen by Mr. Lassell, of Liverpool. Its distance from the primary is 940,000 miles, and its period of revolution is 22 days. It resembles a star of the seventeenth magnitude.

454. *Mass and density of Saturn.*—The distance of a satellite compared with its time of revolution enables us to compare the mass of Saturn, or its quantity of matter, with that of the earth. This mass is thus found to be 100 times that of the earth; but its volume is nearly 800 times that of the earth; hence its density is only about ⅛th that of the earth. Since the density of the earth is 5½ times greater than that of water, the density of Saturn must be about ¾ths that of water. This is the density of the lighter sorts of wood, such as maple and cherry.

URANUS.

455. *Discovery.*—In 1781, the attention of Sir W. Herschel was attracted to an object which he did not find registered in the catalogues of stars, and which, with a high magnifying power, present-

ed a sensible disc; and he soon found that it changed its place among the fixed stars. He first announced this object as a comet; but when it was found to move in an orbit nearly circular, beyond the orbit of Saturn, its proper place among the planets was no longer questioned, and it was proposed to call it the "Georgium Sidus," in compliment to George III. The name Herschel was preferred by Laplace, and was, to some extent, adopted; but the scientific world have at last universally agreed upon the name Uranus.

456. *Former observations of this planet.*—As soon as an approximate orbit of the planet had been obtained, it was possible to compute its place at any past epoch. In this way it was found that the planet Uranus had been observed six times by Flamsteed as a fixed star, twelve times by Lemonnier, and once by Mayer. Thus the planet had been observed as a fixed star at least nineteen times before its real nature was detected by Sir W. Herschel. These observations extend back to 1690, and have proved of the greatest value in accurately determining the planet's orbit.

457. *Distance, period,* etc.—The mean distance of Uranus from the sun is 1828 millions of miles; and, since the eccentricity of its orbit is very small, this distance is increased in aphelion, and diminished in perihelion by less than one twentieth of its entire amount. The plane of its orbit coincides nearly with that of the ecliptic.

The period of one revolution is 84 years; but the interval between two successive oppositions is only 370 days.

458. *Diameter, form of its disc,* etc.—The diameter of this planet is 36,000 miles, being about half that of Saturn, and more than four times that of the earth. Its volume is nearly 100 times that of the earth; and its apparent diameter is about 4″.

The planet may be just discerned by a person gifted with strong sight, without the telescope, in a perfectly dark sky, when its exact position with reference to the surrounding stars is known.

The disc of Uranus appears uniformly bright, and of a pale color, but no appearance of spots or belts has been perceived. For this reason, the time of rotation upon its axis has not been ascertained. Some astronomers think they have detected considerable ellipticity in the form of the planet; but other astronomers, with equally good telescopes, have not succeeded in discovering any difference in the diameters.

Since light moves at the rate of 192,000 miles per second, it would require over 9000 seconds, or 2½ hours, to move from the sun to Uranus. Whatever changes may take place on the surface of the sun, they can not, therefore, be perceived by inhabitants of that planet until 2½ hours after they really take place.

459. *Satellites.*—Soon after the discovery of this planet, Sir W. Herschel announced that it was attended by a system of six satellites, but only four have ever been seen by any other observer. The times of revolution of these four satellites, together with their distances from the primary, have been well determined, and are as follows:

Satellite.	Revolution.	Distance.
1	2 d. 12 h.	120,000 miles.
2	4 3	171,000 "
3	8 17	288,000 "
4	13 11	380,000 "

The third and fourth of these satellites are by far the most conspicuous, and their periods have been ascertained with great accuracy. The existence of the fifth and sixth satellites announced by Herschel must be regarded as quite doubtful.

When the plane of the orbits of the satellites passes through the earth, the orbits appear as straight lines. Such was the case in 1840. When the direction of the earth is at right angles to the line of the nodes, the apparent orbits do not differ sensibly from circles. Such was the case in 1862. In all intermediate positions, the apparent orbits are more or less elliptical.

Contrary to the law which generally prevails in the motions of the planets and their satellites, the orbits of these satellites are inclined to the plane of the ecliptic 79°, being little less than a right angle; and their motions in these orbits are *retrograde*—that is, from east to west. We may, however, consider the motion of the satellites as *direct*, in orbits inclined 101° to the plane of the ecliptic.

NEPTUNE.

460. *Perturbations of the planets.*—If the planets were subject only to the attraction of the sun, they would revolve in exact ellipses, of which the sun would be the common focus; but, since they are also subject to the attraction of each other, they are drawn slightly out of the ellipses which they would otherwise describe. When the masses and distances of the planets are known, these disturbances can be computed with such precision that the exact place of any planet can be determined for any time either past or future.

461. *Irregularities in the motion of Uranus.*—In 1821, Bouvard published a set of tables for computing the place of Uranus. The materials for the construction of these tables consisted of 40 years' regular observations since 1781, and the 19 accidental observations (Art. 456), reaching back almost a century farther. Bouvard was unable to find any elliptic orbit which, combined with the perturbations of known planets, would represent the entire series of observations. He therefore rejected the ancient observations, and founded his tables upon the observations since 1781. These tables represent very well the observations of the 40 years from which they were derived; but soon after 1821 new discrepancies began to appear, which soon increased with great rapidity. This is shown in the following table, which exhibits the differences between the observed places of the planet and those computed from Bouvard's tables:

Year.	Differences.	Year.	Differences.	Year.	Differences.
1690	+43″	1781	−4″	1822	+1″
1712	+76	1785	−1	1825	−1
1715	+44	1790	+3	1828	−8
1750	−67	1795	+2	1831	−22
1753	−60	1800	−4	1834	−38
1756	−66	1805	−1	1837	−61
1764	−49	1810	−1	1840	−82
1769	−32	1815	+1	1843	−107
1771	−15	1820	−1	1846	−128

It is obvious, from an examination of this table, that the discrepancies can not be ascribed to the inaccuracy of the observations, since they evidently follow a definite law. We must, then, conclude that some cause was in operation not taken into account by Bouvard in constructing his tables.

462. *Researches of Le Verrier and Adams.*—In the year 1845, two astronomers, M. Le Verrier, of Paris, and Mr. Adams, of Cambridge, England, independently of each other, attempted to determine the place and magnitude of a planet outside of Uranus, which would account for these irregularities. The problem which they proposed, and which they actually solved, was this: *Given the perturbations produced in Uranus by the action of an unknown planet; it is required to assign the elements of a planet capable of producing these perturbations.*

Le Verrier and Adams, by a most laborious analysis, demonstrated that these irregularities were such as would be caused by an undiscovered planet revolving about the sun at a distance nearly double that of Uranus, and with a mass somewhat greater than that of Uranus; and they pointed out the place in the heavens which this planet ought at present to occupy. Le Verrier was the first to publish to the world the results of his researches, and thus obtained the chief credit for the discovery.

463. *Discovery of the planet at Berlin.*—On the 23d of September, 1846, Dr. Galle, of the Berlin Observatory, received a letter from Le Verrier, announcing the results of his calculations, informing him that the longitude of the unseen planet ought to be 326°, and requesting him to search for it. Dr. Galle did search for it, and found it on the first night. It appeared as a star of the eighth magnitude, having a longitude of 326° 52′, and, consequently, only 52′ from the place assigned by Le Verrier. This planet has been called Neptune.

The orbit of Neptune is smaller than that predicted by either Adams or Le Verrier, and its mass somewhat less; yet its disturbing action upon Uranus is such as perfectly to explain the anomalies which had been observed in the motion of that planet.

464. *Earlier observations of this planet.*—As soon as an approximate orbit of Neptune had been obtained, its place was computed back for several preceding years, and it was found that it had been repeatedly observed as a fixed star. Two such observations were made in 1795, one in 1845, and three in 1846, before it was seen

at Berlin. With the aid of these observations, it was soon possible to obtain a very accurate determination of the orbit of the planet.

465. *Distance, period*, etc.—The mean distance of Neptune from the sun is 2862 millions of miles, and its period of revolution is 164 years. Its apparent diameter is about 2½ seconds, and it resembles a star of the eighth magnitude. Its real diameter is 35,000 miles, which is a little less than that of Uranus.

466. *Bode's law disproved.*—The discovery of Neptune has entirely refuted Bode's law of planetary distances. This law has been stated in Art. 420. The following table shows, first, the true relative distance of each of the planets; second, the distance according to Bode's law; and, third, the error of this law.

	True Distance.	Bode's Law.	Error.
Mercury . . .	3.87	4	0.13
Venus	7.23	7	0.23
Earth	10.00	10	
Mars	15.24	16	0.76
82 Asteroids . .	26.67	28	1.33
Jupiter	52.03	52	0.03
Saturn	95.39	100	4.61
Uranus	191.82	196	4.18
Neptune . . .	300.37	388	87.63

Hence it will be seen that, although this law represents pretty well the distances of the nearer planets, the error is quite large for Saturn and Uranus, and for Neptune the error amounts to more than 800 millions of miles.

467. *Satellite of Neptune.*—Neptune has one satellite, which makes a revolution around the primary in 5d. 21h., at a distance of 236,000 miles, which is about the same as the distance of our moon from the earth. The orbit of this satellite is inclined 29° to the plane of the ecliptic, and its motion in this plane is *retrograde.* This fact is one of great interest, as hitherto the only known instance of retrograde motion among the planets, or their satellites, has been the case of the satellites of Uranus. This satellite is estimated to be equal in brightness to a star of the fourteenth magnitude.

Several observers at first suspected that Neptune was attended by a ring like Saturn, but later observations do not countenance this idea.

468. *Appearance of the solar system as observed from Neptune.*—The apparent diameter of the sun as seen from Neptune is 64″, which but little exceeds the greatest apparent diameter of Venus as seen from the earth. The illuminating effect of the sun at that distance is only about one thousandth part of its effect upon the earth, being about midway between our sunlight and our moonlight.

With reference to Neptune, all the other planets are *inferior*, and most of them never appear to recede many degrees from the sun. The greatest elongation of Uranus is 40°, of Saturn 18°, of Jupiter 10°, of Mars 3°, and of the interior planets still less. Uranus, Saturn, and Jupiter might perhaps therefore be seen by the inhabitants of Neptune as stars of the sixth magnitude, but none of the remaining planets. All the planets, if they could be observed from Neptune, would occasionally appear to travel across the sun's disc, but those which are interior to Jupiter subtend so small an angle that it is doubtful whether they could be seen even with the best telescope; and, on account of the small diameter of the sun, combined with the inclination of the planetary orbits, the transits of the larger planets would be of extremely rare occurrence. A transit of Uranus would not happen oftener than once in 40,000 of our years.

The problem of finding the distance of the fixed stars presents very little difficulty to the Neptunian astronomers, except that which arises from the length of one of their years, required to complete an observation, since they are in possession of a base line thirty times as long as that to which we are confined. See Art. 551.

CHAPTER XVII.

QUANTITY OF MATTER IN THE SUN AND PLANETS.—PLANETARY PERTURBATIONS.

469. *How to determine the mass of a planet.*—By the method employed in Art. 266, we may determine the masses of such of the planets as have satellites. The quantity of matter may also be found in terms of the distance and periodic time of the planet and its satellite.

Let M represent the mass of the sun, R the distance of a planet, and T its periodic time; then, by Art. 248, the central force which retains the planet in its orbit is

$$\frac{4\pi^2 R}{T^2}.$$

But, since the planet is retained in its orbit by the attraction of the sun, and this attraction varies directly as the mass, and inversely as the square of the distance, Art. 256, we shall have

$$\frac{M}{R^2}=\frac{4\pi^2 R}{T^2},$$

or

$$M=\frac{4\pi^2 R^3}{T^2}. \qquad (1)$$

For the same reason, if we put m to represent the mass of a planet, r the distance, and t the periodic time of a satellite revolving around it, we shall have

$$m=\frac{4\pi^2 r^3}{t^2}. \qquad (2)$$

Comparing equations (1) and (2), we find

$$M : m :: \frac{R^3}{T^2} : \frac{r^3}{t^2}.$$

Hence we see that the quantities of matter in the bodies which compose the solar system are directly as the cubes of the mean distances of any bodies which revolve about them, and inversely as the squares of the times in which the revolutions are performed.

Ex. 1. The distance of the earth from the sun is 95,300,000 miles, and its time of revolution 365.256 days. The distance of the moon from the earth is 238,900 miles, and its time of revolution 27.321 days. What is the mass of the sun compared with that of the earth? *Ans.* 355,000 times that of the earth.

Ex. 2. The mean distance of Jupiter from the sun is 495,817,000 miles, and its time of revolution is 103,982 hours; the distance of its fourth satellite is 1,200,000 miles, and its time of revolution 400.53 hours. What is the mass of the sun compared with that of Jupiter? *Ans.* 1047 times that of Jupiter.

Ex. 3. What is the mass of Jupiter compared with that of the earth? *Ans.* 339 times that of the earth.

Ex. 4. The distance of Saturn from the sun is 909,028,000 miles, and its time of revolution 10,759 days; the distance of its outer satellite is 2,268,000 miles, and its time of revolution 79.32 days. What is the mass of the sun compared with that of Saturn? *Ans.* 3500 times that of Saturn.

Ex. 5. What is the mass of Saturn compared with that of the earth? *Ans.* 101 times that of the earth.

Ex. 6. The distance of Uranus from the sun is 1,828,200,000 miles, and its time of revolution 30686.8 days; the distance of its fourth satellite is 380,000 miles, and its time of revolution 13.463 days. What is the mass of the sun compared with that of Uranus? *Ans.* 21,400 times that of Uranus.

Ex. 7. What is the mass of Uranus compared with that of the earth? *Ans.* 16 times that of the earth.

Ex. 8. The distance of Neptune from the sun is 2,862,457,000 miles, and its time of revolution 60126.7 days; the distance of its satellite is 236,000 miles, and its time of revolution 5.87 days. What is the mass of the sun compared with that of Neptune? *Ans.* 17,000 times that of Neptune.

Ex. 9. What is the mass of Neptune compared with that of the earth? *Ans.* 21 times that of the earth.

The masses of those planets which have no satellites have been determined by estimating the force of attraction which they exert in disturbing the motions of other bodies. The mass of Mercury has been determined from the perturbations which it causes in the motions of Encke's comet, which sometimes passes near to that planet. The mass of Venus is determined by the disturbance which it causes in the orbit of the earth; and the mass of Mars is determined in the same manner.

470. *How to determine the density of a planet.*—Having determined the quantity of matter in the sun and planets, and knowing also their volumes, Art. 387, we can compute their densities, for these are proportional to the masses divided by the volumes. Knowing also the specific gravity of the earth, Art. 49, we can compute the specific gravity of each member of the solar system. The following table shows the mass, density, and specific gravity of the principal members of our solar system. The masses are according to Le Verrier, and differ somewhat from the results of the preceding computations.

	Mass.	Density.	Specific Gravity.
Sun	354936	0.25	1.37
Mercury . . .	0.12	2.01	10.97
Venus . . .	0.88	0.97	5.30
Earth. . . .	1.00	1.00	5.46
Mars	0.13	0.72	3.93
Jupiter . . .	338.03	0.24	1.31
Saturn . . .	101.06	0.13	0.71
Uranus . . .	14.79	0.15	0.82
Neptune. . .	24.65	0.27	1.47

Upon comparing the numbers in this table, we do not readily perceive any law connecting the density of a planet with its distance from the sun. The four outer planets have a specific gravity differing but little from that of water, while the other planets increase in density somewhat according to their proximity to the sun.

471. *Problems.*

Prob. 1. How much must the mass of the earth be increased, in order that the moon might revolve about it in the same time as at present, although removed to twice her present distance?

Prob. 2. How much must the mass of the earth be increased to make the moon, at her present distance, revolve once in two days?

Prob. 3. If the earth's mass were 350,000 times as great as at present, in what time would the moon, at her present distance, revolve around it?

Prob. 4. What would be the periodic time of a small body revolving about the moon, at a distance of 5000 miles from the moon's centre, assuming the mass of the moon to be $\frac{1}{80}$th of the mass of the earth?

Prob. 5. What would be the periodic time of a satellite revolving about Jupiter close to the surface of the planet?

Prob. 6. How much faster must Jupiter rotate upon his axis in order that a body on the equator of the planet may lose all its gravity?

PERTURBATIONS OF THE PLANETS.

472. *How to compute the disturbing force of a planet.*—It appears from Art. 470 that the mass of the sun is more than a thousand times greater than the largest planet, and more than a hundred thousand times greater than the smaller planets. Moreover, the difference between the mean distances of the planets is so great, and the eccentricities of their orbits are so small, that, when they approach nearest to each other, the disturbing force exerted by any one upon any other is only a minute fraction of the attraction of the sun. Both the intensity and direction of the disturbing force caused by any one of the planets may be computed in the same manner as was shown in the case of the moon, Art. 269.

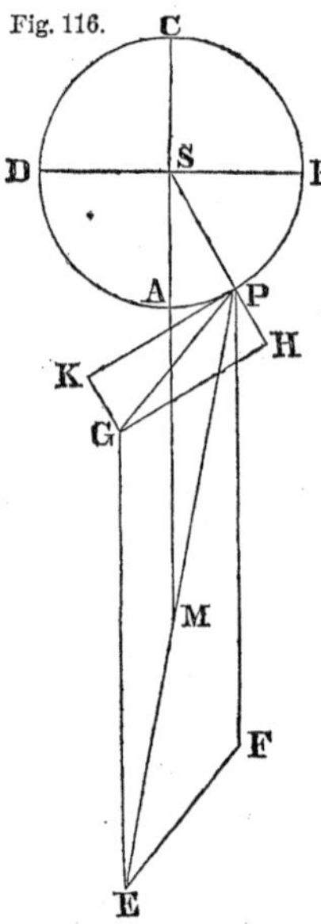

Fig. 116.

Let S represent the sun, P a planet revolving in its orbit ABCD, and let M be another planet which, by its attraction, disturbs the motion of P. Take SM to represent the force with which M attracts S; and in the line PM, produced if necessary, take PE such that $PE : MS :: MS^2 : MP^2$; then PE will represent in quantity and direction the force with which M attracts P. Resolve PE into PF and PG, of which PF is equal and parallel to SM. Then, as in Art. 269, PG represents the disturbing force of M upon P.

The ratio of the line PG to SM may be computed by Trigonometry when we know the distances of the two planets from the sun, and also their relative situations. The disturbing force of M upon P may then be compared with the sun's attraction on P by means of the following proportions:

1. Disturbing force : M's attraction on sun :: PG : SM.
2. M's att. on sun : sun's att. on M :: M's mass : sun's mass.
3. Sun's attraction on M : sun's attraction on P :: $SP^2 : SM^2$.

Compounding these proportions, we have,

Disturbing force : sun's attraction on P :: PG × SP^2 × M's mass : SM^3 × sun's mass.

473. *Disturbing force of Jupiter and Saturn.*—In order to show the application of these principles, we will compute the disturbing force of the two largest planets, Jupiter and Saturn, upon each other in two different positions.

Ex. 1. Compare the disturbing force of Saturn upon Jupiter with the sun's attraction upon Jupiter when the two planets are in conjunction, assuming the distances of Jupiter and Saturn from the sun to be 5.2028 and 9.5388, and the mass of the sun to be 3512 times that of Saturn.

Fig. 117.

S P M

Saturn's att. on Jupiter : Saturn's att. on sun :: $MS^2 : MP^2 :: 9.5388^2 : 4.336^2$ :: 4.8396 : 1.

The force with which Saturn draws Jupiter away from the sun is therefore represented by 3.8396; or,

Disturbing force : Saturn's attraction on sun :: 3.8396 : 1.

Saturn's att. on sun : sun's attraction on Saturn :: 1 : 3512.

Sun's att. on Saturn : sun's att. on Jupiter :: $5.202^2 : 9.538^2$.

By compounding these proportions, we have,

Disturbing force : sun's attraction on Jupiter :: $3.8396 \times 5.202^2 : 3512 \times 9.538^2$ :: 1 : 3075;

that is, *by the disturbing action of Saturn at conjunction, Jupiter's gravity to the sun is diminished by* $\frac{1}{3075}$th *part.*

Ex. 2. Compare the disturbing force of Saturn upon Jupiter with the sun's attraction upon Jupiter when the two planets are in opposition to each other. *Ans.* The ratio is 1 : 20300; that is, *by the disturbing action of Saturn at opposition, Jupiter's gravity to the sun is diminished by* $\frac{1}{20300}$th *part.*

Ex. 3. Compare the disturbing force of Jupiter upon Saturn with the sun's attraction upon Saturn when the two planets are in conjunction, assuming the mass of the sun to be 1050 times that of Jupiter.

Fig. 118.

S M P

Jupiter's att. on Saturn : Jupiter's att. on sun :: $MS^2 : MP^2 :: 5.2028^2 : 4.336^2$ :: 1.4397 : 1.

Hence the force with which Jupiter draws Saturn toward the sun is represented by 2.4397; or,

Disturbing force : Jupiter's attraction on sun :: 2.4397 : 1.

Jupiter's att. on sun : sun's attraction on Jupiter :: 1 : 1050.

Sun's att. on Jupiter : sun's att. on Saturn :: $9.538^2 : 5.202^2$.

By compounding these proportions, we have,

Disturbing force : sun's attraction on Saturn :: $2.4397 \times 9.538^2 :$
$$1050 \times 5.202^2 :: 1 : 128;$$
that is, *by the disturbing action of Jupiter at conjunction, Saturn's gravity to the sun is increased by* $\frac{1}{128}$th *part.*

Ex. 4. Compare the disturbing force of Jupiter upon Saturn with the sun's attraction upon Saturn when the two planets are in opposition to each other. *Ans.* The ratio is 1 : 357; that is, *by the disturbing action of Jupiter at opposition, Saturn's gravity to the sun is diminished by* $\frac{1}{357}$th *part.*

474. *Periodical inequalities of the planets.*—It is thus seen to be possible to compute the direction and intensity of the disturbing force exerted at any time by any planet upon any other planet; we can therefore compute how much each planet will be drawn out of its elliptic path by the disturbing action of the other planets. These disturbances, as stated in Art. 280, are either periodical or secular. The periodical inequalities of the planets are generally small. Those of Mercury can never exceed a quarter of a minute; those of Venus can never exceed half a minute; those of the earth about one minute; and those of Mars about two minutes. Those of Jupiter may amount to 20 minutes, and those of Saturn to 48 minutes, while those of Uranus are less than 3 minutes.

475. *Long inequalities.*—Some of these inequalities are very remarkable for the length of their periods, arising from a near approach to commensurability in the times of revolution. Eight times the period of the earth is nearly equal to thirteen times the period of Venus; or 235 times the period of the earth is almost exactly equal to 382 times the period of Venus, as was shown Art. 409. Hence arises in the motion of both of these planets an inequality having a period of 235 years; amounting, however, to only 2″.9 for Venus, and to 2″.0 for the earth.

476. *Long inequality of Jupiter and Saturn.*—The long inequality in the motion of Jupiter and Saturn is very celebrated in

the history of Astronomy. Five times the period of Jupiter is nearly equal to twice that of Saturn; or 77 revolutions of Jupiter are very nearly equal to 31 of Saturn, corresponding to a period of 913 years. Hence arises in the motion of both of these planets an inequality having a period of over 900 years, amounting, at its maximum, in the case of Jupiter, to about 20′, and in the case of Saturn to 48′.

477. *Long inequality of Uranus and Neptune.*—There is a similar inequality in the motions of Uranus and Neptune. The periodic time of Neptune is nearly double that of Uranus; or, more accurately, 25 revolutions of Neptune correspond to 49 of Uranus. Hence arises in the motions of these planets an inequality having a period of over 4000 years.

478. *Secular inequalities of the planets.*—The secular inequalities of the planets are generally small, but in the lapse of time become important by their continued accumulation. The nodes of all the planetary orbits have a slow motion westward on the ecliptic, amounting, in one case, to 36″ annually. The line of the apsides of their orbits is also in continual motion, that of Mercury moving eastward 5″, that of the earth 12″, and that of Saturn 19″ annually.

The disturbing action of one planet upon another causes the line of the apsides sometimes to progress, and at other times to regress; but in the case of most of the planets the former effect predominates.

479. *Secular variation of the inclination.*—The inclinations of the planetary orbits and their eccentricities are only subject to small periodical variations on each side of a mean, from which they never greatly depart. In no case (excepting the asteroids) does the change of inclination exceed a quarter of a second annually. The inclinations of the orbits of Jupiter and Saturn are closely related to each other; and it has been computed that the inclination of the orbit of Jupiter to the ecliptic must oscillate between the values of 2° 2′ and 1° 17′, while that of Saturn will oscillate between the values of 2° 32′ and 0° 46′, requiring for these changes a period of 50,000 years.

The inclination of the earth's equator to the ecliptic is now 24 minutes less than it was twenty-one centuries ago, and is now decreasing at the rate of half a second annually; but it has been proved that this is a secular inequality of a long period, and, after reaching a minimum, will return in the contrary direction, and thus oscillate back and forth about a mean position. It has been computed that the extent of the deviation on each side of the mean position will be less than 1° 21′, and that the period of this inequality will be about 30,000 years.

480. *Secular variation of the eccentricity.*—The eccentricities of all the planetary orbits are continually changing, but (with the exception of the asteroids) this change in no case exceeds one thousandth part in 300 years. In every instance these changes will always be confined within moderate limits. Those of Jupiter will be confined within the limits of 0.06 and 0.02, while those of Saturn will be confined within the limits of 0.08 and 0.01, the period in each case being 35,000 years.

The eccentricity of the earth's orbit is decreasing at the rate of 0.00004 in a century; but this change will always be confined within the limits of 0.07 and 0.003. The earth's orbit can therefore never become an exact circle. Le Verrier has computed that the eccentricity will continue to diminish for 24,000 years, when its value will be .0033. It will then begin to increase, and at the end of another period of 40,000 years its value will be .02, after which it will again slowly decrease.

This diminution in the eccentricity of the earth's orbit causes an acceleration in the mean motion of the moon amounting to 10″ in a century, Art. 281. This acceleration will continue as long as the earth's orbit is approaching the circular form; but when the eccentricity of the earth's orbit begins to increase, the acceleration of the moon's mean motion will be converted into a retardation.

481. *Secular constancy of the major axes.*—Thus the place of every planet in its orbit is changed by the action of the other planets, and the orbit itself is changed in all its elements but two—in the major axis of the orbit, and the time of the planet's revolution. These two elements of every planetary orbit remain secure against all disturbance. Moreover, all the inequalities in the planetary motions are periodical, and each of them, after a certain period of time, runs again through the same series of changes. Ev-

ery planetary inequality can be expressed by terms of the form Asin. nt, or Acos. nt, where A is a constant coefficient, and n a certain multiplier of t, the time; so that nt is an arc of a circle which increases proportionally to the time. Now, although nt is thus capable of indefinite increase, yet, since sin. nt can never exceed the radius, or unity, the inequality can never exceed A. Accordingly, the value of the term Asin. nt first increases from 0 to A, and then decreases from A to 0; after which it becomes negative, and extends to $-$A, and from thence to 0 again, the period of all these changes depending on n, the multiplier of t. If the value of any of the planetary inequalities contained a term of the form Ant, or Atang. nt, the inequality so expressed would increase without limit. Lagrange, Laplace, and Poisson, in demonstrating that no such terms as these last can enter into the expression of the disturbances of the planets, made known one of the most important truths in physical astronomy. They proved that the planetary system is *stable;* that the planets will neither recede indefinitely from the sun, nor fall into it, but continue to revolve forever in orbits of very nearly the same dimensions as at present, unless there is introduced the action of some external force.

482. *Why the solar system is stable.*—This accurate compensation of the inequalities of the planetary motions depends on certain conditions belonging to the original constitution of the solar system.

1st. It is essential that the mass of the central body should be much greater than that of any of the planets. If the mass of the sun were no greater than that of Jupiter, then the disturbing action of Jupiter upon the nearer planets would be sufficient entirely to change the form of their orbits; but at present the disturbing action of Jupiter upon any one of the planets is small when compared with the attraction of the sun upon the same planet.

2d. It is essential that the distances between the planets, especially of the larger ones, should be considerable when compared with their distances from the sun, otherwise, when in conjunction, their disturbing action upon each other would be so great as entirely to change the form of their orbits. Hence, also,

3d. It is essential that the orbits of the planets, especially of the larger ones, should have but little eccentricity. If the orbit of

Jupiter were as eccentric as that of many of the comets, he might at times approach so near to the earth as entirely to change the form of our orbit. This principle is well illustrated by the case of the comet of 1770. See Art. 521.

4th. It is essential that the planets should revolve around the sun in planes but little inclined to each other. If Jupiter's orbit had great inclination to the ecliptic, he would tend continually to draw the earth out of its present plane of motion, while the axis of the earth would retain a fixed position in space; that is, the obliquity of the ecliptic might change very greatly, and this would involve a change of seasons which might be very unfavorable both to animal and vegetable life.

These four conditions are essential to the stability of any system. In the demonstration by Laplace that the solar system is stable, a fifth condition is required, namely, that the planets all move about the sun in the same direction.

These conditions do not necessarily result from the nature of motion or of gravitation, neither can they be ascribed to chance, for it is improbable that without a cause particularly directed to that object there should be such a conformity in the motions of so many bodies scattered over so vast an extent of space. It seems difficult to avoid the conclusion that all this is the work of intelligence and design, directing the original constitution of the system so as to give stability to the whole.

483. *Effect of commensurability in the periodic times.*—If the periodic times of the planets were commensurable, and could permanently continue thus, it would endanger the stability of the solar system. If, for example, Neptune made *exactly* one revolution while Uranus makes two (as it does very nearly), then the two planets would always come into conjunction in the *same* part of their orbits; the effect which Neptune produces upon Uranus at one conjunction, although small, would be doubled at the second conjunction, and trebled at the third; and, after the lapse of a large number of revolutions, the orbit of Uranus would be entirely changed. But even supposing the periodic times to have been made exactly commensurable, they could not permanently continue so, since any change produced in the periodic time of the disturbed planet is necessarily accompanied by a change in the *opposite* direction in that of the disturbing planet, so that the

periods would become incommensurable by the mere effect of their mutual action. So long as the periodic times of the planets are incommensurable, their conjunctions take place successively upon different parts of the orbit, and their effects compensate each other by mutual opposition. When there is a near approach to commensurability (as in the case already cited), it requires a long period of years for these effects to compensate each other—that is, it gives rise to an inequality of a long period.

CHAPTER XVIII.

COMETS.—COMETARY ORBITS.—SHOOTING STARS.

484. *What is a comet?*—A comet is a nebulous body revolving in an orbit about the sun, sometimes with a bright nucleus and tail, but frequently with neither. The orbits of all known comets are more eccentric than any of the planets. The most eccentric planetary orbit known is that of Polyhymnia, one of the asteroids, whose eccentricity is 0.338; the least eccentric cometary orbit is that of Faye's comet, whose eccentricity is 0.556. In consequence of this eccentricity, and of the faintness of their illumination, all comets, during a part of every revolution, disappear from the effect of distance.

485. *Number of comets.*—The number of comets which have been recorded since the birth of Christ is over 600, and the number of those whose orbits have been computed is 240. Of these, 195 have moved either in parabolic orbits, or in ellipses of such eccentricity that they could not be distinguished from parabolas; there are five whose motions are best represented by an hyperbola, while about 40 have been computed to move in elliptic orbits.

The number of comets belonging to the solar system must amount to many thousands. Eighty comets have been recorded within the past 50 years; and if we admit that the proportion of faint comets was as great before the invention of the telescope as it has been since, we must conclude that more than 4000 comets have approached the sun within the orbit of Mars since the commencement of the Christian era.

Comets are usually named from the year in which they appear. Several of them, however, are known by the name of their first discoverer, or of some astronomer specially connected with their history. Among the most celebrated comets are those of Halley, Encke, Biela, Donati, etc., which are specially described in Arts. 507–524.

486. *Position of cometary orbits.*—The orbits of comets exhibit every possible variety of position. Their inclinations to the ecliptic range from 0° to 90°, and their motion is as frequently retrograde as direct; in other words, their inclinations range from 0° to 180°. Unlike the planets, the comets are seen near the poles of the heavens as well as near the ecliptic.

487. *Period of visibility.*—The duration of a comet's visibility varies from a few days to more than a year, but it most usually happens that it does not exceed two or three months. Only six comets have been observed so long as 8 months. The comet of 1825 was observed nearly 12 months, and that of 1811 was observed 17 months. The period of visibility of a comet depends on its intrinsic brightness, and on its position with reference to the earth and sun.

488. *The coma, nucleus, tail,* etc.—The most splendid comets consist of a roundish, and more or less condensed mass of nebulous matter termed the *head,* from which issues, in a direction opposite to that of the sun, a train of a lighter kind of nebulosity called the *tail.* Within the head is sometimes seen a bright point, like a star or a planet, which is called the *nucleus* of the comet. In most instances the centre of the head exhibits nothing more than a higher degree of condensation of the nebulous matter, which always has a confused appearance in the telescope. The nebulosity which surrounds a highly condensed nucleus is called the *coma.* In most instances the coma is less than 100,000 miles in diameter, and but very rarely exceeds 200,000; but that of the comet of 1811 exceeded a million of miles in diameter.

489. *Dimensions of the nucleus.*—In a few instances the diameter of the nucleus has been computed at 5000 miles; but it seldom exceeds 500 miles; and the majority of comets have no bright nucleus at all.

It is probable that in those cases in which the diameter of the nucleus has been estimated at 5000 miles, the object measured was not a solid body, but simply nebulous matter in a very high degree of condensation. Thus Donati's comet at one time exhibited a bright nucleus whose diameter was computed at over 5000 miles. But as the comet approached the sun and increased in brilliancy, the nucleus steadily decreased; and when the comet was nearest to the sun, the diameter of the nucleus was less than 500 miles. The nuclei of some comets have exactly the appearance of solid bodies; but the true nucleus, apart from the surrounding nebulosity, is probably quite small.

490. *Variations in the dimensions of comets.*—The real dimensions of the nebulosities of comets vary greatly at different dates during their visibility. Many of them *contract* as they approach the sun, and *dilate* on receding from the sun. This has been repeatedly observed in the case of Encke's comet, and the same has been noticed in the case of several other comets. It has been conjectured that this effect may result from the change of temperature to which the comet is exposed.. As the comet approaches the sun, the vapor which composes the nebulous envelope may be converted by intense heat into a transparent and invisible elastic fluid. As it recedes from the sun, the temperature decreasing, this vapor is gradually condensed, and assumes the form of a visible cloud; whence the *visible* volume of the comet is increased, while its *real* volume may perhaps be diminished.

491. *Changes in the nebulosity about the nucleus.*—When comets have a bright nucleus and a splendid train, the nebulosity about the nucleus undergoes remarkable changes as the comet approaches the sun. The nucleus becomes much brighter, and throws out a jet or stream of luminous matter toward the sun. Sometimes two, three, or more jets are thrown out at the same time in different directions. This emission of luminous matter sometimes continues, with occasional interruptions, for several weeks. The form and direction of these luminous streams undergo singular and capricious alterations, so that no two successive nights present the same appearance. These jets, though very bright at their point of emanation from the nucleus, fade away and become diffuse as they expand into the coma, at the same time curving back-

ward, as if impelled against a resisting medium. These streams combined form the outline of a bright parabolic envelope surrounding the nucleus, and this envelope steadily increases in its dimensions, receding from the nucleus. After a few days a second luminous envelope is sometimes formed within the first, the two being separated by a band comparatively dark, and this second envelope steadily increases in its dimensions from day to day. A few days later a third envelope is sometimes formed, and so on for a long series. Donati's comet showed seven such envelopes in succession, each separated from its neighbor by a band comparatively dark, and each steadily receding from the nucleus. See Plate VI., Fig. 3.

These envelopes seem to be formed of substances different from the vapors on the earth's surface, for they do not sensibly refract light. They appear to be driven off from the nucleus by a repulsive force on the side next the sun, somewhat as light particles are thrown off by electric repulsion from an excited conductor; and the dark bands separating the successive envelopes seem to result from a periodical cessation or diminished activity of this repulsive force.

492. *The tail.*—The tail of a comet is but the prolongation of the nebulous envelope surrounding the nucleus. Each particle of matter, as it issues from the nucleus on the side next to the sun, gradually changes its direction by a curved path, until its motion is almost exactly away from the sun. The brightness and extent of the train increase with the brightness and magnitude of the envelopes, the tail appearing to consist exclusively of the matter of the envelopes, driven off by a powerful repulsive force emanating from the sun. On the side of the nucleus opposite to the sun there is no appearance of luminous streams, and hence results a dark stripe in the middle of the tail, dividing it longitudinally into two distinct parts. This stripe was formerly supposed to be the shadow of the head of the comet; but the dark stripe still exists even when the tail is turned obliquely to the sun. The tail is probably a hollow envelope; and when we look at the edges, the visual ray traverses a greater quantity of nebulous particles than when we look at the central line, which circumstance would cause the central line to appear less bright than the sides.

493. *Rapidity of formation of the tail.*—When comets make their first appearance they generally have little or no tail; but, by degrees, the nebulous envelope is formed, the tail soon appears, which increases in length and brightness as the comet approaches perihelion. When the comet is nearest the sun, the tail sometimes increases with immense rapidity. The tail of Donati's comet in 1858 increased in length at the rate of two millions of miles per day; that of the great comet of 1811 increased at the rate of nine millions of miles per day; while that of the great comet of 1843, soon after passing perihelion, increased at the rate of 35 millions of miles per day.

494. *Dimensions of the tail.*—The tails of comets frequently have an immense length. That of 1843 attained a length of 120 millions of miles; that of 1811 had a length of over 100 millions of miles, and a breadth of about 15 millions; and there have been four other comets whose tails attained a length of 50 millions of miles.

The *apparent* length of the tail depends not merely upon its absolute length, but upon the direction of its axis, and its distance from the earth. There are six comets on record whose tails subtended an angle of 90° and upward—that is, whose length would reach from the horizon to the zenith; and there are about a dozen more whose tails subtended an angle of at least 45°.

The tail usually attains its greatest length and splendor a few days after the comet passes its perihelion; and as the comet recedes farther from the sun, the tail fades gradually away, being apparently dissipated in space.

495. *Position of the axis of the tail.*—The axis of the tail, CG, Fig. 119, is not a straight line, and, except near the nucleus, is not directed exactly *from* the sun, but always makes an angle with a radius vector, SC. This angle generally amounts to 10° or 20°, and sometimes even more, the tail always inclining *from* the region toward which the comet proceeds. If the tail were formed by a repulsive force emanating from the sun, which carried particles *instantly* from the comet's head to the extremity of the tail, then the axis of the tail ought to be turned exactly *from* the sun. But, in fact, the particles at the extremity of the tail, as G, are those which were emitted from the nucleus several days previ-

ous, perhaps 20 days, when the head of the comet was at A, and, in consequence of their inertia, they retain the motion in the direction of the orbit which the nucleus had at the time they parted from it. The particles near the middle of the tail, as H, are those which left the nucleus later than the preceding, perhaps 10 days, when the head of the comet was at B, and they retain the motion in the direction of the orbit which the nucleus had at the time when they parted from it.

496. *Probable mode of formation of comet's tails.*—In order to explain the phenomena of comet's tails, it seems necessary to admit the existence of a repulsive force by which certain particles of a comet are driven off from the nucleus, and that these particles are then acted upon by a more powerful repulsive force emanating from the sun.

Fig. 119.

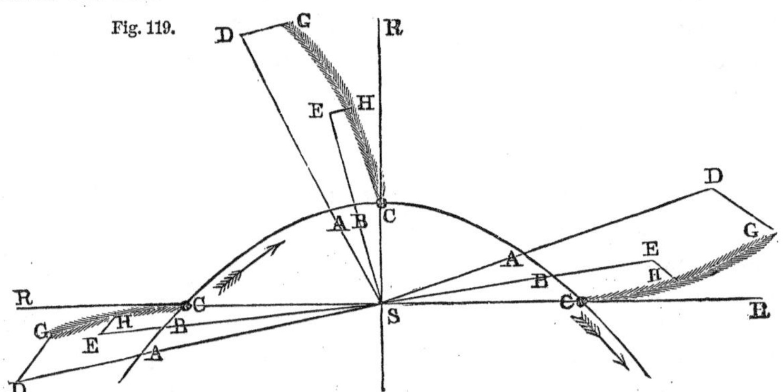

Let S represent the position of the sun, and ABC a portion of a comet's orbit, the comet moving in the direction of the arrows. Suppose, when the nucleus is at A, a particle of matter is expelled from the head of the comet in the direction SAD. This particle will still retain the motion which it had in common with the nucleus, and this motion would carry the particle over the line DG while the head is moving from A to C. When the nucleus reaches B, suppose another particle to be driven off in the direction SBE. This particle will also retain the motion which it had in common with the nucleus, and which would carry it over EH while the head is moving from B to C. Thus, when the nucleus has reached the point C, the particles which were expelled from the head during the time of its motion from A to C will all be situated upon the line CHG. It is evident that this line will be a curve, tangent at C to the radius vector SC produced, and always curving *from* the region toward which the comet proceeds.

497. *Comets with several tails.*—A transverse section of the tail of a comet is not generally a circle, but an oval curve, more or less elongated. In the case of Donati's comet, the greatest diameter of this oval was about four times the least, and in the comet of 1744 the ratio was probably still greater. The longest diameter of the transverse section coincides nearly with the plane of the orbit; in other words, the tail of a comet spreads out like a fan, so that its breadth, measured in the direction of the plane of the orbit, is greater than its breadth measured in a transverse direction.

In order to explain this phenomenon, it seems necessary to admit that the repulsive force of the sun is not the same upon all the particles which form the tail of the comet. Those particles upon which the repulsive force of the sun is very great will form a tail which is turned almost exactly *from* the sun; but those particles upon which the repulsive force of the sun is small will form a tail which falls very much behind the direction of a radius vector. If, then, the head of the comet consists of particles which are *unequally* acted upon by the sun, the comet may have several tails, or perhaps an indefinite number, whose axes occupy somewhat different positions, but all are situated in the plane of the orbit. This theory will enable us to explain the striped appearance of the tail of Donati's comet (see Plate VI., Fig. 1), as well as the faint streamers which extended in a direction very nearly opposite to the sun. It also explains the very remarkable appearance of the comet of 1744, which is commonly said to have had six tails. When near perihelion, the head of the comet being below the horizon, the tail was seen to extend above the horizon, as represented in Plate V., Fig. 5.

498. *Telescopic comets.*—Most comets are not attended by tails. Telescopic comets seldom exhibit this appendage. They, however, generally become elongated as they approach the sun, and the point of greatest brightness does not occupy the centre of the nebulosity.

In many cases, the absence of a tail is probably owing to the smallness of the comet, and the consequent faintness of its light; so that, although a tail is really formed, it entirely escapes observation.

In other cases, it seems probable that, by frequent approaches to the sun, the comet has lost all of that class of particles which are repelled by the sun, and which contribute to form a tail; and such comets exhibit only a slight elongation as they approach the sun.

499. *Quantity of matter in comets.*—The quantity of matter in comets is exceedingly small. Comets have been known to pass near to some of the planets and their satellites, and to have had their own motions much disturbed by the consequent attractions, without producing any sensible disturbance in the motion of the planets or their satellites. Since the quantity of matter in comets is inappreciable in comparison with the satellites, while their volumes are enormously large, the density of the comet's nebulosity must be incalculably small.

The *transparency* of the nebulosities of comets is still more remarkable. Stars of the smallest magnitude have been repeatedly seen through comets of from 50,000 to 100,000 miles in diameter, and, in the majority of cases, not the least perceptible diminution of the star's brightness could be detected.

500. *Do comets exhibit phases?*—Comets exhibit *no* phases like those presented by the moon, and which might be expected from a solid nucleus shining by reflected light. Some have therefore doubted whether comets shine simply by the borrowed light of the sun. The following consideration proves that their light, at least for the most part, depends on their distance from the sun. A self-luminous surface appears of the same brilliancy at all distances as long as it subtends a sensible angle. Thus the surface of the sun, as seen from Uranus, must appear as bright as it does to us, only subtending a smaller angle. If, then, a comet shines by its own light, it should retain its brilliancy as long as its diameter has a sensible magnitude. Such, however, is not the case. Comets gradually become dim as their distance increases; and they vanish simply from loss of light, while they still retain a sensible diameter.

501. *Cometary orbits.*—It was first demonstrated by Newton that a body which revolves under the influence of a central force like gravitation, whose intensity decreases as the square of the distance increases, must move in one of the conic sections; that is, either a parabola, an ellipse, or an hyperbola. Several comets are known to move in ellipses of considerable eccentricity; the orbits of most comets can not be distinguished from parabolas; while a few have been thought to move in hyperbolas. Since the parabola and hyperbola consist of two indefinite branches which diverge from each other, a body moving in either of these curves would not complete a revolution about the sun. It would enter the solar system from an indefinite distance, and, passing through its perihelion, issue in a different direction, moving off to an indefinite distance, never to return. Hence bodies moving in parabolas and hyperbolas are not periodic; but comets moving in elliptic orbits must make successive revolutions like the planets.

It is, however, probable that the orbits which are treated as parabolic are in fact very long ellipses, which differ but little from parabolas in that portion described by the comet while it is visible.

502. *How to deduce the orbit of a comet from the observations.*—The methods explained in Chapter XIV. for determining the orbits of the planets are generally quite inapplicable to the comets, because these methods require observations to be made in particular portions of the orbit, and often involve an interval of several years between the observations; but comets generally continue in sight only a few weeks, and from these few observations it is required to deduce the form and position of the orbit. Sir Isaac Newton first pointed out the method of computing the orbit of a comet from three observations of its position. This method was published in the Principia in 1687, accompanied by a computation of the orbit of the remarkable comet of 1680. This method has since been much simplified, and tables have been prepared by which the computations are greatly facilitated.

In order to determine the orbit of a comet, we must therefore know its direction in the heavens on three different days. These observations may be embraced within an interval of 48 hours; but the longer the interval, the more reliable will be the orbit deduced from the observations. The computations are also much simplified when the interval from the first to the second observation is exactly equal to the interval from the second to the third, although this restriction is by no means a necessary one. We will suppose, then, that we have three observations of a comet's place as seen from the earth. These places are usually denoted by right ascensions and declinations. We begin with converting these places into longitudes and latitudes, which have reference to the ecliptic, because we find it most convenient to refer the comet's motion to the plane of the earth's orbit.

We now take from the Nautical Almanac the longitudes of the sun for the same three instants of observation. These longitudes, increased by 180°, represent the longitudes of the earth as seen from the sun. We then construct a diagram representing the earth's orbit, and set off upon it the places of the earth at the three given dates. From each of these points we draw a line representing the direction in which the comet was seen at the corresponding date. From these data it is required to deduce the orbit in which the comet is moving.

503. *Principles assumed in computing the orbit.*—In computing the orbit, we assume certain laws which have been verified in the case of all the known members of the solar system. These laws are,

1st. The plane of the orbit of the comet must pass through the sun.

2d. The path described by the comet must be a conic section, of which the sun occupies one of the foci; and since we are sure that its orbit is quite eccentric, we know that it can not differ much from a parabola. We therefore assume, in the first case, that the orbit is a parabola.

3d. The motion of every heavenly body in its orbit about the sun is such that the areas described by the radius vector are proportional to the times in which they are described. Hence any area divided by the time gives a quotient which is a constant quantity for the same body, whether it be a planet or a comet.

4th. For different bodies revolving about the sun, the squares of the quotients thus obtained are proportional to the parameters of the orbits.

In applying these principles to determine the orbit of a comet, we may first assume *any* position for the plane of the orbit. If this assumption does not violate any of the preceding principles, we may be sure that we have found the true plane of the orbit. Otherwise we must vary the position of this plane until we have found one which does not conflict with either of these principles. We next compute the comet's places from day to day in the supposed orbit, and compare the observed places with the computed places. These differences must not exceed the unavoidable errors of observation. If they do, we must vary the assumed orbit until the observed places agree with the computed places, within the limits of those errors to which such observations are always liable. We are thus able to decide whether the orbit is truly a parabola, an ellipse, or an hyperbola, independently of any error in our first hypothesis.

504. *Method illustrated by a diagram.*—Let S represent the position of the sun, and A, B, C the positions of the earth in its orbit at the dates of the three observations, the lengths and position of the lines SA, SB, SC being given in the Nautical Almanac. From A draw a line, AD, to represent the direction in which the comet was seen at the first observation; and from B and C, in like manner, draw lines to represent the directions of the comet at the second and third observations. We know that at the date of the first observation the comet was somewhere on the line AD, but we do not know at what point of this line. If we assume that the comet was at G, then, by Principle 1st of Art. 503, its places at the other two dates will not only be on the lines BE, CF, but in a plane passing through S and G. Also, by Principle 3d, Art. 503, the parabolic sectors SHG, SHK, must be equal to each other. If the interval between the observations is only a few days, the parabolic sectors will differ but little from the plane triangles SHG, SHK, which must therefore be nearly equal. These conditions alone will generally enable us to determine the

Fig. 120.

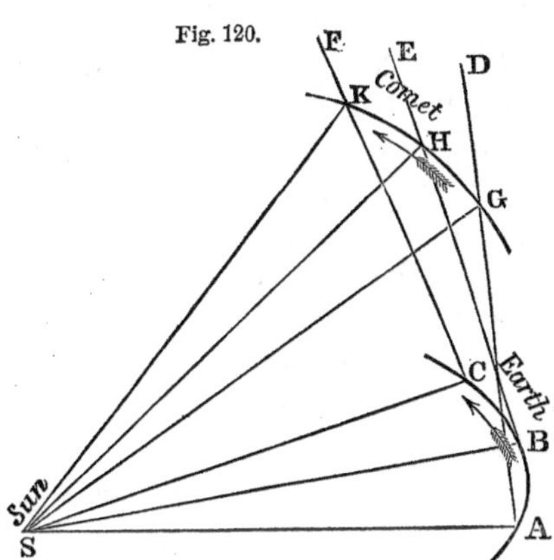

position of the comet's orbit very nearly. If the parabolic sectors are found to be exactly equal, and if the magnitude of these sectors is such as is given by Principle 4th, Art. 503, we may be sure that we have discovered the true orbit. By this method we may, in less than an hour, deduce the approximate orbit of any comet from three observations, embracing an interval of only a few days.

505. *Mode of computing the orbit.*—The actual computation of the elements of a cometary orbit is founded upon the same principles. The geometrical relations here stated are represented by equations, and these equations are solved by successive approximations. Tables have been prepared which greatly facilitate the computation, so that the approximate elements of a cometary orbit can be obtained by the labor of a few hours. The most accurate possible determination of the orbit is only obtained by a careful comparison of all the observations made during the entire period of visibility, and this may involve a labor of several weeks.

The method here indicated is applicable to the determination of the orbit of a planet as well as that of a comet. The motions of planets and comets are governed by exactly the same laws; and there is no essential difference in the mode of computing the orbits, except that we generally assume that a comet moves in a parabola, which is a conic section whose eccentricity is known; in other words, a cometary orbit involves *one less* unknown quantity to be determined than a planetary orbit. The methods of determining the planetary elements explained in Chapter XIV. are applicable to the brighter planets; but when a new asteroid is discovered, we are required to deduce an orbit at once from observations of a few days, and this is accomplished by the method here indicated.

506. *How a comet is known to be periodical.*—Since comets are only seen in that part of their orbit which is nearest to the sun, and since an ellipse, a parabola, and an hyperbola, for a considerable distance from perihelion, do not depart very widely from each other, it is difficult to determine in which of these curves a comet actually moves; but if a comet have an elliptic orbit, it must return to perihelion after completing its revolution. If, then, we find that two comets, visible in different years, moved in

the same path, the presumption is that they were the same body, reappearing after having completed its circuit in an elliptic orbit; and if the comet has been observed at several returns, this evidence may amount to absolute demonstration.

The shortest periodic time of any comet at present known is $3\frac{1}{3}$ years; the longest period which has been positively verified by the return of the comet is 75 years; but there are several comets whose period has been computed to exceed a century; and the periods of some comets probably amount to many centuries. Indeed, if a comet moves in a parabolic orbit, its periodic time must be infinitely long; that is, it could never complete a revolution about the sun.

There are seven comets whose periods have been well established, viz., Halley's, Encke's, Biela's, Faye's, Brorsen's, D'Arrest's, and Winnecke's.

Halley's Comet.

507. Soon after the publication of Newton's Principia, Halley, an eminent English astronomer, computed from recorded observations the elements of a number of comets according to the method furnished by Newton. These elements were published in 1705. On comparing these orbits, he found that a comet, which had been observed by himself and others in 1682, followed a path which coincided very nearly with those of comets which had been observed in 1607 and 1531. This led him to suppose that, instead of three different comets, it might be the same comet revolving in an orbit whose period was 75 or 76 years. He accordingly predicted the reappearance of this body in 1758–9. He observed, however, that as, in the interval between 1607 and 1682, the comet passed near Jupiter, its velocity must have been augmented, and, consequently, its period shortened by the action of that planet, the comet ought not to be expected to appear until the end of 1758, or the beginning of 1759.

508. *Predicted return of Halley's comet.*—As the time approached for the fulfillment of this prediction, two French astronomers, Clairaut and Lalande, undertook to compute the disturbing effect of the planets upon the comet, and thus determine its exact path, with the time of its return to perihelion. The result of these computations was to fix upon April 13, 1759, as the time of perihelion

passage. Clairaut, however, stated that, on account of the small quantities unavoidably neglected in his computations, the time thus assigned might vary from the truth to the extent of a month. The comet passed its perihelion on the 12th of March, just one month before the time announced by Clairaut; and Laplace has shown that if Clairaut had used in his calculations the mass of Saturn as at present received, his prediction would have been in error only 13 days.

Before the comet's next return in 1835, its path was computed by several astronomers, and the most complete computations fixed the time of perihelion at November 14, 1835. It actually passed its perihelion on the 16th of November.

The mean distance of this comet from the sun is about 18 times that of the earth, or a little less than the mean distance of Uranus; but, on account of the eccentricity of its orbit, its distance from the sun at aphelion is considerably greater than that of Neptune.

509. *Physical peculiarities of Halley's comet.*—At its return in 1835, Halley's comet exhibited physical changes remarkable for their magnitude and rapidity. The tail began to be formed about a month before the perihelion passage, and commenced with an emanation of nebulous matter from that part of the comet which was turned toward the sun. This emanation resembled a brush of electric light from a pointed wire in a dark room. As this matter receded from the head, it seemed to encounter a resistance from the sun, by which it was driven back, and carried out to vast distances behind the nucleus, forming the tail. This emanation took place only at intervals; and sometimes the nebulous matter thus emitted presented the appearance of a second tail turned toward the sun. At one time two, and at another time three nebulous streams were observed to issue in diverging directions. See Plate V., Fig. 2. These directions were continually varying, as well as the comparative brightness of the emanations. Sometimes they assumed the form of a swallow-tail. These jets, though very bright at their point of emanation from the nucleus, faded rapidly away, and became diffused as they expanded into the coma.

The tail seemed to be formed entirely of matter thus emitted from the head and repelled by the sun. The velocity with which this matter was driven from the sun was enormous, amounting to not less than two millions of miles per day. On account of the feeble attraction of the nucleus, the matter thus repelled from the head must mostly escape, and be lost in space, never to reunite with the comet. Hence it seems inevitable that at each approach to the sun the comet must lose some of those particles on which the production of the tail depends, so that at each return the dimensions of this appendage must become smaller and smaller. Upon comparing the different descriptions of this comet at its successive returns to perihelion, it has been concluded that it is now much smaller than it was in 1305. But the appearances in 1835 did not indicate any material diminution since 1759, so that we must conclude that, if this comet is actually wasting away, the process is a very gradual one.

Encke's Comet.

510. The periodicity of this comet was discovered in 1819 by Professor Encke, of Berlin, who identified the comet of that year with those that had been observed in 1786, 1795, and 1805, and which had been supposed to be different comets. He found its period to be only about 1207 days, or $3\frac{1}{3}$ years, and he predicted its return in 1822. This prediction was verified, and the comet has been observed at every subsequent return to the sun, making 13 apparitions since 1819, and 17 returns for which we have accurate observations.

At perihelion this comet passes within the orbit of Mercury, while at aphelion its distance from the sun is $\frac{4}{5}$ths that of Jupiter.

511. *Indications of a resisting medium.*—By comparing observations made at the successive returns of this comet, it is found that the periodic time, and, consequently, the mean distance from the sun, is subject to a slow but regular decrease, amounting to about a day in eight revolutions. It also appears that this diminution is not produced by the disturbing action of the planets. In order to explain the observed fact, Encke assumes that the interplanetary spaces are pervaded by an extremely rare medium, which causes no sensible obstruction to the motions of dense bodies like the planets, but which sensibly resists the motion of a mere mass of vapor like a comet. The effect of such a resistance would be a diminution of the comet's orbital velocity, in consequence of which it would be drawn nearer the sun, and perform its revolu-

tion in less time. It appears to follow from this hypothesis that, after the lapse of many ages, not only this comet, but other comets, and the planets also, must be precipitated upon the sun.

512. *Objections to Encke's hypothesis.*—It has been objected to Encke's hypothesis that no indication of a resisting medium has been detected in the motion of other comets. Encke's answer to this objection is that, in order to decide whether a comet is affected by a resisting medium, we require observations at three successive returns of the comet to the sun. Now there are only three comets for which we have such observations, viz., Halley's, Encke's, and Faye's. We do not know whether Halley's comet experiences resistance or not, because the disturbing influence of the planets through an entire revolution has never yet been computed with sufficient precision. Faye's comet, will be described in Arts. 515, 516.

Biela's Comet.

513. In 1826, Captain Biela, an Austrian officer, discovered a comet, which was afterward observed by other astronomers. The path which it pursued was found to be similar to that of comets which had appeared in 1772 and 1805; and Biela concluded that this body revolved in an elliptic orbit, with a period of about $6\frac{2}{3}$ years.

This comet has since been observed at three returns in conformity with prediction, viz., in 1832, 1846, and 1853. The returns which must have taken place in 1839 and 1859 escaped observation, owing to the unfavorable position and extreme faintness of the comet.

At perihelion the distance of this comet from the sun is a little less than that of the earth, while at aphelion its distance somewhat exceeds that of Jupiter.

The orbit of this comet approaches the earth's orbit within a distance less than the sum of the semi-diameters of the earth and comet. The earth passes this point of its orbit on the 30th of November. If Biela's comet should ever arrive at the same point on the 30th of November, the earth must penetrate a portion of the comet. In 1832, the comet passed this point on the 29th of October—a circumstance which created no little alarm.

514. *Division of Biela's comet into two comets.*—In 1846, this comet presented the singular phenomenon of a double comet, or two distinct comets moving through space side by side. See Plate V., Fig. 3. At first, one portion was extremely faint as compared with the other, but the fainter gradually increased, and by the middle of February they were nearly equal in brightness; after which the variable comet began to diminish, and in about a month disappeared, while the other continued visible several weeks longer as a single comet. The orbits of these two bodies were found to be ellipses entirely independent of each other; and during their entire visibility in 1846, their distance apart was about 200,000 miles.

Biela's comet reappeared in August, 1852, and continued visible about four weeks. The changes of relative brilliancy of the two comets were similar to those observed in 1846. At first one body was fainter than the other; subsequently the fainter became the brightest; and, a few days later, it again became the fainter of the two. The distance of the two bodies from each other in 1852 was about 1,500,000 miles. By following both bodies backward in their orbits, we find that about the last of September, 1844, their distance from each other was only 15,000 miles, which is less than the radius of one of these bodies. Their separation, then, probably took place in the latter part of the year 1844. This separation may have been caused by the operation of some internal force like that which causes the tails of comets, or perhaps by collision with some other body like one of the asteroids.

Faye's Comet.

515. In 1843, M. Faye, of the Paris Observatory, discovered a comet, and determined its orbit to be an ellipse, with a period of only 7½ years. Le Verrier computed its orbit with great care, and predicted its succeeding return to perihelion for April 3, 1851. The comet was first seen November 28, 1850, very nearly in the place assigned it by Le Verrier; and it reached its perihelion within about a day of the time predicted.

The distance of Faye's comet from the sun at perihelion is 161 millions of miles, and at aphelion 565 millions. This comet is remarkable as having an orbit more closely resembling in form the orbits of the planets than any other cometary orbit known, its eccentricity being only 0.55.

516. *Does Faye's comet afford evidence of a resisting medium?*—The observations of Faye's comet at its first appearance embraced a period of nearly six months, and enabled astronomers to compute the orbit with uncommon precision. At its second appearance, the observations embraced a period of more than three months, and at its third appearance, in 1858, they embraced a period of more than one month. All these observations may be very accurately represented by an elliptic orbit, without supposing that the comet has experienced any resistance from an assumed ether. The comet made its fourth appearance in 1865, and its observed positions agreed almost exactly with those which had been predicted, showing that this body does not encounter any appreciable resistance.

Encke's comet is therefore the only body at present known which requires us to admit the existence of a resisting medium; and according to Professor Encke, this resistance is not appreciable beyond the orbit of Venus, and the density of the medium is assumed to vary inversely as the square of the distance from the sun. A resisting medium must produce an effect upon the motion of a comet, quite different from that which some persons would anticipate. Such a medium would diminish the comet's tangential velocity; that is, it would diminish its centrifugal force; in consequence of which, the comet must be drawn nearer to the sun, so that it would describe a *smaller* orbit. But, according to Art. 245, when the orbit diminishes, the absolute velocity increases. Hence we conclude, that the absolute velocity of a planetary or cometary body is increased by encountering a resisting medium.

Brorsen's Comet.

517. In 1846, Mr. Brorsen, of Denmark, discovered a telescopic comet, which has been found to revolve around the sun in about 5½ years. The date of its next arrival at perihelion was fixed for September, 1851. Its position at that time was very unfavorable for observations, and the comet was not found. It was, however, seen at its subsequent return to perihelion in 1857. It was discovered at Berlin March 18th, and passed its perihelion March 29th, 1857.

The distance of this comet from the sun at perihelion is 62 millions of miles, being less than the distance of Venus; and at aphelion 538 millions, which is somewhat greater than the distance of Jupiter. Its periodic time is 2031 days. The orbit of this comet, when projected on the ecliptic, is included wholly within that of Biela.

This comet should have returned again to the sun in September, 1862, but, on account of its unfavorable position, it passed unobserved.

D'Arrest's Comet.

518. In 1851, Dr. D'Arrest, of Leipsic, discovered a faint telescopic comet, whose orbit was computed to be an ellipse, having a period of 6.4 years. It was accordingly predicted that it would return again to the sun about the last of November, 1857. On account of its great southern declination, this comet was not visible in the northern hemisphere, but it was discovered at the Cape of Good Hope in December, 1857, and followed until the middle of January. It passed the perihelion November 28th, and pursued almost exactly the path predicted for it in 1851. Its distance from the sun at perihelion is 111 millions of miles, and at aphelion 546 millions.

Winnecke's Comet.

519. In 1819, M. Pons, at Marseilles, discovered a comet, which he continued to observe for 38 days. Its orbit was computed by Encke to be an ellipse, with a period of 5.6 years. This comet was not seen again until 1858, when it was rediscovered by Dr. Winnecke, at Bonn, having made seven revolutions since its apparition in 1819, making the time of one revolution 5.54 years. Its distance from the sun at perihelion is 73 millions of miles, and at aphelion 526 millions.

The orbits of six of the periodical comets here described (all except Halley's) bear a striking resemblance to each other. The direction of their motion about the sun is the same as that of the planets, and they move in planes not more inclined to the ecliptic than the orbits of the asteroids. In the dimensions of their orbits, and in the degree of their eccentricity, as well as in the position of their orbits, there is a family resemblance almost as decided as that between the different individuals of the group of asteroids.

The Comet of 1744.

520. The comet of 1744 was the most splendid comet of the 18th century. Its distance from the sun at perihelion was only about one fifth that of the earth, or a little more than one half the mean distance of Mercury. Three weeks before the perihelion passage, its light was equal to that of Jupiter at his greatest brilliancy, and a fortnight before perihelion its light was little inferior to Venus. On the day of perihelion passage the head was seen with a telescope at noonday, and many persons followed it with the naked eye some time after the sun had risen.

The tail of this comet attained a length of 19 millions of miles. A fortnight before perihelion the tail appeared divided into two branches, one 7° and the other 24° long. On the day before perihelion the tail exhibited remarkable curvature, being nearly in the form of a semi-parabola. Then followed a week of cloudy weather, during which the comet could not be observed; but six days after perihelion, about two hours before sunrise, when the head of the comet was far below the horizon, the extremity of the tail rose above the horizon, and appeared spread out like a fan, as shown in Plate V., Fig. 5. This portion presented the appearance of six tails, extending from 30° to 44° from the head of the comet.

The Comet of 1770.

521. The comet of 1770 is remarkable for its near approach to the earth and Jupiter, and the consequent changes in the form of its orbit. This comet was found to describe an elliptic orbit, with a periodic time of about 5½ years. By tracing back the comet's path, it was found that early in 1767 it was very near to Jupiter, the distance between the two bodies being at one time only $\frac{1}{58}$th of the comet's distance from the sun, in which position the influence of the planet must have been three times greater than that of the sun. For, by Art. 256, $G : g :: \frac{M}{D^2} : \frac{m}{d^2} :: \frac{1050}{58^2} : 1 :: 1 : 3$. The motion of the comet at this part of the orbit being nearly in the same direction as that of Jupiter, it was subjected for several months to a powerful disturbance from that planet; and the small ellipse in which the comet was seen to move in 1770 was the result of Jupiter's attraction. Previous to that time it had been moving in an orbit requiring 48 years for a revolution, and its

perihelion distance was about 300 millions of miles, at which distance it could never be seen from the earth.

This comet has not been seen since 1770. Its observation on its first return in 1776 was rendered impossible by its great distance from the earth, and before another revolution could be accomplished it again passed very near to Jupiter. In August, 1779, the distance of the comet from Jupiter was only $\frac{1}{492}$ of its distance from the sun, in which position the action of the planet must have exceeded that of the sun 230 times. For $G : g :: \frac{1050}{492^2} : 1 :: 1 : 230$. In consequence of this attraction, the orbit was so changed that the time of revolution became 16 years, and its perihelion distance again became about 300 millions of miles, at which distance the comet can not be seen from the earth. Thus this comet has been entirely invisible from the earth both before and since the year 1770. The annexed diagram shows the form of the orbit of this comet in 1770, and its relation to the orbits of the earth and Jupiter.

Fig. 121.

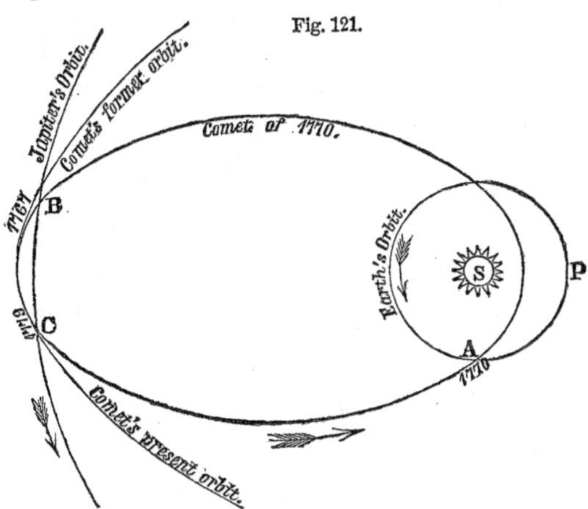

522. *Mass of this comet.*—In July, 1770, this comet made a nearer approach to the earth than any other comet on record, its distance at one time being only 1,400,000 miles. In this position, the nebulosity surrounding the nucleus subtended an angle of 2° 23′, or nearly five times that of the moon. Laplace has computed that if the mass of this comet had been equal to that of the earth, it would have changed the earth's orbit to such an extent as to have

increased the length of the year by 2h. 48m. But it is proved from astronomical observations that the length of the year has not been increased by a quantity so large as two seconds, from which it is inferred that the mass of the comet can not have been so great as $\frac{1}{5000}$th of the mass of the earth.

The mass of the comet must indeed have been smaller than this, for, although the comet approached Jupiter within a distance less than that of his fourth satellite, the motions of the satellites suffered no perceptible derangement.

The Great Comet of 1843.

523. One of the most brilliant comets of the present century was the great comet of 1843. It was seen in many parts of the world on the 28th of February, at midday, close to the sun; and soon after this, it became visible as a very conspicuous object in the evening twilight. The apparent length of its tail varied from 50 to 70 degrees, and its greatest real length was about 120 millions of miles. At perihelion this comet came almost in contact with the sun's disc, and it has been computed that it must have become 2000 times hotter than red-hot iron. For several days after perihelion the tail exhibited a decided fiery appearance. The heat to which it had been subjected was doubtless the cause of its extraordinary tail, which not merely attained an enormous length, but was formed with astonishing rapidity.

This comet moved in a very elongated ellipse. Attempts have been made to identify it with comets which appeared in 1668 and 1689; but the most careful computations indicate that its period amounts to about 170 years.

Donati's Comet of 1858.

524. This comet was discovered at Florence by Donati in June, 1858, and for two months remained a faint object, not discernible by the unaided eye. During the latter part of August, traces of a tail were noticed. The comet passed through perihelion on September 29th, and was at its least distance from the earth on October 10th. The tail continually increased until October 10th, when it had attained a length of 50 millions of miles, and subtended an angle of 60°. The nucleus of the comet was uncommonly large, and was intensely brilliant. It was not seen in Europe after the end of October, but in the southern hemisphere it

was followed till March, 1859. This comet undoubtedly revolves in an elliptic orbit, but the period can not be less than 1600 years, and is probably about 2100 years. This comet is remarkable for the changes which were noticed in the number and dimensions of its nebulous envelopes, which were similar to those described in Art. 491.

525. *Is it possible for a comet to strike the earth?*—Since comets move through the planetary spaces in every direction, it is quite possible that in the lapse of time the earth may come in collision with one of them. The comet of 1770 approached within 1,400,000 miles of the earth. In 1832, Biela's comet approached the earth's orbit so near that a portion of the orbit must have been included within the nebulosity of the comet; the earth was, however, at that time distant many millions of miles from the comet. The first comet of 1864 also approached within 600,000 miles of the earth's orbit. The consequences which would result from a collision between the earth and a comet would depend mainly upon the mass of the comet. If the comet had no solid nucleus, it is probable that it would be entirely arrested by the earth's atmosphere, and no portion of it might reach the earth's surface.

That the earth may some time pass through the tail of a comet is highly probable; and, indeed, we know of several cases in which the earth has passed very near to the tail of a comet, if it has not been actually enveloped in the nebulosity.

Shooting Stars.

526. Shooting stars are those small luminous bodies which at night are frequently seen to shoot rapidly across the heavens, and suddenly disappear. They may be seen on every clear night, and at times follow each other so rapidly that it is quite impossible to count them. They generally increase in frequency from the evening twilight throughout the night until the morning twilight; and, when the light of day does not interfere, they are most numerous about 6 A.M.

527. *Height, velocity*, etc.—By means of simultaneous observations made at two or more stations at suitable distances from each other, we may determine their height above the earth's surface, the length of their paths, and the velocity of their motion. It is

found that they begin to be visible at an average height of 74 miles, and they disappear at an average elevation of 50 miles. The average length of their visible paths is 42 miles. The average velocity relative to the earth's surface for the brighter class of shooting stars amounts to 29 miles per second; and they come in the greatest numbers from that point of space toward which the earth is moving in its annual course around the sun.

528. *The meteors of August and November.*—Shooting stars are most numerous in the month of August; and about the 10th of August the number is five times as great as the average for the entire year. The paths of most of them then diverge from the constellation Perseus, a region about 40° north of that point toward which the earth at that time is moving.

In the year 1833, shooting stars appeared in extraordinary numbers on the morning of November 13th. It was estimated that the number visible at a single station could not have been less than 200,000. They seemed to emanate chiefly from a point in the constellation Leo, which is about 10° north of that point in the heavens toward which the earth at that time was moving. A similar exhibition took place on the 12th of November, 1799, as also on several other years about the same day of November.

Unusual numbers of shooting stars have been noticed on other days of the year, particularly about the 21st of April of certain years.

529. *Meteoric orbits*, etc.—Having determined the velocity and direction of a meteor's path with reference to the earth, we can compute the direction and velocity of the motion with reference to the sun. In this manner it has been shown that these bodies, before they approached the earth, were revolving about the sun in ellipses of considerable eccentricity. In some instances the velocity has been so great as to indicate that the path differed little from a parabola.

Thus we see that ordinary shooting stars are bodies moving through space in paths similar to the comets; and it is probable that they do not differ materially from comets except in their dimensions, and perhaps also in their density.

Their light probably results from the heat generated by the compression of the air before them. It has been objected that at the height of 50 miles the atmosphere is too rare to develop so much heat. But we know that the motion of a large body moving about 30 miles per second is entirely lost in a second or two, and this motion, communicated to the particles of the surrounding air, must be sufficient to develop an enormous amount of heat and light.

We also conclude that shooting stars are not distributed uniformly through space, but many of them are grouped together, forming complete or incomplete rings of minute bodies revolving together around the sun. These rings are so situated that the earth encounters one of them annually on the 10th of August, and another occasionally on the 12th of November, furnishing meteoric displays of unusual splendor. The plane of the August zone appears to be nearly perpendicular to the plane of the earth's orbit; and this is the reason why at that time the radiant point is found so far distant from the ecliptic.

Detonating Meteors.

530. Ordinary shooting stars are not accompanied by any audible *sound*, though sometimes seen to break into pieces. Occasionally meteors of extraordinary brilliancy are succeeded by an *explosive* noise. These have been called detonating meteors. On the morning of November 15th, 1859, a meteor passed over the southern part of New Jersey, and was so brilliant that its flash attracted attention in the presence of an unclouded sun. Soon after the flash, there was heard a series of terrific explosions, which were compared to the discharge of a thousand cannon. From a comparison of numerous observations, it was computed that the height of this meteor when first seen was over 60 miles; and when it exploded its height was 20 miles. The length of its visible path was more than 40 miles. Its velocity relative to the earth was at least 20 miles per second; but its velocity relative to the sun was about 28 miles per second, indicating that it was moving about the sun in a very eccentric ellipse, or perhaps a parabola.

On the 2d of August, 1860, in the evening, a magnificent fireball was seen throughout the whole region from Pittsburg to New Orleans, and from Charleston to St. Louis. A few minutes after its disappearance there was heard a tremendous explosion like the sound of distant cannon. The length of its visible path was about 240 miles, and its time of flight was 8 seconds, showing a velocity of 30 miles per second. Its velocity relative to the sun was 24 miles per second.

531. *Number, velocity,* etc.—The number of detonating meteors recorded in scientific journals is over 800. Their average height at the first instant of apparition is 92 miles, and at the instant of vanishing is 32 miles. Their average velocity is 19 miles per second.

Comparing these results with those derived from the ordinary shooting stars, we conclude that the two classes of bodies do not probably differ much from each other except in size and density. The noise which succeeds the appearance of a detonating meteor is perhaps due to the collapse of the air rushing into the vacuum which is left behind the advancing meteor. No audible sound proceeds from ordinary shooting stars, because they are small bodies, of feeble density, and are generally consumed while yet at an elevation of 50 miles above the earth's surface.

Aerolites.

532. There is no evidence that any thing coming from ordinary shooting stars ever reaches the earth's surface; but occasionally solid bodies descend to the earth's surface from beyond the earth's atmosphere. These are called aerolites. In December, 1807, a meteor of great brilliancy passed over the southern part of Connecticut, and soon after its disappearance there were heard three loud explosions like those of a cannon, and there fell a shower of meteoric stones. The entire weight of all the fragments discovered was at least 300 pounds. The specific gravity of these stones was 3.6; their composition was one half silex, one third oxyd of iron, and the remainder chiefly magnesia. The length of the visible path of this meteor was at least 100 miles, and its velocity several miles per second.

In May, 1860, an aerolite exploded over Eastern Ohio, and from it descended a shower of stones whose entire weight was estimated at 700 pounds. Their specific gravity was 3.54, and their composition very similar to that of the meteor of 1807. There are 20 well-authenticated cases in which aerolites have fallen in the United States since 1807. The specific gravity of 16 of these meteors ranged from 3 to 3.66.

In July, 1847, an aerolite exploded over Bohemia, and from it there were seen to descend two masses of iron, which together weighed 72 pounds. Its specific gravity was 7.71. Its composition was 92 per cent. of iron, 5 per cent. nickel, with a small quantity of cobalt, etc. There are one or two other cases in which iron meteors have been known to fall to the earth; and there have been found over 100 other similar masses believed to be aerolites, although the date of their fall is unknown.

The elements of which aerolites consist are the same as those found in the crust of the earth; yet the manner in which these elements are combined is peculiar, so that the general aspect of aerolites is sufficient to distinguish them from all terrestrial minerals.

533. *Origin of aerolites.*—Various hypotheses have been proposed to account for the origin of aerolites.

1st. It has been conjectured that they are formed in the atmosphere like rain or hail. This supposition is inadmissible, because, allowing the aerolite to be once formed, there is no known cause which could impel it in a direction nearly horizontal with a velocity of several miles per second.

2d. It has been conjectured that aerolites are masses ejected from terrestrial *volcanoes.* This supposition is inadmissible, because the greatest velocity with which stones have ever been ejected from volcanoes is less than two miles per second, and the direction of this motion must be nearly vertical; while aerolites frequently move in a direction nearly horizontal, and with a velocity of several miles per second.

3d. It has been conjectured that aerolites have been ejected from *volcanoes in the moon* with a velocity sufficient to carry them out of the sphere of the moon's attraction into that of the earth's attraction. This supposition is unsatisfactory, because the lunar volcanoes are at present entirely extinct. If, then, aerolites once belonged to the moon, they must have been projected from its surface many years ago. Since that time they must have been moving in orbits around some larger body, such as the earth or the sun; that is, whatever may have been the first source of aerolites, they must now be regarded as satellites of the earth or the sun.

534. *Orbits of aerolites.*—The facts which have been established respecting shooting stars and detonating meteors can leave but little doubt that aerolites are bodies revolving about the sun like the planets and comets, and are encountered by the earth in its annual motion around the sun; and it is probable that the chief difference between these three classes of bodies depends upon their size and density.

We hence conclude that the interplanetary spaces, instead of being absolutely void, are filled with a countless number of minute bodies, whose aggregate mass must be very great. The comets, like the earth, must encounter an immense number of these bodies, and a part of their motion must be thereby destroyed. This effect may be appreciable in the case of the periodic comets, although it is thus far inappreciable in the case of the earth and the other planets.

CHAPTER XIX.

THE FIXED STARS—THEIR LIGHT, THEIR DISTANCE, AND THEIR MOTIONS.

535. *What is a fixed star?*—The fixed stars are so called because from century to century they preserve almost exactly the same positions with respect to each other. Many of the stars form groups which are so peculiar that they are easily identified; and the relative positions of these stars are nearly the same now as they were two thousand years ago. Accurate observations, however, made with telescopes, have proved that many, and probably all of the so-called fixed stars, have a real motion. There are, however, only about 30 stars whose motion is as great as one second in a year, and generally the motion is only a few seconds in a century.

536. *How the fixed stars are classified.*—The stars are divided into classes according to their different degrees of apparent brightness. The most conspicuous are termed stars of the *first* magnitude; those which are next in order of brightness are called stars of the *second* magnitude, and so on, the first six magnitudes embracing all which can be distinctly located by the naked eye.

Telescopic stars are classified in a similar manner down to the twelfth, and even smaller magnitudes.

The distribution of stars into magnitudes is arbitrary, and astronomers have differed in the magnitude they have assigned to the same star. According to the best authority, the number of stars of the first magnitude is 20; of the second magnitude, 34; third, 141; fourth, 327; fifth, 959; and sixth, 4424; making 5905 stars visible to the naked eye. Of these only about one half can be above the horizon at one time; and it is only on the most favorable nights that stars of the sixth magnitude can be clearly distinguished by the naked eye. Even then, only the brighter stars can be seen near the horizon.

The number of stars of the seventh magnitude is estimated at 13,000; eighth magnitude, 40,000; and ninth magnitude, 142,000; making about 200,000 stars from the first to the ninth magnitude. It is estimated that the number of stars visible in Herschel's reflecting telescope of 18 inches aperture was more than 20 millions; and the number visible in more powerful telescopes is still greater.

537. *Comparison of the brightness of the stars.*—Sir W. Herschel estimated that if an average star of the sixth magnitude be taken as unity, the light emitted by an average star of the fifth magnitude will be represented by 2; one of the fourth magnitude by 6; of the third magnitude by 12; the second magnitude by 25; and the first magnitude by 100. There is, however, considerable variety in the brightness of stars that are classed as of the same magnitude. The light of Sirius, the brightest star in the heavens, is from 5 to 10 times as great as some of the stars of the first magnitude, and more than 300 times as great as an average star of the sixth magnitude.

538. *Cause of this diversity of brightness.*—It is probable that these varieties of magnitude are chiefly caused by difference of distance rather than by difference of intrinsic splendor among the objects themselves. Those stars which are placed immediately about our solar system appear bright in consequence of their proximity, and are called stars of the first magnitude; those which lie beyond are more numerous, and appear less bright, and are called stars of the second magnitude; and thus, as the distance

of the stars increases, their apparent brightness diminishes, until at a certain distance they become invisible to the naked eye.

Some deviations from this general rule are to be expected. In fact, some of the fainter stars are among those which are nearest to us.

539. *Have the fixed stars a sensible disc?*—When a telescope is directed to a planet, the planet appears with a distinct disc, like that which the moon presents to the naked eye. But it is different even with the brightest of the fixed stars. The telescope, instead of magnifying, actually diminishes them. A star viewed by the naked eye appears surrounded by a radiation, and the appearance may be represented by a dot with rays diverging from every side of it. The telescope cuts off this radiation, and exhibits the star as a lucid point of very small diameter, even when the highest magnifying powers are employed. With a power of 6000, the apparent diameter of the stars seems *less* than with lower powers.

The brighter stars, when viewed with the best telescope, do, however, exhibit a small disc; but this disc is spurious, and probably arises from the dispersion of light in passing through the earth's atmosphere. That these discs are not real is proved by the fact that they are not magnified by an increase of telescopic power, and also by the fact that, in the occultation of a bright star by the moon, its extinction is absolutely instantaneous, not the smallest trace of gradual diminution of light being perceptible.

540. *How do the stars become visible to us?*—The term *magnitude* applied to the stars is therefore used to designate simply their relative *brightness*. None of the stars have any measurable magnitude at all. There is, however, reason to believe that the absolute diameters of the stars are very great; hence we are compelled to conclude that the distance of these bodies is so enormous that their apparent diameter seen from the earth is 6000 times less than any angle which the naked eye is capable of appreciating.

Stars, then, become sensible to the eye, not by subtending an appreciable angle, but from the intensity of the light which they emit. The quantity of light which the eye receives from a star varies inversely as the square of its distance. At a certain distance this light is insufficient to produce sensation, and the star becomes invisible.

When a star becomes invisible to the naked eye, the telescope may render it visible by uniting in the image as many rays as can enter the aperture of the object-glass. The increase of illumination from the use of a telescope will depend upon the ratio of the area of the aperture of the object-glass to that of the pupil of the eye. By augmenting the aperture of the telescope we may therefore increase the apparent brightness of an object, so that a star of the sixth magnitude may appear as bright as a star of the first magnitude does to the naked eye.

541. *Twinkling of the stars.*—The scintillation, or twinkling of the stars, which contrasts so strongly with the steady light of the principal planets, is an optical phenomenon, supposed to be due to what is termed the interference of light. Humboldt, the celebrated traveler, states that in the pure air of Cumana, in South America, the stars do not twinkle after they have attained an elevation, on the average, of 15° above the horizon.

542. *Division into constellations.*—For the sake of more readily distinguishing the stars, they have been divided into groups called *constellations.* These constellations are represented under the forms of various animals, such as bears, lions, goats, serpents, and so on. In some instances we may easily imagine that the arrangement of the stars bears some resemblance to the object from which the constellation is named, as, for example, the Swan and the Scorpion; in other instances no such resemblance can be traced. This fanciful mode of grouping the stars is of very ancient date, and is continued by modern astronomers chiefly for the sake of avoiding the confusion that might arise from an alteration in the old system.

543. *Names of the constellations.*—There are twelve constellations lying upon the zodiac, and hence called the *zodiacal constellations,* viz., Aries, Taurus, Gemini, Cancer, Leo, Virgo, Libra, Scorpio, Sagittarius, Capricornus, Aquarius, and Pisces. These are also the names of the twelve divisions of 30° each into which the ecliptic has been divided; but the effect of precession, which throws back the place of the equinox among the stars 50″ a year,

has caused a displacement of the signs of the zodiac with respect to the corresponding constellations. The sign Taurus at present occupies the constellation Aries, the sign Gemini the constellation Taurus, and so on, the signs having retreated among the stars 30° since the present division of the zodiac was adopted.

The principal constellations in the northern half of the heavens, in addition to such of the zodiacal ones as lie north of the celestial equator, are:

Andromeda.	Cassiopeia.	Draco.	Perseus.
Aquila.	Cepheus.	Hercules.	Ursa Major.
Auriga.	Corona Borealis.	Lyra.	Ursa Minor.
Bootes.	Cygnus.	Pegasus.	

The principal constellations situated on the south side of the equator, exclusive of the six southern zodiacal ones, are:

Argo Navis.	Cetus.	Ophiuchus.
Canis Major.	Crux.	Orion.
Canis Minor.	Eridanus.	Piscis Australis.
Centaurus.	Monoceros.	

Others will be found upon celestial globes and charts, raising the total number of constellations at present recognized by astronomers to about eighty.

544. *How particular stars are designated.*—Many of the brighter stars had proper names assigned them at a very early date, as Sirius, Arcturus, Rigel, Aldebaran, etc., and by these names they are still commonly distinguished.

It was the custom in former times to indicate the locality of a star by its position in the constellation to which it belonged; but this method was found to be extremely tedious, besides being frequently liable to misconception. Bayer, a German astronomer, in 1604 published a series of maps of the heavens, in which the stars of each constellation were distinguished by the letters of the Greek and Roman alphabets, the brightest being called α, the next β, and so on. Thus α Lyræ denotes the brightest star in the constellation Lyra, β Lyræ the second star, and so on.

In consequence either of a want of proper care in assigning letters to the stars, or perhaps from a real change of brightness of the stars since the time of Bayer, we sometimes find that the brightness of the stars in a constellation does not follow the order of the letters by which they are distinguished. Thus α Draconis is not so bright as either β or γ of the same constellation. Flamsteed, the first astronomer royal at Greenwich, distinguished the stars of each constellation by the numerals 1, 2, 3, etc., and stars are often referred to by these numbers. In large catalogues of stars, the stars are usually numbered continuously from beginning to end in the order of their right ascensions.

545. *Remarkable constellations enumerated.*—One of the most conspicuous constellations in the northern firmament is Ursa Major, or the Great Bear, in which we find seven stars which may easily be conceived to form the outline of a dipper, of which the two brightest are nearly in a straight line with the pole star, and are hence called the *pointers.* They are not far from the zenith at New Haven at 10 o'clock in the evening in the month of April.

The constellation Cassiopeia presents six stars which may be conceived to form the outline of a chair. It is not far from the zenith at 10 o'clock in the evening in the month of October.

The constellation Ursa Minor contains seven stars which may also be conceived to form the outline of a dipper, the pole star forming the extremity of the handle. The principal stars of these three constellations are represented in Fig. 2, page 14.

The constellation Orion is one of the most magnificent in the heavens, and with some imagination may be conceived to resemble a great giant. It is in the south at 10 o'clock in the evening in the month of January. To the left of Orion, and a little below it, is then seen the star Sirius, which far surpasses all others in brilliancy.

The square of Pegasus is formed by four moderately bright stars, which appear at a considerable altitude above the horizon in the southern quarter of the sky about 10 in the evening in the middle of October.

The Pleiades form a group of stars in the constellation Taurus. The naked eye discovers six or seven, but in the telescope upward of two hundred are revealed. This group passes the meridian at 10 o'clock in the evening in the month of December. A little below, and to the left of the Pleiades, is a wedge of stars called the *Hyades*, of which Aldebaran is the conspicuous member.

546. *Catalogues of stars.*—Various catalogues of stars have been formed, in which are indicated their right ascensions and declina-

tions for a certain epoch. Hipparchus is believed to have been the first who undertook such a compilation, 128 years before the Christian era. His catalogue included 1022 stars, and has been preserved to us in the Almagest of Claudius Ptolemy. Some modern catalogues contain a much larger number of stars. The British Association catalogue contains 8377 stars; the catalogue of Lalande contains 47,390 stars; Cooper's catalogue contains 60,066 stars near the ecliptic; and the entire number tabulated at the present time amounts to several hundreds of thousands.

547. *Periodic stars.*—Some stars exhibit periodical changes in their brightness, and are therefore called *periodic* stars. One of the most remarkable of this class is the star Omicron Ceti, often termed *Mira*, or the *wonderful* star. This star retains its greatest brightness for about 14 days, being then usually equal to a star of the second magnitude. It then decreases, and in about two months ceases to be visible to the naked eye. After remaining thus invisible for six or seven months, it reappears, and increases gradually for two months, when it recovers its maximum splendor. It goes through all its changes in 332 days, and in 1867 its maximum brilliancy occurred on the 17th of January. At the times of the least light, it becomes reduced to the tenth or twelfth magnitude.

Another remarkable periodic star is Algol, in the constellation Perseus. It generally appears of the second magnitude, and continues thus for about 61 hours. It then diminishes in brightness, and in less than four hours is reduced to a star of the fourth magnitude, and thus remains about twenty minutes. It then increases, and in about four hours more it recovers its original splendor. The exact period in which all these variations are performed is 2d. 20h. 48m. 55s. It was at its minimum of brightness in 1867, December 10th, at 7h. 44m. in the evening, New Haven time, from which data the time of any other minimum can be computed.

There are more than 100 stars known to be variable to a greater or less extent. The periods of these changes vary from a few days to many years. The star 34 Cygni varies from the third to the sixth magnitude in a period of about 18 years. The bright star Capella, in the constellation Auriga, is believed to have increased in lustre during the present century, while within the same period one of the seven bright stars (δ) in Ursa Major has probably diminished. Many instances of a similar kind might be mentioned.

548. *Cause of this periodicity.*—These phenomena have been explained, 1st, by supposing that a dark, opaque body may revolve about the variable star, and at certain times intercept a portion of its light; or, 2d, that a nebulous body of great extent may revolve round the star, and intercept a portion of its light when interposed between us and the star. 3d. The stars themselves may not be uniformly luminous all over their surfaces, but occasionally, from their axial rotations, present toward the earth a disc partially covered with dark spots, thereby shining with a dimmer light. 4th. Some stars may have the form of thin flat discs, and by rotation present to us alternately their edge and their flat side, producing corresponding changes of brightness. It is certain that these variations are entirely independent of any effect which the earth's atmosphere could produce.

The first-mentioned hypothesis will not explain the long-continued obscuration of Omicron Ceti. Neither of the last two hypotheses will explain the sudden changes in the brightness of Algol. The second hypothesis is believed to furnish the most plausible explanation of the phenomena which has yet been proposed.

549. *Temporary stars.*—Several instances are recorded of stars suddenly appearing where none had before been observed, sometimes surpassing the light of stars of the first magnitude, remaining thus for a short time, and then gradually fading away. The first on record was observed by Hipparchus 125 B.C., the disappearance of which is said to have led that astronomer to compile the star-catalogue bearing his name. In the year 389 A.D. a star blazed forth near *α* Aquilæ, which shone for three weeks, appearing as splendid as the planet Venus, after which it disappeared, and has never since been seen. In the autumn of 1572 a new star suddenly appeared in the constellation Cassiopeia. When first noticed it was as bright as Sirius, the brightest star in our firmament; and it finally attained such splendor that it was distinctly visible at midday. In about a month it began to diminish, and in sixteen months it entirely disappeared.

Another temporary star became suddenly visible in Ophiuchus in 1604, and exceeded Jupiter in splendor. It remained visible till 1606, and then disappeared.

In 1848 a star of the fourth magnitude was seen in the constellation Ophiuchus, in a place where no star had ever been observed before. After a few weeks it declined in brightness, and has now faded away to the twelfth magnitude, so that it can not be seen without a superior telescope. It is possible that the temporary stars do not differ from the periodic stars except in the length of their periods.

550. *Distance of the fixed stars.*—That the distance of the fixed stars from the earth is immense is proved by the following considerations. The earth, in its annual course around the sun, revolves in an orbit whose diameter is 190 millions of miles. The station from which we observe the stars on the 1st of January is distant 190 millions of miles from the station from which we view them on the 1st of July; yet from these two remote points the stars present the same appearance, proving that the diameter of the earth's orbit must be a mere point compared with the distance of the nearest stars.

551. *Annual parallax.*—The greatest angle which the radius of the earth's orbit subtends at a fixed star is called its annual parallax. Numerous attempts have been made to measure the amount of this parallax. Suppose a star to be situated at the pole of the ecliptic, and that it is near enough to the earth to have a sensible parallax. Then, while the earth travels round the sun, the star, as projected on the distant firmament, will appear to describe a small circle, ABCD, whose centre, S, is on the line joining the sun and star; and the diameter of this circle will diminish as the distance of the star from the earth increases.

Fig. 122.

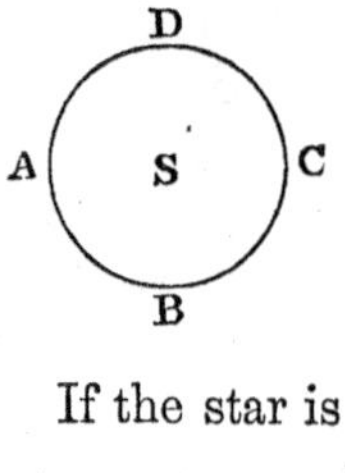

If the star is situated in the plane of the ecliptic, then, for three months of the year, it will appear to move a little to the east of its mean position, and in the next three months it will return to its first position. It will then appear to move a little to the west of its mean position, and afterward return to its first position, its apparent motion being confined to a straight line, AC.

Fig. 123.

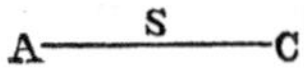

If the star is situated between the ecliptic and its pole, the motion of the earth about the sun will give to the star an apparent motion in an ellipse, ABCD, whose eccentricity will increase as the star's latitude decreases.

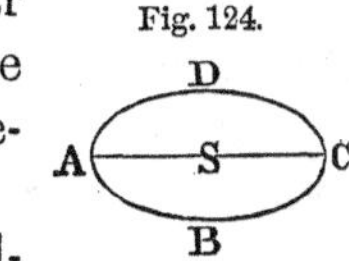

Fig. 124.

If, then, a fixed star had any considerable parallax, it would be easy to discover it by measuring accurately its position from one season to another; but, among the many thousand stars which have been carefully observed by astronomers, not one has been found which exhibits a parallax exceeding one second.

552. *Parallax of Alpha Centauri.*—Observations made upon the star Alpha Centauri, one of the brightest stars of the southern hemisphere, indicate an annual parallax of $\frac{92}{100}$ths of a second. Having determined the parallax, we can compute the distance of the star by the proportion

sin. $0''.92 : 1 :: 95$ millions of miles : the distance of the star, which is found to be *twenty millions of millions* of miles. This distance is so immense that a ray of light, moving at the rate of 192,000 miles per second, requires $3\frac{1}{2}$ years to travel from this star to the earth. We do not see the star as it actually is, but it shines with the light emitted $3\frac{1}{2}$ years ago. Hence, if it were obliterated from the heavens, we should continue to see it for more than three years after its destruction; yet Alpha Centauri is probably our nearest neighbor among the fixed stars.

553. *How differences of parallax may be detected.*—Since the best astronomical observations are liable to minute errors, which render it difficult to determine a star's absolute place with the accuracy required for the measurement of parallax, astronomers have sought for some method of detecting parallax which shall be free from the errors of ordinary observations. The following method has been proposed for this purpose. Let S be a star which we will suppose to have a visible parallax, and let ABCD be the small ellipse which it appears to describe in consequence of the motion of the earth about the sun. Let S′ and S″ be two other

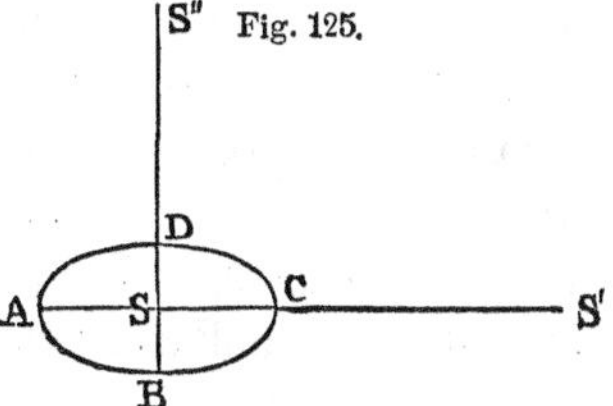

Fig. 125.

stars, so distant from the earth as to have *no* sensible parallax, and situated on the axes of the ellipse ABCD, and suppose the three stars to be included in the same field of the telescope. The apparent distance of the star S from S′ will change during the year from AS′ to CS′, and its distance from S″ will change from BS″ to DS″. These distances can be measured with great exactness by means of a micrometer; and these measurements are independent of the errors which affect the determination of the absolute places of the stars. If the stars S′ and S″ have a small parallax, then these observations will determine the *difference* of parallax between the star S and the stars of comparison. When we wish to select stars which have no appreciable parallax, we choose those of the smallest magnitude, which for that reason are presumed to be at the greatest distance from the earth.

554. *Parallax of* 61 *Cygni.*—By the method here indicated, the parallax of the star 61 Cygni was determined by the great astronomer Bessel, of Königsberg, to be 0″.35. The observations of Mr. Johnson, at Oxford, make the parallax of this star 0″.40; Struve, at Pulkova, makes the parallax 0″.51; and Auwers, at Königsberg, makes the parallax 0″.56. The mean of these four determinations is 0″.45, indicating a distance of 44 millions of millions of miles, a space which light would not traverse in less than 7¼ years.

555. *Parallax of other stars.*—No other star has yet been found whose parallax exceeds about one quarter of a second. Sirius and Alpha Lyræ have apparently a parallax of nearly a quarter of a second, and observations have indicated about an equal parallax in two other small stars. All the other stars of our firmament are apparently at a greater distance from us; and if the distance of the nearest stars is so great, we must conclude that those faint stars which are barely discernible in powerful telescopes are much more distant. Hence we conclude that we do not see them as they now are, but as they were years ago; perhaps, in some instances, with the rays which proceeded from them several thousands of years ago; and it is possible that they may have changed their appearance, or have been entirely annihilated years ago, although we actually see them at the present moment.

556. *Light of the sun compared with that of the fixed stars.*—The fixed stars must be *self-luminous*, for no light reflected from our sun could render them visible at the enormous distances at which they are situated from us. Indeed, it is demonstrable that many of the fixed stars actually give out as much light as our sun. It is estimated that the light of our sun is 450,000 times greater than that of the full moon; and it has been proved that the light of the full moon is 13,000 times greater than that of Sirius; that is, the light of the sun is about 6000 million times greater than that of Sirius. Since the quantity of light which the eye receives from a star varies inversely as the square of its distance, and since the distance of Sirius is 800,000 times that of the sun, it follows that, if Sirius were brought as near to us as the sun, its light would be 640,000 million times as great as it appears at present; that is, the light emitted by Sirius is a hundred times that of our sun. Many other fixed stars probably emit as much light as Sirius; in other words, the fixed stars belong to the same class of bodies as our sun, in respect of the amount of light which they emit; and it is probable that many of them are bodies of at least equal dimensions, otherwise the intensity of their illumination must be very much greater than that of our sun.

557. *Proper motion of the stars.*—The changes in the position of the stars due to aberration and nutation are merely apparent movements, and their exact amounts can be readily calculated for any star. The effects of precession can be determined with equal facility. It is found by observation, however, that most stars exhibit a slow motion in the heavens which can not be thus accounted for. After due allowance has been made for precession, aberration, and nutation, there still remain very appreciable changes of position. These are not such periodical motions to and fro as would be produced by parallax; on the contrary, they are uniformly progressive from year to year. A star in Ursa Major (known as 1830 of Groombridge's catalogue) travels at the rate of seven seconds in a year; 61 Cygni, whose parallax, as already mentioned, has been determined, is moving at the rate of five seconds annually. The star Alpha Centauri has a proper motion of nearly four seconds annually, and most of the brighter stars of the firmament have a sensible proper motion. The result of this motion is a slow but constant change in the figure of the

constellations. In the case of several of the stars, this change in 2000 years has become quite sensible to the naked eye. The proper motion of Arcturus in 2000 years has amounted to more than one degree; that of Sirius and Procyon to three quarters of a degree.

558. *Cause of this proper motion.*—There are two ways in which such movements may be explained. Either the star itself may be supposed to have a real motion through space, or the sun, attended by the planets, may have a real motion in a contrary direction to that of the star's apparent one. On extending the inquiry to a great number of stars, it appears beyond doubt that both causes must be in existence, certain stars having really an independent motion in the heavens, which, to distinguish it from merely apparent displacements, is termed the *proper motion*, while the solar system itself *travels through space.*

559. *How could a motion of the solar system be detected?*—If we suppose the sun, attended by the planets, to be moving through space, we ought to be able to detect this motion by an apparent motion of the stars in a contrary direction, as, when an observer moves through a forest of trees, his own motion imparts an apparent motion to the trees in a contrary direction. All the stars would not be equally affected by such a motion of the solar system. The nearest stars would appear to have the greatest motion, but all the changes of position would appear to take place in the same direction. The stars would all appear to recede from

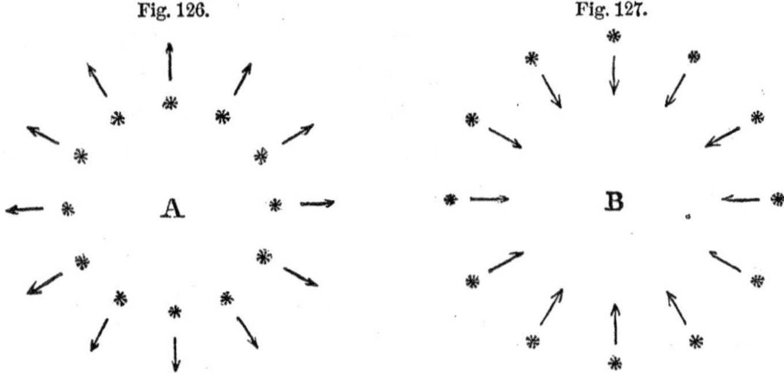

A, that point of the heavens *toward* which the sun is moving, while in the opposite quarter, B, the stars would become crowded more closely together.

560. *Direction of the sun's motion.*—In 1783, Sir William Herschel announced that the proper motion of a large portion of the stars might be explained by supposing that the sun has a motion toward a point in the constellation Hercules. More recent and extensive investigations have not only established the fact of the solar motion, but likewise indicated a direction very nearly coincident with that assigned by Herschel, viz., nearly toward the star ρ Herculis. The average displacement of the stars, as estimated by Struve, indicates that the motion of the sun in one year is about 150 millions of miles, which is about one fourth of the velocity of the earth in its orbit, or five miles per second; but, according to the estimate of Airy, the motion of our solar system is about twenty-seven miles per second.

561. *Is the sun's motion rectilinear?*—It is probable that the solar system does not advance from age to age in a straight line, but that it revolves about the centre of gravity of the group of stars of which it forms a member. It is also probable that this centre of gravity is situated nearly in the plane of the Milky Way; and if the orbit of the sun is nearly circular, this centre must be about 90° distant from ρ Herculis, the point toward which the solar system is moving. Maedler conjectured that the brightest star in the Pleiades was the central sun of the universe, but without sufficient reason. The orbit of the solar system is probably so large that ages may elapse before it will be possible to detect any change in the direction of the sun's motion.

CHAPTER XX.

DOUBLE STARS.—CLUSTERS OF STARS.—NEBULÆ.

562. *Double stars.*—Many stars which to the naked eye, or with telescopes of small power, appear to be single, when examined with telescopes of greater power are found to consist of two stars placed close together. These are called *double stars.* Some of these are resolved into separate stars by a telescope of moderate power, as Castor, which consists of two stars at the distance of 5″ from each other, each being of the third or fourth magnitude. Many of them, however, for their separation, require the most

powerful telescope. Some stars, which to ordinary telescopes appear only double, when seen through more powerful instruments are found to consist of three stars, forming a triple star; and there are also combinations of four, five, or more stars, lying within small distances from each other, thus forming *quadruple*, *quintuple*, and *multiple* stars. Only four double stars were known until the time of Sir W. Herschel, who discovered upward of 500, and subsequent observers have extended this number to 6000.

563. *Classification of double stars.* — Herschel divided double stars into four classes, according to the angular distance between the two components. The first class comprised those only in which the distance between the two components does not exceed 4″; the second class those in which it exceeds 4″, but falls short of 8″; the third class extends from 8″ to 16″; and the fourth class extends from 16″ to 32″. Struve has subdivided some of Herschel's classes, making thus eight classes instead of four. When the distance between two stars exceeds 32″, they are not generally admitted into the catalogue of double stars.

In some instances the components of a double star are of equal brilliancy, but it more frequently happens that one star is brighter than the other. Occasionally the inequality of light is so great that the smaller star is almost lost in the refulgence of its brighter neighbor.

564. *Colored stars.*—Many stars shine with a colored light, as red, blue, green, or yellow. These colors are exhibited in striking contrast in many of the double stars. Combinations of blue and yellow, or green and yellow, are not infrequent, while in fewer cases we find one star white and the other purple, or one white and the other red. In several instances each star has a rosy light. The colors of the two components are sometimes complementary to each other—that is, if combined, they would form white light. In such cases, if one star is much smaller than the other, we may attribute the difference of color to the effect of contrast only. Thus, if the larger one be yellow, the companion may incline to blue; or, if the former have a greenish light, the latter may be tinged with crimson. Yet it can hardly be doubted that in many cases the light of the stars is actually of different colors; that there exist in the universe numbers of yellow, blue,

green, and crimson suns, whose refulgence must produce the most beautiful effects upon the planets which circulate around them.

Single stars of a fiery red or deep orange color are not uncommon, but there is no instance of an isolated deep blue or green star; these colors are apparently confined to the compound stars.

Below the constellation Orion there is a star of the seventh magnitude of a blood-red color, and near it is another star of similar brightness, but presenting a pure white light.

The following are a few of the most interesting colored double stars:

Name of star.	Color of larger one.	Color of smaller one.
γ Andromedæ	Orange	Sea-green.
α Piscium	Pale green	Blue.
β Cygni	Yellow	Sapphire-blue.
σ Cassiopeæ	Greenish	Fine blue.
A star in Argo	Pale rose	Greenish-blue.
A star in Centaurus	Scarlet	Scarlet.

565. *Stars optically double.*—If two stars be very nearly in the same line of vision, though one may be vastly more distant than the other, they will form a star *optically* double, or one whose components are only apparently connected by the near coincidence of their directions as viewed from the earth. Thus the two stars A and B, seen from the earth at E, will appear in close juxtaposition, although they may be separated by an interval greater than the distance of the nearest from the earth. The chances, however, are greatly against there being a large number of stars thus optically joined together. If the stars down to the seventh magnitude were scattered fortuitously over the entire firmament, the chances against any two of them having a position so close to each other as 4″ would be 9000 to 1. But more than 100 such cases of juxtaposition are known to exist.

Fig. 128.

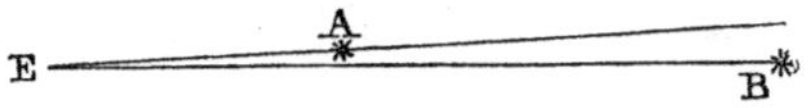

E A B

566. *Binary stars.*—In the year 1780, Sir William Herschel undertook an extensive series of observations of double stars, recording the relative position of the components, and the distance by which they were separated. By this means he hoped to be able to detect a parallax. He found that the distance and relative po-

sition of the components of a double star were subject to change, but the period of this change had no relation to the earth's motion about the sun. After twenty years of observations, he ascertained and announced that these apparent changes of position were due to real motions of the stars themselves; that the components of several of the double stars moved in orbits in the same manner as the planets move around the sun; that there exist sidereal systems consisting of two stars revolving about each other, or, rather, both revolving round their common centre of gravity. These stars are termed *physically* double, or *binary stars*, to distinguish them from other double stars in which no such periodic change of position has been discovered.

567. *The star Gamma Virginis.*—One of the most remarkable of the binary stars is γ Virginis. This is a star of the fourth magnitude, and its components are almost exactly equal. It has been known to consist of two stars since 1718, their distance being then 7″; and since 1780 they have been regularly observed. In 1836, their distance from each other was less than half a second, so that no telescope, unless of a very superior quality, could show them otherwise than as a single star. At present their distance from each other is about 4″. During the interval of 146 years, the direction of one of the components, as seen from the other, has changed by nearly 360°. The entire series of observations is well represented by supposing each of the stars to revolve about their common centre of gravity in an ellipse whose major axis is 7″, and in a period of 169 years. In the annexed figure, the dotted line represents the apparent orbit of one of the stars about the other, while the black line represents the form of the actual orbit as computed. The orbit, as viewed from the earth, is seen somewhat obliquely, and the apparent length of the major axis is thereby somewhat reduced.

Fig. 129.

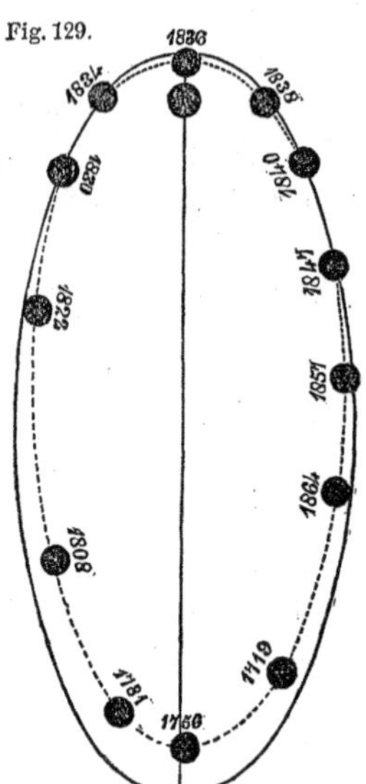

568. *The star* 70 *Ophiuchi.*—The star 70 Ophiuchi consists of

two components, one of the fourth, the other of the seventh magnitude. Since 1779, one has made nearly a complete revolution about the other in an ellipse whose major axis is about 8″, and the period is about 92 years. In the annexed figure, the dotted line represents the apparent orbit of one of the stars about the other, while the black line represents the form of the actual orbit as computed.

Fig. 130.

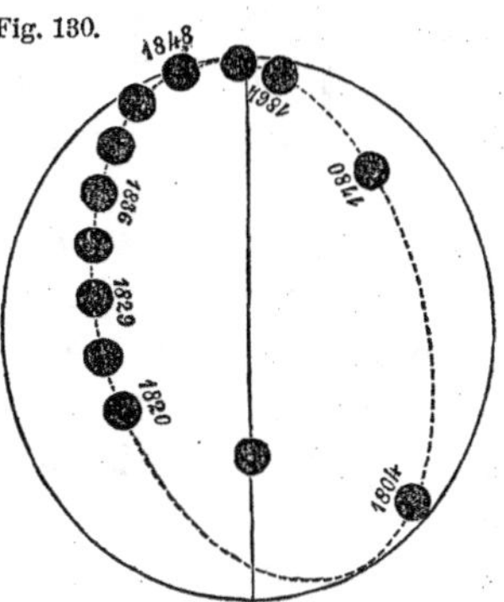

569. *The star Xi Ursæ Majoris*, etc.—The star ξ Ursæ Majoris consists of two components, one of the fourth, the other of the fifth magnitude. Since 1780, one has completed an entire revolution about the other, and has entered upon a second period. The major axis of the orbit is about 5″, and the time of revolution 61 years. The annexed figure represents both the apparent and the real orbit.

Fig. 131.

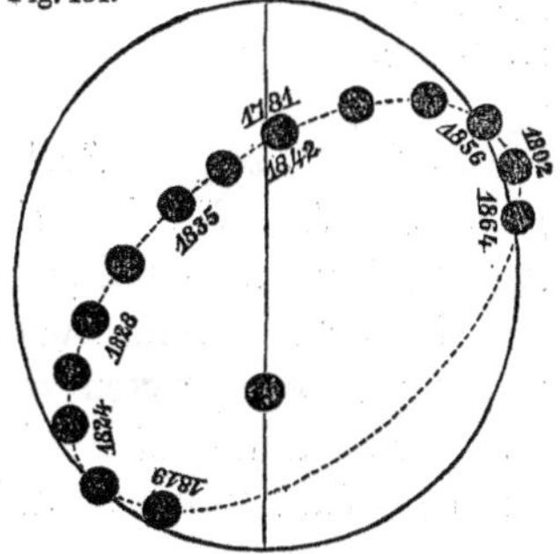

The star ζ Herculis consists of two components, one of the third, the other of the sixth magnitude. Since 1782, one has completed two entire revolutions about the other in an ellipse whose major axis is $2\frac{1}{2}''$, and the period of a revolution is 36 years.

570. *The star Alpha Centauri.*—The star α Centauri consists of two components, one of the first, the other of the second magnitude. These two stars were observed by Lacaille in 1751; and since 1826 their positions have been frequently and carefully observed. The annexed figure represents the apparent path from

Fig. 132.

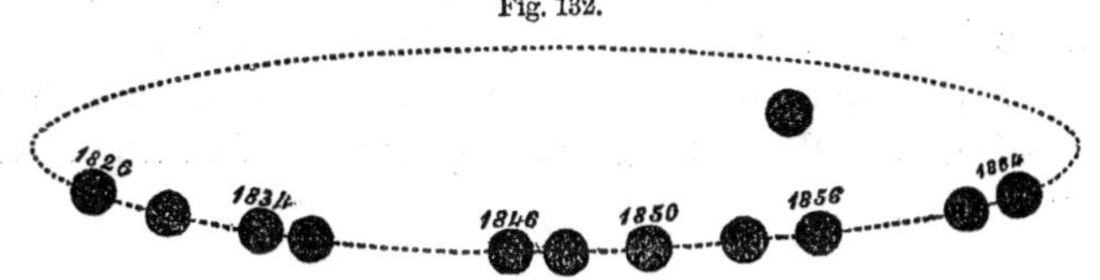

1826 to 1864, as well as the remaining portion of the probable apparent orbit. The major axis of this orbit is about 30″, and the time of one revolution about 80 years.

571. *Number of the binary stars.*—There are 467 double stars in which observations have indicated a change in their relative positions, and which are therefore believed to form binary systems. Of these there are only eight whose periods are less than a century; there are 142 whose periods are less than a thousand years; while the periods of 325 apparently exceed a thousand years. When the motion is so slow, it must require observations extending over a long interval of time to determine with accuracy the period of a complete revolution. It is probable that a large majority of the double stars will in time be proved to be physically connected.

572. *The law of gravitation extends to the fixed stars.*—It has been proved, in Art. 249, that if a body revolve in an ellipse by an attractive force directed toward the focus, that force must vary inversely as the square of the distance. But several of the binary stars have been proved to revolve in ellipses; hence it is inferred that the same law of gravitation which prevails in the solar system prevails among the sidereal systems.

573. *Absolute dimensions of the orbit of a binary star.*—If we knew the distance of a binary star from the earth, we could compute the absolute dimensions of the orbit described. Now α Centauri and 61 Cygni are both binary stars, and their distances are tolerably well determined. The distance of α Centauri is 224,000 times the radius of the earth's orbit. Hence we shall have the proportion

$$R : 224{,}000 :: \text{tang. } 15'' : 16;$$

that is, the radius of the orbit described by the components of α Centauri is 16 times the radius of the earth's orbit, or about four fifths the distance of Uranus from the sun. In a similar manner it has been computed that the radius of the orbit described by the components of 61 Cygni is 44 times the radius of the earth's orbit, which is considerably greater than the orbit of Neptune.

574. *Mass of a binary star computed.*—Since the relation between the dimensions of the orbit and the time of revolution determines the relative masses of the central bodies, we are enabled to compare the mass of a binary star with that of our sun, when we know the absolute radius of the orbit, and the periodic time of the star. In Art. 469 we found

$$\mathrm{M} : m :: \frac{\mathrm{R}^3}{\mathrm{T}^2} : \frac{r^3}{t^2}.$$

If m represent the mass of our sun, r the radius of the earth's orbit, and t the time of the earth's revolution, or one year, then

$$\mathrm{M} = \frac{\mathrm{R}^3}{\mathrm{T}^2} \cdot m.$$

In the case of α Centauri, R has been found equal to 16, and T=80 years. Hence

$$\mathrm{M} = \frac{16^3}{80^2} \cdot m = \tfrac{3}{5} m;$$

that is, the mass of the double star α Centauri is about three fifths that of our sun. In a similar manner it has been computed that each of the stars which compose 61 Cygni is about one third of our sun.

From observations of 70 Ophiuchi continued through a period of several years, it has been concluded that its parallax amounts to 0″.16, indicating that its distance from us is 1,289,000 times the radius of the earth's orbit. Hence it has been computed that the major axis of its orbit is 30½ times the radius of the earth's orbit, from which it follows that the mass of this system is three times that of our sun.

575. *The fixed stars are suns.*—We thus see that the stars are bodies essentially like our sun. Some of them have a power of attraction nearly equal to that of our sun, and it is probable that others have a greater power of attraction. Some of them emit more light than our sun. The stars are therefore self-luminous bodies of vast size, and are entitled to be called *suns.* In the binary stars, then, we have examples, not of planets revolving round a sun, as in our solar system, but of *sun* revolving round *sun.*

576. *Triple stars.*—Besides the binary stars, there are some triple stars which are proved to be physically connected. Of these the

most remarkable is Zeta Cancri. It consists of three components, one being of the sixth, and the other two of the ninth magnitude. Since 1781, one of these components has made a complete revolution about one of the others in an ellipse whose major axis is 2″, and the period of revolution is 58 years. During the same period the other component has advanced more than 30 degrees in its orbit, from which it is estimated that its period of revolution must be about 500 years. The annexed diagram represents the orbit of the nearest component, and a portion of the orbit of the more remote component.

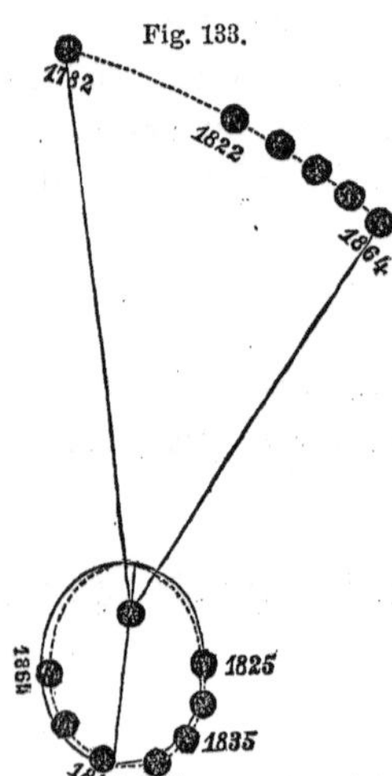

Fig. 133.

The star 51 Libræ is a triple star, two of whose components are of the fifth magnitude, and the other of the seventh. It has been computed that the nearer component makes a revolution in 105 years, and the more remote component in 600 or 700 years.

577. *Quadruple and quintuple stars.*—ε Lyræ furnishes an instance of a quadruple star, in which all the components are believed to be physically connected. Three of the components are of the fifth, and the other of the sixth magnitude. The movement of these stars is extremely slow, and, at the present rate of motion, it will require nearly a thousand years for the nearest component to complete one revolution, and many thousand years for the most remote component.

Theta Orionis is a quintuple star, in which one of the components is of the sixth magnitude, two are of the seventh, one of the eighth, and the other is of the fifteenth magnitude. In the best telescopes, a sixth star of extreme faintness may also be seen. The relative position of these stars has not sensibly changed since they were first observed by Herschel, and it is uncertain whether they are physically connected.

578. *Clusters of stars.*—In many parts of the heavens we find stars crowded together in clusters, frequently in such numbers as to defy all attempts to count them. Some of these clusters are visible to the naked eye. In the cluster called the Pleiades,

six stars are readily perceived by the naked eye, and we obtain glimpses of many more. With a telescope of moderate power 188 stars can be counted.

In the constellation Cancer is a luminous spot called Præsepe, or the bee-hive, which a telescope of moderate power resolves entirely into stars. There is a remarkable group in the sword-handle of Perseus, in which the stars are readily seen with a common night-glass, though the whole have a blurred aspect to the naked eye.

One of the most magnificent clusters in the northern hemisphere occurs in the constellation Hercules, between the stars η and ξ. It is visible to the naked eye on clear nights as a hazy-looking object, and the stars composing it are readily seen with a telescope of moderate power. When examined in a powerful instrument, its aspect is grand beyond description; the stars, which are coarsely scattered at the borders, come up to a perfect blaze in the centre.

The richest cluster in the entire heavens is situated in the constellation Centaurus, which belongs to the southern hemisphere, and is called ω Centauri. To the naked eye it appears like a nebulous or hazy star of the fourth magnitude, while in the telescope it is found to cover a space two thirds of the apparent diameter of the moon, over which the stars are congregated in countless numbers. See Plate VII., Fig. 2.

We can not doubt that most of the stars in such a cluster as ω Centauri are near enough to each other to feel each other's attraction. They must therefore be in motion, and we must regard this cluster as a magnificent *astral system*, consisting of a countless number of suns, each revolving in an orbit about the common centre of gravity.

579. *Nebulæ.*—In various parts of the firmament we discover with a telescope dim patches of light, presenting a hazy, undefined, or cloud-like appearance. These objects are called *nebulæ.* A large proportion are either round or oval, brighter toward their centres than at their borders, and when viewed with small optical power very much resemble comets, for which they are often mistaken. In more powerful instruments, such as those brought into use by Sir William Herschel, a considerable number are readily resolved into *clusters of stars*, like Præsepe, or the group in Per-

seus above mentioned; some hundreds, or even thousands of stars are wedged together within the space of two or three minutes of arc or less. Many others present a mottled, glittering aspect when thus viewed, which shows that they are similarly constituted, but too distant for our telescopes to separate them into stars; while, as might be expected, there are also very many that the most powerful optical means hitherto devised have altogether failed to exhibit otherwise than as faint, cloud-like objects. Very many of them have the same form and general appearance as the resolvable nebulæ seen in common telescopes, and hence there is reason for supposing them to be similar clusters of stars, but situated at far greater distances from the earth. About 5000 nebulæ have been observed, and their places are assigned in catalogues. The following are among the most remarkable of this class of objects.

580. *The great nebula in Andromeda.*—This is a very conspicuous nebula, distinctly visible without a telescope, and is often mistaken for a comet. It was discovered nearly 1000 years ago, though not much noticed until attention was directed to its singular appearance by Simon Marius in 1612. It is of an oval shape, and has been described as resembling the light of a candle shining through horn. When observed with the best telescopes, its boundaries appear greatly extended, its extreme length being 90 minutes, and its breadth 15 minutes. The great telescope at Cambridge Observatory exhibits two dark bands or canals nearly straight and parallel, about one degree in length, running in the direction of the longer axis of the nebula. Till very recently, this nebula defied all the optical power that could be brought to bear upon it to resolve it into stars, or even to afford any symptoms of its stellar character. But within the last few years, decisive evidence of its consisting of stars has been obtained with the Cambridge telescope. Plate VII., Fig. 1, gives a representation of the appearance of this nebula in Herschel's telescope.

581. *The great nebula in the sword-handle of Orion.*—This nebula was first discovered and figured by Huygens in 1659. It consists of irregular nebulous patches, extending over a surface about 40 minutes square, its superficial magnitude being more than twice that of the moon's disc. The brightest portion of the nebula re-

sembles the head and jaws of some monstrous animal with an enormous proboscis. The nebulosity in this vicinity is flocculent, and of a greenish-white tinge. It was irresolvable until the completion of Lord Rosse's telescope; but in this instrument there are strong indications of its being composed of a vast multitude of stars, far removed from us in the profundity of space. A comparison of the earlier with the more recent representations of this nebula might lead to the conclusion that it had changed its form within two hundred years, but no such conclusion can be safely drawn on account of the imperfection of the telescopes with which the early observations were made. Plate VII., Fig. 3, gives a representation of the appearance of this nebula.

582. *The spiral nebula.*—This nebula is situated near the extremity of the tail of the Great Bear. It is a double nebula, with two centres about five minutes apart. From one of the centres proceed several luminous streams, which wind spirally round the nucleus, suggesting the idea of a body not in a state of permanent equilibrium. Though not clearly resolved into stars with Lord Rosse's telescope, some evidence is thereby afforded that it is so composed. Other nebulæ have similar spiral coils, but less distinctly marked than in the one above. Plate VIII., Fig. 4, gives a representation of this nebula.

583. *The dumb-bell nebula.*—This nebula is situated between the constellations Swan and Eagle. In a small telescope it exhibits two centres, connected by a nebulous band, its entire diameter being 7 or 8 minutes. Sir John Herschel compared its appearance to that of a dumb-bell. In Lord Rosse's telescope the form appears less regular, but its general outline is elliptical. Plate VIII., Fig. 1, gives a representation of this nebula.

584. *The crab nebula.*—This nebula is situated near the star ζ in the southern horn of Taurus. In an ordinary telescope it appears of an oval form, but in Lord Rosse's telescope it is seen as a densely-crowded cluster, with branches streaming off from the oval boundary like claws, so as to give it an appearance that in a measure justifies the name of the crab nebula by which it is often distinguished. Plate VIII., Fig. 2, gives a representation of this nebula.

585. *The annular nebula in Lyra.*—This nebula is situated in the constellation Lyra, between the stars β and γ. In Sir J. Herschel's telescope it appeared like a ring of light of a somewhat oval form. The centre was not entirely black, but filled with a faint nebulous light. In Lord Rosse's telescope are seen fringes extending from each side of the annulus, and also stripes crossing the central portion. Though apparently a small nebula, its actual dimensions must be enormous. Even supposing it no farther from us than 61 Cygni, the diameter of the ring would be 20,000 millions of miles, and it is not improbable that its real distance is incomparably greater than that of the above star. Plate VIII., Fig. 3, gives a representation of this nebula.

586. *Planetary nebulæ.*—Planetary nebulæ exhibit discs of uniform brightness throughout, often very sharply defined at the borders, or only a little curdled or furred, as the edges of a planet frequently appear when the night is unfavorable for telescopic observation. They are called planetary nebulæ from the great resemblance they offer to the discs of planets. Not far from the star β in Ursa Major is a fine nebula of this kind. It is circular, nearly 3 minutes in diameter, and of equable light on its whole surface, and, after a long inspection, looks like a condensed mass of attenuated light, seemingly of the size of Jupiter. Supposing it placed at a distance from us not more than that of 61 Cygni, it would have a linear diameter seven times greater than that of the orbit of Neptune. About twenty planetary nebulæ have been observed. They can not be globular clusters of stars, otherwise they would be brighter in the middle than at the borders. It has been conjectured that they may be hollow spherical shells, or circular flat discs, whose planes are nearly at right angles to our line of vision.

587. *Nebulous stars.*—Nebulous stars are stars surrounded by a faint nebulosity, usually of a circular form, and sometimes several minutes in diameter. In some cases the nebulosity is sharply defined at the borders, in others it gradually fades away to darkness. The stars thus attended have nothing in their appearance to distinguish them from ordinary stars, nor does the nebulosity in which they are situated offer the slightest indications of resolvability into stars with any telescopes hitherto constructed. As

instances of nebulous stars may be mentioned one of the fifth magnitude, numbered 55 in Andromeda; and another of the same brightness, numbered 8 in Canes Venatici.

588. *Distribution of the nebulæ.*—The nebulæ are not distributed uniformly over all parts of the heavens. From certain regions they are wholly absent; in others they are rarely found; while in other regions they are crowded in amazing profusion. They are most numerous in the constellations Leo, Virgo, and Ursa Major.

589. *Are all nebulæ resolvable into stars?*—Clusters of stars exhibit all gradations of closeness, from the Pleiades down to those which resemble the diffuse light of a comet. Many clusters, in which, with ordinary telescopes, the component stars are undistinguishable, if seen through more powerful telescopes are resolved wholly into masses of stars, so that some have concluded that all nebulæ are but clusters of stars too remote for the individual stars to be separately seen. In other nebulæ the most powerful telescopes resolve certain portions into masses of stars, while other portions still retain the nebulous appearance. This result may sometimes be ascribed to difference of distance, while in other cases certain portions of the nebulæ may consist of stars having actually a less magnitude, and crowded more closely together.

The telescope which has been most successful in resolving difficult nebulæ is that of Lord Rosse. This is a reflecting telescope, having a clear aperture of 6 feet, and a focal length of 52 feet. Objects which are quite faint in ordinary telescopes, when seen with this instrument appear of dazzling brilliancy. Hitherto every increase of power of the telescope has augmented the number of nebulæ which are resolved into clusters; still it would be unsafe to infer that all nebulosity is but the glare of stars too remote to be separated by the utmost power of our instruments. While Lord Rosse's telescope has shown certain nebulæ to contain an immense number of stars, it has also revealed to us new nebulous appendages of extreme faintness, which we must regard either as not composed of stars, or as composed of stars of very small absolute dimensions.

590. *Have any nebulæ changed their forms?*—The forms of many

of the nebulæ are so peculiar that it is difficult to regard them as having attained a condition of permanent equilibrium, and it has been supposed that we see them now in the state of transition toward stable forms. A comparison of the present appearance of many nebulæ with the representations of those furnished by former astronomers would lead to the conclusion that they had sensibly changed their form within 100 years. Such a conclusion may be premature, but it is probable that future astronomers will discover changes that are incontestable. If any of the nebulæ consist wholly of vaporous matter, they are probably in a state of gradual condensation; and if they all consist of clusters of stars, then these stars are doubtless in motion, forming astral systems of wonderful complexity.

591. *Variations in the brightness of nebulæ.*—Some of the nebulæ have exhibited decided changes of brightness. A nebula, situated near ε in Taurus, at the date of its discovery in 1852 was easily seen with a good telescope, whereas in 1862 it was invisible with instruments of far greater power. A small star close to this nebula likewise faded within the same lapse of time. Another nebula, situated near the Pleiades, in 1859 could be seen with a three-inch telescope, whereas in 1862 it could only be seen with difficulty through the largest telescope. Five or six cases of this kind have been noticed. It is not improbable that these variations of brightness are due to the same cause as the changes of the variable stars.

592. *The Via Lactea, or Milky Way.* — The Galaxy, or Milky Way, is that whitish luminous band of irregular form which is seen on a clear night stretching across the expanse of heaven from one side of the horizon to the other. To the naked eye it presents merely a diffused milky light, stronger in some parts than in others; but when examined in a powerful telescope, it is found to consist of myriads of stars so small that no one of them singly produces a sensible impression on the unassisted eye.

The general course of the Milky Way is in a great circle, inclined about 63° to the celestial equator, and intersecting it near the constellations Orion and Ophiuchus.

The distribution of the telescopic stars within its limits is far from uniform. In some regions several thousands are crowded together within the space of one square degree; in others only a few glittering points are scattered upon the black ground of the heavens. In some parts it presents to the naked eye a bright glow of light from the closeness of the constituent stars; in others there are dark spaces containing scarcely a single star. Such vacancies occur in the constellations Scorpio and Ophiuchus.

593. *Law of distribution of the stars.*—In order to decide whether the stars are distributed over the surface of the heavens according to any general law, Sir W. Herschel undertook a rigorous telescopic survey of the heavens, counting the number of the stars visible in the field of his telescope when directed to different parts of space. He thus discovered that around the poles of the Milky Way the stars are more thinly scattered than elsewhere; that as we advance toward the Milky Way the number of stars included in the field of view of the telescope increases, at first slowly, but afterward more rapidly; and that along the Milky Way the stars are crowded so closely together that it becomes in many cases impossible to count them.

594. *Hypothesis of Sir William Herschel.*—In 1784, Sir W. Herschel advanced the following hypothesis respecting the Milky Way: The stars of our firmament, instead of being scattered in all directions promiscuously through space, constitute a cluster with definite limits, in the form of a stratum, of which the thickness is small in comparison with its length and breadth, and in which the earth occupies a position somewhere about the middle of its thickness. For if we suppose the stars to be scattered pretty uniformly through space, the number of stars visible in the field of a telescope ought to be about the same in every direction, provided the stars extend in all directions to an equal distance. But if the stars about us compose a stratum whose thickness is small in comparison with its length and breadth, then the number of stars visible in the different directions will lead us to a knowledge both of the exterior *form* of this stratum, and of the place occupied by the observer. For example, if within a certain circle of the heavens we count ten stars, and in a circle of the same diameter, taken in a different direction, we count eighty stars with the same telescope, the lengths of the two visual rays will be in the ratio of 1 to 2, or the cube roots of 1 and 8.

595. *This hypothesis is untenable.*—This hypothesis assumes, 1st, that the stars are uniformly distributed through space; and, 2d, that Herschel was able with his telescope to penetrate to the limits of our stratum.

At a later period of his life, Herschel abandoned each of these hypotheses. Every increase in the power of his telescopes disclosed new stars which before had been invisible, and he was compelled to admit that with his telescope of 40 feet the Milky Way was entirely *fathomless;* and instead of the stars being distributed uniformly through space, he admitted that there is a great and sudden condensation of stars in the neighborhood of the Milky Way.

In every part of the heavens the stars seem to extend to a distance beyond the reach of the most powerful telescope hitherto constructed, and hence the *shape* of that portion of space which the stars occupy must be entirely unknown to us; that is, the material universe appears to us to be *boundless.*

596. *Mädler's hypothesis respecting the Milky Way.*—Mädler supposes that the stars of the Milky Way are grouped together in the form of an immense ring, or perhaps a system of detached but concentric star-rings of unequal thickness and various dimensions, but all situated nearly in the same plane. To an observer situated in the centre of such a system of rings, the inner ring would seem to cover the exterior ones; that is, the stars would seem to form but a single ring, and this ring would be a great circle of the sphere. The Milky Way, in fact, divides our firmament into two portions, whose areas are to each other in the ratio of about 8 to 9, from which it is concluded that the solar system is not situated exactly in the plane of the Milky Way, but somewhat toward the south, or in the direction of the constellation Virgo.

The division of the Milky Way throughout a considerable portion of its extent into two separate branches indicates that in this part of the firmament the star-rings do not cover each other, which Mädler explains by supposing that we are eccentrically situated, being nearer to the southern than to the northern part of the rings. This supposition would also explain the greater brilliancy of the Milky Way in the neighborhood of the south pole.

597. *Original condition of the universe.*—The question naturally

arises, Was the universe created substantially as we now see it, or has it been brought to its present condition by a succession of gradual changes under the operation of general laws? We find in our solar system several remarkable coincidences which we can not well suppose to be fortuitous, and which naturally suggest the idea of some grand and comprehensive law.

1st. All the planets (now 90 in number) revolve about the sun from west to east, and, with slight exceptions, nearly in the same plane, viz., the plane of the sun's equator. There are only 4 planets (and these are minute asteroids) whose orbits are inclined to the ecliptic as much as 20°.

2d. The sun rotates on an axis in the same direction as that in which the planets revolve around him.

3d. All the major planets (except perhaps Uranus and Neptune) rotate on their axes in the same direction as that in which they move around the sun.

4th. The satellites (as far as known) revolve around their primaries in the same direction in which the latter turn on their axes.

5th. The orbits of all the larger planets and their satellites have small eccentricity. Only seven of the asteroids have an eccentricity as great as one quarter.

6th. The planets, upon the whole, increase in density as they are found nearer the sun.

7th. The orbits of the comets have usually great eccentricity, and have every variety of inclination to the ecliptic.

These coincidences are not a consequence of the law of universal gravitation, yet it is highly improbable that they should be the result of chance. They seem rather to indicate the operation of some uniform law. Can we discover any law from which these coincidences would necessarily result?

598. *Conclusions from geological phenomena.*—An examination of the condition and structure of the earth has led geologists to conclude that our entire globe was once liquid from heat, and that it has gradually cooled upon its surface, while a large portion of the interior still retains much of its primitive heat. The shape of the mountains in the moon seems to indicate that that body has at some former time been in a state of fusion. But if the earth and moon were ever subjected to such a heat, it is proba-

ble that the other members of the solar system were in a like condition, perhaps at a temperature sufficient to volatilize every solid and liquid body, constituting perhaps a single nebulous mass of the smallest density.

599. *The nebular hypothesis stated.*—Let us suppose, then, that the matter composing the entire solar system once existed in the condition of a single nebulous mass, extending beyond the orbit of the most remote planet. Suppose that this nebula has a slow rotation upon an axis, and that by radiation it gradually cools, thereby contracting in its dimensions. As it contracts in its dimensions, its velocity of rotation, according to the principles of Mechanics, must necessarily increase, and the centrifugal force thus generated in the exterior portion of the nebula would at length become equal to the attraction of the central mass.

This exterior portion would thus become detached, and revolve independently as an immense zone or ring. As the central mass continued to cool and contract in its dimensions, other zones would in the same manner become detached, while the central mass continually decreases in size and increases in density.

The zones thus successively detached would generally break up into separate masses revolving independently about the sun; and if their velocities were slightly unequal, the matter of each zone would ultimately collect in a single planetary, but still gaseous mass, having a spheroidal form, and also a motion of rotation about an axis.

As each of these planetary masses became still farther cooled, it would pass through a succession of changes similar to those of the first solar nebula; rings of matter would be formed surrounding the planetary nucleus, and these rings, if they broke up into separate masses, would ultimately form satellites revolving about their primaries.

600. *Phenomena explained by this hypothesis.*—The planet Saturn affords the only instance in the solar system in which these rings have preserved their unbroken form; and the group of asteroids between Mars and Jupiter presents a case in which a ring broke up into a large number of small fragments, which continued to revolve in independent orbits about the sun.

The first six of the phenomena mentioned in Art. 597 are obvious consequences of this theory. The eccentricity of some of the orbits, and their inclination to the sun's equator, must be ascribed to the accumulated effect of the disturbing action of the planets upon each other.

601. *Apparent anomalies explained.*—The planets thus formed would all have a motion of rotation, but they would not all necessarily rotate in the same direction as the motion of revolution. The outer planets might rotate in the contrary direction, but the satellites must in all cases revolve in their orbits in the same direction as the rotation of the primary. The satellites of Uranus and Neptune have a retrograde motion; and if it shall be discovered that these planets rotate upon their axes in the same direction, these movements would all be consistent with the nebular hypothesis.

Comets may consist of nebulous matter encountered by the solar system in its motion through space, and thus brought within the attractive influence of the sun. They are thus compelled to move in orbits around the sun, and these orbits may become so modified by the attraction of the planets that they may sometimes become permanent members of our solar system. Some of the comets may perhaps consist of small portions of nebulous matter which became detached in the breaking up of the planetary rings, and continued to revolve independently about the sun.

602. *How this hypothesis may be tested.*—It has been attempted to subject this hypothesis to a rigorous test in the following manner. The time of revolution of each of the planets ought to be equal to the time of rotation of the solar mass at the period when its surface extended to the given planet. It remains, then, to compute what should be the time of rotation of the solar mass when its surface extended to each of the planets. It has been found that if we suppose the sun's mass to be expanded until its surface extends to each of the planets in succession, its time of rotation at each of these instants would be very nearly equal to the actual time of revolution of the corresponding planet; and the time of rotation of each primary planet corresponds in like manner with the time of revolution of its different satellites.

The nebular hypothesis must therefore be regarded as possessing considerable probability, since it accounts for a large number of circumstances which hitherto had remained unexplained.

Miscellaneous Problems.

1. At what hour does the sun rise at Havana, Lat. 23° 9′, at the time of the winter solstice?

2. What is the greatest, and also the least meridian altitude of the sun at Chicago, Lat. 41° 52′?

3. What is the least latitude in which twilight lasts all night at the time of the summer solstice?

4. In what azimuth does the sun rise at Boston, Lat. 42° 21′, on the 10th of May, when his declination is 17° 45′ N.?

5. At what hour of the day is the sun due east at New York, Lat. 40° 42′, on the 10th of August, when his declination is 15° 26′ N.?

6. Find the duration of twilight at Cincinnati, Lat. 39° 6′, on the 21st of January, when the sun's declination is 20° S.

7. Find the latitude of the place where the sun's centre remains above the horizon for a hundred successive days.

8. At Washington, Lat. 38° 54′, on the 1st of May, when the sun's declination is 15° 14′ N., the length of the shadow cast by a tower at noon on a horizontal plane is m feet; determine the height of the tower.

9. At New Haven, Lat. 41° 18′, on the 20th of May, when the sun's declination is 20° 6′ N., at what hour of the day will a man's shadow be double his height?

10. Find the altitude of the sun at Philadelphia, Lat. 39° 57′, on the day of the equinox at 9 o'clock in the morning.

11. Find the time of sunrise on the longest day at a place in Lat. 45°.

12. Determine the latitude of the place in which the longest day contains 16 hours.

13. Find the sun's altitude at 6 o'clock in terms of the latitude of the place, and declination of the sun.

14. Find the sun's altitude when on the prime vertical in terms of the latitude and declination.

15. The sun's altitude at 6 o'clock was 14°, and its altitude when due east was 23°; required the latitude of the place.

16. Determine the declination of the sun that it may set in the S.W. point at a place whose latitude is 65° N.

17. Determine the latitude of the place where the sun rises in

the N.E. point, and also the time of its rising, the sun's declination being 20° N.

18. The sun's meridian altitude is 66°, and his depression below the horizon at midnight is 30°; required the sun's declination and the latitude of the place.

19. The longitude of Sirius on the 1st of January, 1864, was 101° 1′ 10″; what was its longitude at the commencement of the Christian era, allowing 50″.24 for mean amount of precession?

20. In the year 1852 there were five Sundays in the month of February; when will a similar case happen again?

21. How much faster than at present must the earth rotate upon its axis in order that bodies on its surface at the equator may lose half their weight?

22. How much faster than at present must the earth rotate upon its axis in order that bodies on its surface, in Lat. 60°, may lose all their gravity?

23. Determine the latitude of the place where the longest day is 6 hours and 12 minutes longer than the shortest day.

24. Determine the latitude of the place at which the sun sets at 10 o'clock on the longest day, and also find the latitude of the place where it sets at 3 o'clock on the shortest day.

25. Aldebaran (Dec. 16° 14′ N.) was observed when on the prime vertical both east and west, and the intervening time was 9h. 20m.; required the latitude of the place.

26. Determine the latitude of the place at which the sun rises in the N.N.E. point at the summer solstice.

27. At a place in Lat. 38°, when the sun's declination was 20° N., the sun was observed to rise at a point E. by N. according to a surveyor's compass; required the variation of the needle.

28. The horizontal refraction being 34′ 54″, find how much the rising of the sun is accelerated by it at New Haven at the time of the summer solstice.

29. Prove that the sun's rising is least accelerated by refraction at the time of the equinoxes.

30. Supposing the quantity of matter in the sun to be increased nine times, and the orbits of the planets to continue the same, how would the periodic times be altered?

31. If the mass of a planet be 4 times that of the earth, and the distance of its satellite 16 times that of the moon from the earth, in what time will the satellite make one revolution?

32. If the periodic time of Mercury be to that of the earth as 4 to 17, determine the time of one synodic revolution.

33. What must be the relation of the distances from the sun of a superior and an inferior planet that their synodic revolutions may be equal?

34. Determine when Saturn will appear stationary, assuming his distance from the sun to be to that of the earth from the sun as 19 to 2, and the orbits to be circles.

35. How high must a man be elevated above the surface of the earth at New York, Lat. 40° 42′, to see the sun at midnight at the time of the summer solstice?

36. A place in Lat. 42° has its horizon so surrounded with mountains that the sun is not visible until it is 10° above the rational horizon in the morning, and it again disappears when 10° above the rational horizon in the evening; how much is the longest day shortened by this circumstance?

37. At a place in Lat. 35° N., Aldebaran (R. A. 67° 2′, Dec. 16° 14′ N.) was seen in the same vertical plane with Sirius (R. A. 99° 47′, Dec. 16° 32′ S.); required the azimuth.

38. At a place in Lat. 35° N., find the hour at which Aldebaran and Sirius will be in the same azimuth on the 1st of January, when the sun's R. A. is 18h. 45m.

39. Aldebaran and Sirius were found to set at the same instant; required the latitude of the place of observation.

40. Find the azimuth of a star when its change of altitude in a given time is a maximum.

41. Find at what time on the longest day of the year, the variation of the sun's altitude at New Haven is the most rapid.

42. Given the sun's apparent diameter, and the latitude of the place, it is required to determine his declination when the time of rising of the sun's disc is a minimum.

43. Find the time when the apparent diurnal motion of *α* Ursæ Majoris (Dec. 62° 29′) is perpendicular to the horizon at New Haven.

44. Compare the times during which *α* Ursæ Majoris moves eastward and westward at New Haven.

45. Find the sun's longitude, or the day of the year, when Sirius rises with the sun at a place in Lat. 42°.

46. Find the day of the year when Sirius sets with the sun at a place in Lat. 42°.

47. When the sun's declination was 15° N., his altitude was found to be 20°, and after one hour's interval his altitude was found to be 31°; required the latitude of the place of observation.

48. If the length of the day be to that of the night as 3 to 2, and the altitude of the sun at noon double his depression at midnight, determine the latitude of the place, and the sun's declination.

49. Determine at what place and at what time of the year, day breaks at 2 o'clock, and the sun rises at half past four.

50. Find the sun's declination when the twilight is shortest at New York.

THE following Alphabet is given in order to facilitate, to the student who is unacquainted with it, the reading of those parts in which the Greek letters are used:

Letters.		Names.	Letters.		Names.
Α	α	Alpha.	Ν	ν	Nu.
Β	β	Bēta.	Ξ	ξ	Xi.
Γ	γ	Gamma.	Ο	ο	Omicron.
Δ	δ	Delta.	Π	ϖ π	Pi.
Ε	ε	Epsilon.	Ρ	ρ	Rho.
Ζ	ζ	Zēta.	Σ	σ ς	Sigma.
Η	η	Eta.	Τ	τ	Tau.
Θ	ϑ θ	Thēta.	Υ	υ	Upsilon.
Ι	ι	Iōta.	Φ	φ	Phi.
Κ	κ	Kappa.	Χ	χ	Chi.
Λ	λ	Lambda.	Ψ	ψ	Psi.
Μ	μ	Mu.	Ω	ω	Omega.

Name.	Symbol.	Distance from the Sun. Mean.	Greatest.	Least.	Eccentricity.	Sidereal Revolution in Days.
Mercury..	☿	0.38710	0.46669	0.30750	0.20562	87.969
Venus ...	♀	0.72333	0.72826	0.71840	.00683	224.701
Earth	♁	1.00000	1.01678	0.98322	.01677	365.256
Mars.....	♂	1.52369	1.66578	1.38160	.09326	686.980
Jupiter...	♃	5.20280	5.45378	4.95182	.04824	4332.585
Saturn ...	♄	9.53885	10.07328	9.00442	.05600	10759.220
Uranus...	♅	19.18264	20.07612	18.28916	.04658	30686.821
Neptune..	♆	30.03697	30.29888	29.77506	.00872	60126.722

Name.	Synodical Revolution in Days.	Mean daily Motion. ′ ″	Hourly Motion in Miles.	Inclination of Orbit. ° ′ ″	Light at Perihelion.	Light at Aphelion.	Compression.
Mercury..	115.877	245 32.6	109795	7 0 8	10.576	4.592	$\frac{1}{150}$
Venus ...	583.921	96 7.8	80320	3 23 31	1.938	1.885	
Earth		59 8.3	68311		1.034	0.967	$\frac{1}{294}$
Mars.....	779.936	31 26.7	55341	1 51 5	0.524	0.360	$\frac{1}{50}$
Jupiter...	398.884	4 59.3	29948	1 18 40	0.041	0.034	$\frac{1}{17}$
Saturn ...	378.092	2 0.6	22118	2 29 28	0.012	0.010	$\frac{1}{10}$
Uranus...	369.656	42.4	15597	0 46 30	0.003	0.002	
Neptune..	367.489	21.6	12464	1 46 59	0.001	0.001	

Name.	Time of Rotation. h. m. s.	Equatorial Diameter. Apparent. ″	In Miles.	Volume.	Mass.	Density.	Gravity.	Bodies fall in one Second. Feet.
Sun	600	1923.6	888000	1416000	354936	0.25	28.56	459.2
Mercury..	24 5 28	6.7	3000	0.059	0.118	2.01	0.53	8.5
Venus ...	23 21 21	16.6	7700	0.912	0.883	0.97	0.92	14.8
Earth	23 56 4		7926	1.000	1.000	1.00	1.00	16.1
Mars.....	24 37 22	7.3	4500	0.183	0.132	0.72	0.52	8.4
Jupiter...	9 55 26	38.3	92000	1412.0	338.034	0.24	2.70	43.4
Saturn ...	10 29 17	17.0	75000	770.0	101.064	0.13	1.19	19.1
Uranus...		4.1	36000	95.9	14.789	0.15	0.75	12.1
Neptune..		2.6	35000	89.5	24.648	0.27	0.85	13.7

No.	Name.	Date of Discovery.	Discoverer.	Mean Distance.	Sidereal Revolution in Days.	Eccentricity.
1	Ceres	1801, Jan. 1	Piazzi	2.7667	1680.9	0.0802
2	Pallas	1802, March 28	Olbers	2.7696	1683.5	.2400
3	Juno	1804, Sept. 1	Harding	2.6680	1591.7	.2573
4	Vesta	1807, March 29	Olbers	2.3613	1325.3	.0898
5	Astræa	1845, Dec. 8	Hencke	2.5765	1510.6	.1900
6	Hebe	1847, July 1	Hencke	2.4251	1379.4	.2030
7	Iris	1847, Aug. 13	Hind	2.3863	1346.5	.2313
8	Flora	1847, Oct. 18	Hind	2.2014	1192.9	.1567
9	Metis	1848, April 25	Graham	2.3858	1346.0	.1241
10	Hygeia	1849, April 12	Gasparis	3.1511	2043.1	.1000
11	Parthenope	1850, May 11	Luther	2.4525	1402.9	.0993
12	Victoria	1850, Sept. 13	Hind	2.3342	1302.6	.2189
13	Egeria	1850, Nov. 2	Gasparis	2.5763	1510.4	.0867
14	Irene	1851, May 19	Hind	2.5895	1522.0	.1652
15	Eunomia	1851, July 29	Gasparis	2.6437	1570.0	.1872
16	Psyche	1852, March 17	Gasparis	2.9264	1828.5	.1341
17	Thetis	1852, April 17	Luther	2.4733	1420.7	.1276
18	Melpomene	1852, June 24	Hind	2.2956	1270.4	.2177
19	Fortuna	1852, Aug. 22	Hind	2.4416	1393.5	.1572
20	Massilia	1852, Sept. 19	Gasparis	2.4089	1365.6	.1443
21	Lutetia	1852, Nov. 15	Goldschmidt	2.4354	1388.2	.1620
22	Calliope	1852, Nov. 16	Hind	2.9092	1812.4	.1010
23	Thalia	1852, Dec. 15	Hind	2·6271	1555.3	.2323
24	Themis	1853, April 5	Gasparis	3.1420	2034.2	.1170
25	Phocæa	1853, April 7	Chacornac	2.4008	1358.8	.2546
26	Proserpina	1853, May 5	Luther	2.6561	1581.1	.0875
27	Euterpe	1853, Nov. 8	Hind	2.3468	1313.2	.1735
28	Bellona	1854, March 1	Luther	2.7785	1691.6	.1501
29	Amphitrite	1854, March 1	Marth	2.5544	1491.2	.0739
30	Urania	1854, July 22	Hind	2.3663	1329.6	.1270
31	Euphrosyne	1854, Sept. 1	Ferguson	3.1571	2049.0	.2181
32	Pomona	1854, Oct. 26	Goldschmidt	2.5868	1519.6	.0824
33	Polyhymnia	1854, Oct. 28	Chacornac	2.8651	1771.3	.3382
34	Circe	1855, April 6	Chacornac	2.6871	1608.9	.1056
35	Leucothea	1855, April 19	Luther	3.0059	1903.5	.2141
36	Atalanta	1855, Oct. 5	Goldschmidt	2.7450	1661.1	.3007
37	Fides	1855, Oct. 5	Luther	2.6413	1568.0	.1767
38	Leda	1856, Jan. 12	Chacornac	2.7401	1656.8	.1554
39	Lætitia	1856, Feb. 8	Chacornac	2.7700	1683.9	.1111
40	Harmonia	1856, March 31	Goldschmidt	2.2677	1247.3	.0463
41	Daphne	1856, May 22	Goldschmidt	2.7674	1681.5	.2703
42	Isis	1856, May 23	Pogson	2.4400	1392.2	.2256
43	Ariadne	1857, April 15	Pogson	2.2034	1194.6	.1676
44	Nysa	1857, May 27	Goldschmidt	2.4234	1378.0	.1503
45	Eugenia	1857, June 27	Goldschmidt	2.7205	1639.0	.0822

No.	Name.	Date of Discovery.	Discoverer.	Mean Distance.	Sidereal Revolution in Days.	Eccentricity.
46	Hestia	1857, Aug. 16	Pogson	2.5261	1466.5	0.1647
47	Aglaia	1857, Sept. 15	Luther	2.8802	1785.4	.1324
48	Doris	1857, Sept. 19	Goldschmidt	3.1095	2002.8	.0766
49	Pales	1857, Sept. 19	Goldschmidt	3.0840	1978.2	.2374
50	Virginia	1857, Oct. 4	Ferguson	2.6497	1575.4	.2873
51	Nemausa	1858, Jan. 22	Laurent	2.3655	1328.9	.0667
52	Europa	1858, Feb. 6	Goldschmidt	3.1000	1993.6	.1014
53	Calypso	1858, April 4	Luther	2.6188	1547.9	.2037
54	Alexandra	1858, Sept. 10	Goldschmidt	2.7123	1631.6	.1969
55	Pandora	1858, Sept. 10	Searle	2.7590	1673.9	.1447
56	Melete	1857, Sept. 9	Goldschmidt	2.5971	1528.7	.2370
57	Mnemosyne	1859, Sept. 22	Luther	3.1573	2049.1	.1041
58	Concordia	1860, March 24	Luther	2.6950	1616.0	.0403
59	Elpis	1860, Sept. 12	Chacornac	2.7132	1632.4	.1171
60	Echo	1860, Sept. 15	Ferguson	2.3931	1352.2	.1847
61	Danaë	1860, Sept. 19	Goldschmidt	2.9854	1884.1	.1651
62	Erato	1860, Oct.	Förster	3.1287	2021.4	.1702
63	Ausonia	1861, Feb. 10	Gasparis	2.3957	1354.4	.1254
64	Angelina	1861, March 4	Tempel	2.6807	1603.2	.1295
65	Cybele	1861, March 8	Tempel	3.4209	2311.0	.1204
66	Maia	1861, April 9	Tuttle	2.6635	1587.8	.1339
67	Asia	1861, April 17	Pogson	2.4209	1375.8	.1844
68	Leto	1861, April 29	Luther	2.7822	1695.6	.1875
69	Hesperia	1861, April 29	Schiaparelli	2.9950	1893.1	.1745
70	Panopæa	1861, May 5	Goldschmidt	2.6129	1542.7	.1830
71	Feronia	1861, May 29	Peters	2.2654	1245.4	.1195
72	Niobe	1861, Aug. 13	Luther	2.7562	1671.3	.1737
73	Clytie	1862, April 7	Tuttle	2.6655	1589.6	.0454
74	Galatea	1862, Aug. 29	Tempel	2.7785	1691.7	.2384
75	Eurydice	1862, Sept. 22	Peters	2.6708	1594.3	.3067
76	Freia	1862, Oct. 21	D'Arrest	3.1890	2080.0	.0302
77	Frigga	1862, Nov. 12	Peters	2.6738	1596.9	.1358
78	Diana	1863, March 15	Luther	2.6263	1554.6	.2067
79	Eurynome	1863, Sept. 14	Watson	2.4478	1398.8	.1932
80	Sappho	1864, May 2	Pogson	2.2971	1271.6	.2005
81	Terpsichore	1864, Sept. 30	Tempel	2.7800	1693.0	.1313
82	Alcmene	1864, Nov. 27	Luther	2.7427	1659.1	.1981
83	Beatrix	1865, April 26	Gasparis	2.4288	1382.5	.0842
84	Clio	1865, Aug. 25	Luther	2.3600	1324.2	.2346
85	Io	1865, Sept. 19	Peters	2.6536	1578.9	.1907
86	Semele	1866, Jan. 4	Tietjen	3.0908	1984.8	.2049
87	Sylvia	1866, May 16	Pogson	3.4927	2384.1	.0811
88	Thisbe	1866, June 15	Peters	2.6733	1596.5	.1888
89	Julia.	1866, Aug. 6	Stephan	2.5475	1485.1	.1672
90	Antiope	1866, Oct. 1	Luther	3.1188	2011.8	.1477

Elements of the Moon.

Mean distance from the earth ....238900 miles.
Mean sidereal revolution ....27d. 7h. 43m. 11.46s.
Mean synodical revolution ....29d. 12h. 44m. 2.87s.
Mean revolution of nodes.....................18yrs. 218d. 21h. 22m. 46s.
Mean revolution of perigee ....8yrs. 310d. 13h. 48m. 53s.
Mean inclination of orbit ....5° 8′ 48″.
Eccentricity of the orbit ....0.054908.
Diameter of the Moon.........................2160 miles.
Density of the Moon, that of the earth being 1..0.5657.
Mass of the Moon, that of the earth being 1....$\frac{1}{81}$.

Elements of the Satellites of Jupiter.

Sat.	Sidereal Revolution.	Distance in Radii of Jupiter.	Distance in Miles.	Orbit inclined to Jupiter's Equator.	Diameter.		Mass, that of Jupiter being 1.
					Apparent.	In Miles.	
	d. h. m. s.			° ′ ″	″		
1	1 18 27 33	6.049	278542	0 0 7	1.0	2436	.0000173
2	3 13 14 36	9.623	442904	0 1 6	0.9	2187	.0000232
3	7 3 42 33	15.350	706714	0 5 3	1.5	3573	.0000885
4	16 16 31 0	26.998	1200000	0 0 24	1.3	3057	.0000427

Elements of the Satellites of Saturn.

Sat.	Date of Discovery.	Sidereal Revolution.	Distance in Radii of Saturn.	Mean apparent Distance.	Mean Distance in Miles.	Eccentricity of Orbit.	Diameter in Miles.
		d. h. m. s.		″			
1	1789	0 22 37 5	3.361	26.8	118000	0.0689	
2	1787	1 8 53 7	4.312	34.4	152000		
3	1684	1 21 18 26	5.340	42.6	188000	0.0051	500
4	1684	2 17 41 9	6.840	54.5	240000	.02	500
5	1672	4 12 25 11	9.550	76.2	336000	.0227	1200
6	1655	15 22 41 25	22.140	176.5	778000	.0292	2850
7	1848	22 12	28.	222.1	940000	.115	
8	1671	79 7 53 40	64.359	514.5	2268000	.025	1800

Elements of the Satellites of Uranus.

Sat.	Date of Discovery.	Sidereal Revolution.	Distance in Radii of Uranus.	Mean apparent Distance.	Mean Distance in Miles.
		d. h. m.		″	
1	1847	2 12 17	6.94	13.5	119994
2	1847	4 3 28	9.72	19.3	170863
3	1787	8 16 56	15.89	33.9	288600
4	1787	13 11 7	21.27	45.2	380000

Elements of the Satellite of Neptune.

Sidereal revolution..................................5d. 20h. 50m. 45s.
Apparent mean distance.............................16″.98.
Mean distance in miles ....236000.
Orbit inclined to the plane of the ecliptic...............151°.

Table IV.—Eclipses of the Sun from 1865 to 1900.

The following is a list of all the subsequent solar eclipses that will be visible in the city of Boston during the present century. The dates are given in mean time for the meridian of Boston, reckoned astronomically.

1. 1865, *October 18th and 19th.*

Beginning	18th,	21h.	9m.	55s.
Greatest obscuration		22	44	58
Apparent conjunction		22	46	4
End	19th,	0	25	7

Magnitude of the eclipse (sun's diameter=1) 0.692, on sun's south limb.

This eclipse will be annular in the States of North and South Carolina; at Charleston the ring will last 6¾ minutes.

This is the third return of the eclipse of September, 1811, which was annular in Virginia.

2. 1866, *October 7th and 8th.*

Beginning	7th,	23h.	11m.	33s.
Apparent conjunction		23	33	50
Greatest obscuration		23	41	25
End	8th,	0	10	34

Magnitude of the eclipse 0.043, on sun's north limb.

South of Connecticut there will be no eclipse, and no central eclipse in any part of the earth.

3. 1869, *August 7th.*

Beginning	5h.	21m.	17s.
Apparent conjunction	6	16	7
Greatest obscuration	6	16	40
End	7	7	28

Magnitude of the eclipse 0.853, on sun's south limb.

This eclipse will be total in North Carolina and Virginia.

4. 1873, *May 25th.*

The sun and moon will be in contact at sunrise; but the sun will be eclipsed to places at a greater distance from the equator, and in less longitude from Greenwich.

5. 1875, *September 28th.*

Sun rises eclipsed	17h.	56m.	0s.
Formation of the ring	18	20	21
Apparent conjunction	18	21	28
Nearest approach of centres	18	21	37
Rupture of the ring	18	22	52
End of the eclipse	19	30	43

Magnitude of the eclipse at sunrise, 0.603; at nearest approach, 0.951.

This eclipse will be annular in Boston, and in some part of Maine, New Hampshire, Massachusetts, and Vermont.

6. 1876, *March 25th.*

Beginning	4h.	11m.	29s.
Greatest obscuration	5	2	39
Apparent conjunction	5	7	42
End	5	48	24

Magnitude of the eclipse 0.276, on sun's north limb.

7. 1878, *July 29th.*

Beginning	4h.	56m.	10s.
Greatest obscuration	5	50	1
Apparent conjunction	5	53	57
End	6	39	8

Magnitude of the eclipse 0.615, on sun's south limb.

This eclipse will not be total in any part of the United States, but probably will be so in the island of Cuba.

This is the fourth return of the total eclipse of June 16th, 1806.

8. 1880, *December 30th.*

Sun rises eclipsed	19h.	30m.	0s.
Greatest obscuration	20	12	50
Apparent conjunction	20	12	59
End	21	11	37

Magnitude of the eclipse at sunrise, 0.230; at greatest obscuration, 0.457, on sun's north limb. This eclipse can not be central in any place.

At the time of this eclipse the sun and moon are very nearly at their least possible distance from the earth.

9. 1885, *March 16th.*

Beginning	0h.	35m.	0s.
Greatest obscuration	1	55	55
Apparent conjunction	1	57	22
End	3	10	49

Magnitude of the eclipse 0.537, on sun's north limb.

10. 1886, *August 28th.*

Beginning	18h.	30m.	22s.
Apparent conjunction	18	33	15
Greatest obscuration	18	40	1
End	18	51	52

Magnitude of the eclipse 0.018, on sun's south limb.

North of Massachusetts there will be no eclipse.

11. 1892, *October 20th.*

Beginning	0h.	18m.	39s.
Apparent conjunction	1	45	11
Greatest obscuration	1	51	8
End	3	20	8

Magnitude of the eclipse 0.682, on sun's north limb.

The sun will probably be centrally eclipsed in the Canadas and Labrador.

12. 1897, *July 28th.*

Beginning	21h.	7m.	35s.
Greatest obscuration	22	15	35
Apparent conjunction	22	24	56
End	23	23	59

Magnitude of the eclipse 0.369, on sun's south limb.

13. 1900, *May 27th.*

Beginning	20h.	8m.	41s.
Apparent conjunction	21	22	50
Greatest obscuration	21	23	6
End	22	45	32

Magnitude of the eclipse 0.918, on sun's south limb.

The sun will be totally eclipsed in the State of Virginia.

Moon..........	1768, June 29.6, Mag. 1.21.	1786, July 10.9, Mag. 1.06.
Sun	July 13.6, Total.	July 24.9, Total.
Sun	Dec. 8.8, Partial.	Dec. 20.2, Partial.
Moon..........	Dec. 23.2, Mag. 1.75.	1787, Jan. 3.5, Mag. 1.74.
Sun	1769, Jan. 7.6, Partial.	Jan. 18.9, Partial.
Sun	June 3.6, Total.	June 15.2, Total.
Moon..........	June 18.8, Mag. 1.04.	June 30.1, Mag. 1.20.
Sun	Nov. 27.8, Total.	Dec. 9.2, Total.
Moon..........	Dec. 12.7, Mag. 0.77.	Dec. 24.1, Mag. 0.77.
Moon..........	1804, July 22.2, Mag. 0.91.	1822, Aug. 2.5, Mag. 0.75.
Sun	Aug. 5.2, Total.	Aug. 16.5, Total.
Sun	Dec. 31.6, Partial.	1823, Jan. 11.9, Partial.
Moon..........	1805, Jan. 14.8, Mag. 1.74.	Jan. 26.2, Mag. 1.73.
Sun	Jan. 30.3, Partial.	Feb. 10.6, Partial.
Sun	June 26.5, Partial.	July 7.7, Partial.
Moon..........	July 11.3, Mag. 1.37.	July 22.6, Mag. 1.51.
Sun	July 25.7, Partial.	Aug. 6.1, Partial.
Sun	Dec. 20.5, Total.	Dec. 31.8, Annular.
Moon..........	1806, Jan. 4.5, Mag. 0.78.	1824, Jan. 15.8, Mag. 0.78.
Moon..........	1840, Aug. 12.7, Mag. 0.61.	1858, Aug. 24.1, Mag. 0.47.
Sun	Aug. 26.6, Total.	Sept. 7.1, Total.
Sun	1841, Jan. 22.3, Partial.	1859, Feb. 2.5, Partial.
Moon..........	Feb. 5.6, Mag. 1.72.	Feb. 16.9, Mag. 1.69.
Sun	Feb. 20.9, Partial.	Mar. 4.3, Partial.
Sun	July 18.1, Partial.	July 29.4, Partial.
Moon..........	Aug. 1.9, Mag. 1.67.	Aug. 13.2, Mag. 1.81.
Sun	Aug. 16.4, Partial.	Aug. 27.7, Partial.
Sun	1842, Jan. 11.1, Annular.	1860, Jan. 22.5, Annular.
Moon..........	Jan. 26.2, Mag. 0.79.	Feb. 6.6, Mag. 0.81.
Moon..........	1876, Sept. 3.4, Mag. 0.33.	1894, Sept. 15.7, Mag. 0.21.
Sun	Sept. 17.4, Total.	Sept. 29.7, Total.
Sun	1877, Feb. 12.8, Partial.	1895, Feb. 24.1, Partial.
Moon..........	Feb. 27.3, Mag. 1.62.	Mar. 11.7, Mag. 1.56.
Sun	Mar. 15.6, Partial.	Mar. 26.9, Partial.
Sun	Aug. 9.7, Partial.	Aug. 20.0, Partial.
Moon..........	Aug. 23.5, Mag. 1.66.	Sept. 4.7, Mag. 1.54.
Sun	Sept. 8.0, Partial.	Sept. 13.3, Partial.
Sun	1878, Feb. 1.8, Total.	1896, Feb. 13.2, Total.
Moon..........	Feb. 17.9, Mag. 0.82.	Feb. 28.4, Mag. 0.83.

TABLE VI.—TRANSITS OF MERCURY OVER THE SUN'S DISC.

Date.	Duration.	Date.	Duration.	Date.	Duration.	Date.	Duration.
	h. m.		h. m.		h. m.		h. m.
1631, Nov. 6	5 23	1707, May 5	7 54	1776, Nov. 2	1 13	1835, Nov. 7	5 8
1644, Nov. 8	3 57	1710, Nov. 6	5 24	1782, Nov. 12	1 15	1845, May 8	6 45
1651, Nov. 2	3 31	1723, Nov. 9	4 58	1786, May 3	5 28	1848, Nov. 9	5 23
1661, May 3	7 36	1736, Nov. 10	2 42	1789, Nov. 5	4 52	1861, Nov. 11	4 1
1664, Nov. 4	5 17	1740, May 2	3 0	1799, May 7	7 25	1868, Nov. 4	3 31
1674, May 6	4 30	1743, Nov. 4	4 32	1802, Nov. 8	5 27	1878, May 6	7 47
1677, Nov. 7	5 13	1753, May 5	7 47	1815, Nov. 11	4 28	1881, Nov. 7	5 18
1690, Nov. 9	3 36	1756, Nov. 6	5 25	1822, Nov. 4	2 43	1891, May 9	5 8
1697, Nov. 2	3 56	1769, Nov. 9	4 47	1832, May 5	6 56	1894, Nov. 10	5 15

App. Altitude.	Mean Refraction.	Difference for 10′.	App. Altitude.	Mean Refraction.	Difference.	App. Altitude.	Mean Refraction.	Difference.	External Therm.	Factor T.	External Therm.	Factor T.
° ′	′ ″	″	° ′	′ ″	″	°	′ ″	″	°		°	
0 0	34 54.1	124.9	8 30	6 8.4	6.6	27	1 52.8	4.6	—10	1.130	42	1.013
10	32 49.2	116.9	40	6 1.8	6.4	28	1 48.2	4.4	9	1.128	43	1.011
20	30 52.3	108.8	50	5 55.4	6.1	29	1 43.8	4.1	8	1.125	44	1.009
30	29 3.5	100.8	9 0	5 49.3	6.0	30	1 39.7	3.9	7	1.123	45	1.007
40	27 22.7	92.9	10	5 43.3	5.7	31	1 35.8	3.7	6	1.120	46	1.005
50	25 49.8	85.2	20	5 37.6	5.6	32	1 32.1	3.4	5	1.118	47	1.003
1 0	24 24.6	77.9	30	5 32.0	5.5	33	1 28.7	3.3	4	1.115	48	1.001
10	23 6.7	71.1	40	5 26.5	5.2	34	1 25.4	3.1	3	1.113	49	1.000
20	21 55.6	64.7	50	5 21.3	5.1	35	1 22.3	3.0	2	1.110	50	.998
30	20 50.9	59.0	10 0	5 16.2	5.0	36	1 19.3	2.8	— 1	1.108	51	.996
40	19 51.9	53.9	10	5 11.2	4.8	37	1 16.5	2.7	0	1.106	52	.994
50	18 58.0	49.4	20	5 6.4	4.7	38	1 13.8	2.6	+ 1	1.103	53	.992
2 0	18 8.6	45.6	30	5 1.7	4.5	39	1 11.2	2.5	2	1.101	54	.990
10	17 23.0	42.3	40	4 57.2	4.4	40	1 8.7	2.4	3	1.098	55	.988
20	16 40.7	39.8	50	4 52.8	4.3	41	1 6.3	2.3	4	1.096	56	.986
30	16 0.9	37.5	11 0	4 48.5	4.2	42	1 4.0	2.2	5	1.094	57	.984
40	15 23.4	35.6	10	4 44.3	4.1	43	1 1.8	2.1	6	1.091	58	.982
50	14 47.8	33.2	20	4 40.2	3.9	44	59.7	2.0	7	1.089	59	.980
3 0	14 14.6	30.9	30	4 36.3	3.9	45	57.7	2.0	8	1.087	60	.978
10	13 43.7	28.7	40	4 32.4	3.7	46	55.7	1.9	9	1.084	61	.977
20	13 15.0	26.7	50	4 28.7	3.7	47	53.8	1.9	10	1.082	62	.975
30	12 48.3	24.6	12 0	4 25.0	3.6	48	51.9	1.7	11	1.080	63	.973
40	12 23.7	23.0	10	4 21.4	3.4	49	50.2	1.8	12	1.078	64	.971
50	12 0.7	21.8	20	4 18.0	3.4	50	48.4	1.7	13	1.075	65	.969
4 0	11 38.9	20.6	30	4 14.6	3.3	51	46.7	1.6	14	1.073	66	.967
10	11 18.3	19.7	40	4 11.3	3.3	52	45.1	1.6	15	1.071	67	.965
20	10 58.6	19.0	50	4 8.0	3.1	53	43.5	1.6	16	1.069	68	.964
30	10 39.6	18.4	13 0	4 4.9	3.1	54	41.9	1.5	17	1.066	69	.962
40	10 21.2	17.9	10	4 1.8	3.0	55	40.4	1.5	18	1.064	70	.960
50	10 3.3	16.8	20	3 58.8	2.9	56	38.9	1.4	19	1.062	71	.958
5 0	9 46.5	15.6	30	3 55.9	2.9	57	37.5	1.4	20	1.060	72	.956
10	9 30.9	14.9	40	3 53.0	2.8	58	36.1	1.4	21	1.057	73	.955
20	9 16.0	14.1	50	3 50.2	2.8	59	34.7	1.4	22	1.055	74	.953
30	9 1.9	13.5	14 0	3 47.4	2.7	60	33.3	1.3	23	1.053	75	.951
40	8 48.4	12.8	10	3 44.7	2.6	61	32.0	1.3	24	1.051	76	.949
50	8 35.6	12.3	20	3 42.1	2.6	62	30.7	1.3	25	1.049	77	.948
6 0	8 23.3	11.7	30	3 39.5	2.5	63	29.4	1.2	26	1.047	78	.946
10	8 11.6	11.3	40	3 37.0	2.5	64	28.2	1.3	27	1.044	79	.944
20	8 0.3	10.8	50	3 34.5	2.4	65	26.9	1.2	28	1.042	80	.942
30	7 49.5	10.3	15 0	3 32.1	13.5	66	25.7	1.2	29	1.040	81	.941
40	7 39.2	10.0	16 0	3 18.6	12.0	67	24.5	1.2	30	1.038	82	.939
50	7 29.2	9.5	17 0	3 6.6	10.8	68	23.3	1.1	31	1.036	83	.937
7 0	7 19.7	9.2	18 0	2 55.8	9.7	69	22.2	1.2	32	1.034	84	.935
10	7 10.5	8.8	19 0	2 46.1	8.8	70	21.0	1.1	33	1.032	85	.934
20	7 1.7	8.4	20 0	2 37.3	8.0	71	19.9	1.1	34	1.030	86	.932
30	6 53.3	8.2	21 0	2 29.3	7.4	72	18.8	1.1	35	1.028	87	.930
40	6 45.1	7.9	22 0	2 21.9	6.7	73	17.7	1.1	36	1.026	88	.929
50	6 37.2	7.6	23 0	2 15.2	6.3	74	16.6	1.1	37	1.024	89	.927
8 0	6 29.6	7.3	24 0	2 8.9	5.7	75	15.5	5.3	38	1.022	90	.925
10	6 22.3	7.1	25 0	2 3.2	5.4	80	10.2	5.1	39	1.019	91	.924
20	6 15.2	6.8	26 0	1 57.8	5.0	85	5.1	5.1	40	1.017	92	.922
30	6 8.4		27 0	1 52.8		90	0.0		41	1.015	93	.920

TABLE VII.—ASTRONOMICAL REFRACTIONS.

Barom.	Factor B.	Barom.	Factor B.	Barom.	Factor B.	Attached Therm.	Factor t.	Attached Therm.	Factor t.
inches.		inches.		inches.		°		°	
27.9	0.943	29.0	0.980	30.1	1.017	— 15	1.004	40	0.999
28.0	.946	29.1	.983	30.2	1.020	— 10	1.004	45	0.999
28.1	.949	29.2	.987	30.3	1.024	— 5	1.003	50	0.998
28.2	.953	29.3	.990	30.4	1.027	0	1.003	55	0.998
28.3	.956	29.4	.993	30.5	1.031	+ 5	1.003	60	0.997
28.4	.960	29.5	.997	30.6	1.034	10	1.002	65	0.997
28.5	.963	29.6	1.000	30.7	1.037	15	1.002	70	0.997
28.6	.966	29.7	1.003	30.8	1.041	20	1.001	75	0.996
28.7	.970	29.8	1.007	30.9	1.044	25	1.001	80	0.996
28.8	.973	29.9	1.010	31.0	1.047	30	1.000	85	0.995
28.9	.976	30.0	1.014	31.1	1.050	35	1.000	90	0.995

TABLE VIII.—REDUCTION OF THE MOON'S EQUATORIAL PARALLAX.

Latitude.	Moon's Equatorial Parallax.			Latitude.	Moon's Equatorial Parallax.			Latitude.	Moon's Equatorial Parallax.		
	53′	57′	61′		53′	57′	61′		53′	57′	61′
	″	″	″		″	″	″		″	″	″
2	0.0	0.0	0.0	32	3.0	3.2	3.4	62	8.3	8.9	9.5
4	0.1	0.1	0.1	34	3.3	3.6	3.8	64	8.6	9.2	9.9
6	0.1	0.1	0.1	36	3.6	3.9	4.2	66	8.9	9.5	10.2
8	0.2	0.2	0.2	38	4.0	4.3	4.6	68	9.1	9.8	10.5
10	0.3	0.3	0.4	40	4.4	4.7	5.0	70	9.4	10.1	10.8
12	0.5	0.5	0.5	42	4.7	5.1	5.4	72	9.6	10.3	11.1
14	0.6	0.7	0.7	44	5.1	5.5	5.9	74	9.8	10.6	11.3
16	0.8	0.9	0.9	46	5.5	5.9	6.3	76	10.0	10.8	11.5
18	1.0	1.1	1.2	48	5.8	6.3	6.7	78	10.2	10.9	11.7
20	1.2	1.3	1.4	50	6.2	6.7	7.1	80	10.3	11.1	11.9
22	1.5	1.6	1.7	52	6.6	7.1	7.6	82	10.4	11.2	12.0
24	1.7	1.9	2.0	54	6.9	7.5	8.0	84	10.5	11.3	12.1
26	2.0	2.2	2.3	56	7.3	7.8	8.4	86	10.6	11.4	12.2
28	2.3	2.5	2.7	58	7.6	8.2	8.8	88	10.6	11.4	12.2
30	2.6	2.8	3.0	60	8.0	8.6	9.2	90	10.6	11.4	12.2

TABLE IX.—ELEMENTS OF PERIODICAL COMETS.

Name.	Time of Perihelion Passage.	Longitude of Perihelion.	Ascending Node.	Inclination of Orbit.	Semi-major Axis.	Period in Days.	Eccentricity.	Motion.
		° ′	° ′	° ′				
Halley's...	1835, Nov. 16	304 32	55 10	17 45	17.9885	27867	0.9674	R.
Faye's	1858, Oct. 1	49 52	209 40	11 22	3.8118	2718	.5560	D.
Biela's	1852, Sept. 23	109 6	245 57	12 34	3.5018	2394	.7555	D.
D'Arrest's.	1857, Nov. 28	323 0	148 27	13 56	3.4411	2332	.6609	D.
Brorsen's..	1857, March 29	115 44	101 46	29 49	3.1392	2032	.8023	D.
Winnecke's	1858, May 2	275 40	113 31	10 48	3.1388	2031	.7550	D.
Encke's ...	1862, Feb. 6	158 1	334 31	13 5	2.2173	1205	.8474	D.

TABLE X.—ALTITUDES OF THE PRINCIPAL LUNAR MOUNTAINS.

Name.	Altitude in Feet.	Selenographic Position.		Name.	Altitude in Feet.	Selenographic Position.	
		Longitude.	Latitude.			Longitude.	Latitude.
		°	°			°	°
Newton	23800	16 E.	77 S.	Tycho......	17800	12 E.	43 S.
Curtius.....	22300	3 W.	67 S.	Kircher	17600	43 E.	67 S.
Casatus	20800	35 E.	74 S.	Pythagoras .	16900	60 E.	63 N.
Calippus ...	20400	10 W.	39 N.	Clavius.....	16800	15 E.	58 S.
Posidonius..	19800	29 W.	31 N.	Endymion ..	16700	55 W.	53 N.
Short	18700	10 E.	74 S.	Catharina ..	16400	23 W.	17 S.
Moretus	18400	7 E.	70 S.	Theophilus .	15900	26 W.	11 S.
Mutus......	18300	30 W.	63 S.	Harpalus ...	15800	44 E.	53 N.
Huyghens ..	18000	2 E.	20 N.	Eratosthenes	15600	11 E.	14 N.
Blancanus ..	18000	21 E.	63 S.	Werner	15600	3 W.	28 S.

TABLE XI.—DIAMETERS OF SOME OF THE ANNULAR MOUNTAINS ON THE MOON.

Name.	Diameter in English Miles.	Selenographic Position.		Name.	Diameter in English Miles.	Selenographic Position.	
		Longitude.	Latitude.			Longitude.	Latitude.
		°	°			°	°
Clavius.....	143	15 E.	58 S.	Scheiner ...	70	26 E.	60 S.
Schikard ...	134	55 E.	44 S.	Posidonius..	62	29 W.	31 N.
Ptolemy....	115	3 E.	9 S.	Plato	60	9 E.	51 N.
Schiller	113	38 E.	52 S.	Flamsteed ..	60	44 E.	5 S.
Gauss	111	75 W.	37 N.	Piccolomini .	58	31 W.	29 S.
Riccioli	106	75 E.	2 S.	Copernicus..	55	20 E.	9 N.
Hipparchus .	97	5 W.	6 S.	Fabricius ...	55	41 W.	42 S.
Boussingault	92	55 W.	68 S.	Tycho......	54	12 E.	43 S.
Cleomedes..	78	55 W.	27 N.	Aristarchus .	28	47 E.	23 N.
Hevelius ...	70	67 E.	2 N.	Kepler	22	38 E.	8 N.

TABLE XII.—TRANSITS OF VENUS OVER THE SUN'S DISC.

Year.	Greenwich Mean Time of Conjunction.				Duration of Transit.			Least Distance of Venus from Sun's Centre.	
		h.	m.	s.	h.	m.	s.	′	″
1639	Dec. 4,	6	0	20	6	34	0	9	0 S.
1761	June 5,	17	35	14	6	16	0	9	30 S.
1769	June 3,	9	58	34	5	59	46	10	10 N.
1874	Dec. 8,	16	8	24	4	9	22	13	51 N.
1882	Dec. 6,	4	16	24	6	3	26	10	29 S.
2004	June 7,	20	51	24	5	29	40	11	19 S.
2012	June 5,	13	17	40	6	41	30	8	20 N.

TABLE XIII.—SCHWABE'S OBSERVATIONS OF THE SOLAR SPOTS.

Year.	Number of observing Days.	Groups of Spots observed.	Days on which the Sun was free from Spots.	Year.	Number of observing Days.	Groups of Spots observed.	Days on which the Sun was free from Spots.
1826	277	118	22	1846	314	157	1
1827	273	161	2	1847	276	257	0
1828	282	225 max.	0	1848	278	330 max.	0
1829	244	199	0	1849	285	238	0
1830	217	190	1	1850	308	186	2
1831	239	149	3	1851	308	151	0
1832	270	84	49	1852	337	125	2
1833	267	33 min.	139	1853	299	91	3
1834	273	51	120	1854	334	67	65
1835	244	173	18	1855	313	79	146
1836	200	272	0	1856	321	34 min.	193
1837	168	333 max.	0	1857	324	98	52
1838	202	282	0	1858	335	188	0
1839	205	162	0	1859	343	205	0
1840	263	152	3	1860	332	210 max.	0
1841	283	102	15	1861	322	204	0
1842	307	68	64	1862	317	160	3
1843	324	34 min.	149	1863	330	124	2
1844	321	52	111	1864	325	130	4
1845	332	114	29	1865	307	93	26

TABLE XIV.—PARALLAX OF FIXED STARS.

Star.	Magnitude.	Parallax.	Distance (Sun's Distance = 1).	Light Interval in Years.	Name of Observer.
		″			
α Centauri	1	0.92	224201	3.54	Henderson and Maclear.
61 Cygni	6	0.45	458366	7.24	Bessel, Johnson, etc.
21258 Lalande	8	0.26	793326	12.53	Auwers.
17415 Oeltzen	9	0.24	859437	13.57	Krüger.
Sirius	1	0.23	896803	14.16	Henderson and Maclear.
1830 Groombridge	7	0.16	1289055	20.36	Struve, Peters, etc.
α Lyræ	1	0.16	1289055	20.36	Struve and Peters.
70 Ophiuchi	5	0.16	1289055	20.36	Krüger.
ι Ursæ Majoris	3	0.13	1586653	25.06	Peters.
Arcturus	1	0.13	1586653	25.06	Peters.
Procyon	1	0.12	1718873	27.15	Auwers.

TABLE XV.—ELEMENTS OF THE ORBITS OF BINARY STARS.

Name of Star.	A. R.	Dec.	Semi-major Axis.	Eccentricity.	Period of Revolution in Years.	By whom computed.
	° ′	° ′	″			
ζ Herculis	248 54	+31 52	1.25	0.448	36	Villarceau.
ζ Cancri	120 54	+18 5	0.89	0.443	58	Mädler.
ξ Ursæ Majoris	167 32	+32 22	2.29	0.403	61	Mädler.
η Coronæ	229 15	+30 49	1.20	0.404	67	Villarceau.
α Centauri	217 22	−60 12	15.50	0.950	80	Jacob.
τ Ophiuchi	268 43	− 8 10	0.82	0.037	87	Mädler.
70 Ophiuchi	269 28	+ 2 32	4.50	0.480	92	Mädler.
λ Ophiuchi	245 50	+ 2 18	0.84	0.477	95	Hind.
ξ Libræ	239 1	−10 57	1.29	0.000	105	Mädler.
3210 Cassiopeæ	359 36	+57 36	1.00	0.575	146	Mädler.
ξ Bootis	221 7	+19 43	5.59	0.454	160	Mädler.
γ Virginis	188 31	− 0 37	3.86	0.880	169	Mädler.
δ Cygni	295 4	+44 46	1.81	0.606	178	Hind.

TABLE XVI.—VARIABLE STARS.

No.	Star.	R. A. 1870.	Declination. 1870.	Period. Days.	Change of Magnitude.	Discoverer.
		h. m. s.	° ′			
1	o Ceti	2 12 47	− 3 34.1	331.34	2 to 12	Fabricius, 1596
2	β Persei	2 59 43	+40 27.2	2.867	2.5 " 4	Montanari, 1669
3	ε Aurigæ	4 52 38	+43 37.7	350	3.5 " 4.5	Heis, 1846
4	R Leonis	9 40 34	+12 1.8	312	5 " 11.5	Koch, 1782
5	η Argûs	10 40 2	−59 0.1	46 years	1 " 4	Burchell, 1827
6	R Hydræ	13 22 37	−22 36.4	449	4 " 10	Maraldi, 1704
7	30 Herculis	16 24 22	+42 10.1	106	5 " 6	Baxendell, 1857
8	Nova Ophiuchi	16 52 13	−12 41.4		4.5 " 13.5	Hind, 1848
9	κ Coronæ Aust.	18 24 24	−38 48.9		3 " 6	Halley, 1676
10	R Scuti Sobies.	18 40 33	− 5 50.5	71.75	5 " 9	Pigott, 1795
11	β Lyræ	18 45 17	+33 12.7	12.91	3.5 " 4.5	Goodricke, 1784
12	χ Cygni	19 45 34	+32 35.2	406	5 " 13	Kirch, 1687
13	η Aquilæ	19 45 51	+ 0 40.4	7.176	3.5 " 4.5	Pigott, 1784
14	34 Cygni	20 13 0	+37 37.8	18 years	3 " 6	Jansen, 1600
15	24 Cephei	20 23 41	+88 44.0	73 years	5 " 11	Pogson, 1856
16	μ Cephei	21 39 31	+58 11.1	5 or 6 y.	4 " 6	W. Herschel, 1782
17	δ Cephei	22 24 21	+57 45.0	5.366	3.7 " 4.8	Goodricke, 1784

EXPLANATION OF THE TABLES.

Table I., page 321, contains the principal elements of the planetary system, with the exception of the minor planets. These elements have been taken from Le Verrier's Annales de l'Observatoire, tome second, p. 58–61, as far as they are there given; other numbers depending upon these have been derived from them by computation; and the remainder of the Table has been derived from various sources, but chiefly from Hind.

Table II., pages 322 and 323, contains the elements of the minor planets. These elements have been derived from the Berlin Astronomisches Jahrbuch for 1866, with the exception of the last three, which were derived from recent periodicals.

Table III., page 324, contains the elements of the satellites of the primary planets. These elements were derived from a comparison of various authorities, such as Herschel's Astronomy, Chambers's Hand-book of Astronomy, Hind's Solar System, and Chazallon's Annuaire des Marées pour 1860.

Table IV., pages 325 and 326, contains a catalogue of all the eclipses of the sun that will be visible in the city of Boston from 1865 to 1900. It is copied from the American Almanac for 1831, and was computed by Mr. R. T. Paine.

Table V., page 327, contains a catalogue of eclipses designed to illustrate several important principles. It shows, first, that seven eclipses may occur in one year; and, second, it illustrates the principle of the Saros. The data for the past eclipses were derived from the English Nautical Almanac; and those for future eclipses were derived chiefly from Chambers's Hand-book of Astronomy.

Table VI., page 327, contains a complete catalogue of the transits of Mercury over the sun's disc from 1631 (the first transit observed) to the close of the present century. It is derived from Delambre's Astronomie, t. ii., p. 518.

Table VII., pages 328 and 329, contains Bessel's Astronomical Refractions in an abridged form. It requires, in addition to the observed apparent altitude, an observation of the height of the barometer, upon which depends the factor B; of the thermometer attached to the barometer, upon which depends the factor t; and of the temperature of the external air, upon which depends the factor T.

Take the mean refraction corresponding to the observed altitude; take the factor B corresponding to the height of the barometer; also the factor t corresponding to the attached thermometer; and the factor T corresponding to the external thermometer. Multiply these four numbers together, and the product will be the true refraction.

Example. The observed apparent altitude of a star was 34° 11′ 15″; the barometer, 28.856 inches; the external and the attached thermometers both stood at +19°.6 Fahr. It is required to compute the refraction.

Mean refraction for 34° 11′ 15″ - - - - - - 1′ 24″.8.
Barometer, 28.856. Factor B, 0.975.
Thermometer, 19°.6 { Factor t, 1.001.
Factor T, 1.061. }
Product, $0.975 \times 1.001 \times 1.061 = 1.0355$.
True refraction = 84″.8 × 1.0355 = 1′ 27″.8.

For small altitudes, when great accuracy is required, the computation is most conveniently performed by logarithms. A Table, which furnishes the logarithms of all these factors, is given in my Practical Astronomy, pages 364–5.

Table VIII., page 329, shows the quantity by which the moon's equatorial horizontal parallax must be diminished, to obtain the horizontal parallax belonging to any other latitude. This reduction is given for three values of the moon's equatorial parallax, viz., 53′, 57′, and 61′; and for any other value, the equatorial parallax may be easily found by interpolation.

Table IX., page 329, contains the elements of the seven comets whose periods have been well established. These elements have been derived chiefly from the Astronomische Nachrichten.

Table X., page 330, exhibits the altitude in English feet of the principal lunar mountains according to the observations of Beer and Mädler.

Table XI., page 330, exhibits the breadths in English miles of some of the larger craters, or annular mountains on the moon's surface, according to the observations of Beer and Mädler.

Table XII., page 330, contains a catalogue of all the transits of Venus over the sun's disc from 1639 (the first ever observed) to the end of the 21st century. It is derived from Delambre's Astronomie, t. ii., p. 473.

Table XIII., page 331, exhibits the results of 39 years of observations of the solar spots by M. Schwabe, of Dessau, in Germany. Column 2 shows the number of days in each year upon which observations were made; column 3 shows the number of groups of spots observed; and column 4 shows the number of days on each year upon which the sun was free from spots. These observations decidedly indicate a periodicity in the number of the solar spots, a maximum recurring at an interval of from 9 to 12 years.

Table XIV., page 331, exhibits the results of the best observations hitherto made for determining the parallax of some of the fixed stars. Several of the results here given are the averages of the determinations by two or more astronomers. The results for the two stars first mentioned are entitled to considerable confidence; all the others are to be regarded as quite doubtful, except as indicating that the parallax can not much exceed the quantities here given.

Table XV., page 331, furnishes the elements of those binary stars whose periods are less than two centuries.

EXPLANATION OF THE PLATES.

Plate I. is a chart of the world with cotidal lines marked upon it. The numerals upon the cotidal lines denote the hour, in Greenwich time, of high water on the day of new moon or full moon. The map is mainly copied from Professor Airy's chart in the Encyclopædia Metropolitana, Article Tides, with modifications suggested by the observations of the United States Coast Survey, and other recent observations in the Pacific Ocean.

Plate II., Fig. 2, is a representation of the appearance of the full moon, copied from the engraving of Beer and Mädler, modified according to a photographic picture taken at the Cambridge (Massachusetts) Observatory.

Fig. 1 is a representation of a small portion of the moon's surface as seen with a powerful telescope near the time of first quarter. This figure is derived from Mitchel's Sidereal Messenger, vol. i., p. 32.

Plate III., Fig. 1, is a representation of the total solar eclipse of July 18th, 1860, as observed in the northern part of Spain. The figure is copied from a photograph taken by De la Rue one minute after total obscuration. Fig. 2 is copied from a photograph taken immediately previous to the reappearance of the sun. In Fig. 1 the luminous protuberances are almost entirely on the left-hand side of the sun's disc, while the right side is almost entirely free from them. In Fig. 2 protuberances had come into view on the right-hand side, while those on the left hand have mostly disappeared, showing conclusively that these protuberances are attached to the disc of the sun, and not to that of the moon.

Plate IV. contains representations of the planets Venus, Mars, Jupiter, and Saturn. The figure of Venus is copied from a drawing by Schröter, representing the planet near its inferior conjunction. The figure of Mars is copied from a drawing by Secchi, published with the Observations of the Roman Observatory for 1856. The figure of Jupiter is copied from a drawing in the Sidereal Messenger, vol. i., p. 72; and the figure of Saturn is copied from a drawing by Dawes in the Astronomische Nachrichten, vol. xxxv., p. 395. (*See Frontispiece.*)

Plate V. contains representations of several comets. Fig. 1 is Encke's comet, from a drawing by Struve in 1828; Fig. 2 is a representation of the head of Halley's comet as observed in October, 1835, by Bessel, showing the luminous jets which emanated from the nucleus; Fig. 3 is a representation of Biela's comet as observed in February, 1846, by Struve, showing the division into two comets; Fig. 4 is a representation of the great comet of 1843 as seen by the naked eye; and Fig. 5 is a representation of the remarkable comet of 1744 as seen March 8th, at Geneva, by Cheseaux.

Plate VI. also contains representations of comets. Fig. 2 is a representation of Donati's comet as it appeared to the naked eye October 10, 1858, according to a drawing by Professor Bond; Fig. 1 is a representation of Halley's comet as it appeared to the naked eye October 29, 1835, according to Struve; and Fig. 3 is a telescopic view of the head of Donati's comet as it appeared October 2, 1858, according to a drawing by Professor Bond.

Plate VII., Fig. 1, is a representation of the great nebula in Andromeda, copied from Herschel's Astronomy, Plate II.; Fig. 2 is a representation of the great cluster ω Centauri, copied from Herschel's Cape of Good Hope Observations, Plate V.; and Fig. 3 is a representation of the great nebula in Orion, copied from Herschel.

Plate VIII., Fig. 1, is a representation of the dumb-bell nebula; Fig. 3, the annular nebula in Lyra; Fig. 2, the crab nebula; and Fig. 4, the spiral nebula—all copied from figures in Nichol's System of the World.

INDEX.

PLATE I.

CHART OF COTIDAL LINES.
ASIA
RUSSIAN POSSESSIONS
NORTH AMERICA
BRITISH POSSESSIONS
UNITED STATES
MEXICO
GREENLAND
COLOMBIA
PERU
BRAZIL
CHILI
SOUTH AMERICA
NORWAY
LAPLAND
RUSSIAN EMPIRE
EUROPE
TURKEY
TARTARY
IRKUTSK
ASIA
PERSIA
ARABIA
CHINESE EMPIRE
HINDOSTAN
BARBARY
SAHARA
AFRICA
EGYPT
DARFUR
GUINEA
ADEL
LOANGO
C. COLONY
AUSTRALIA

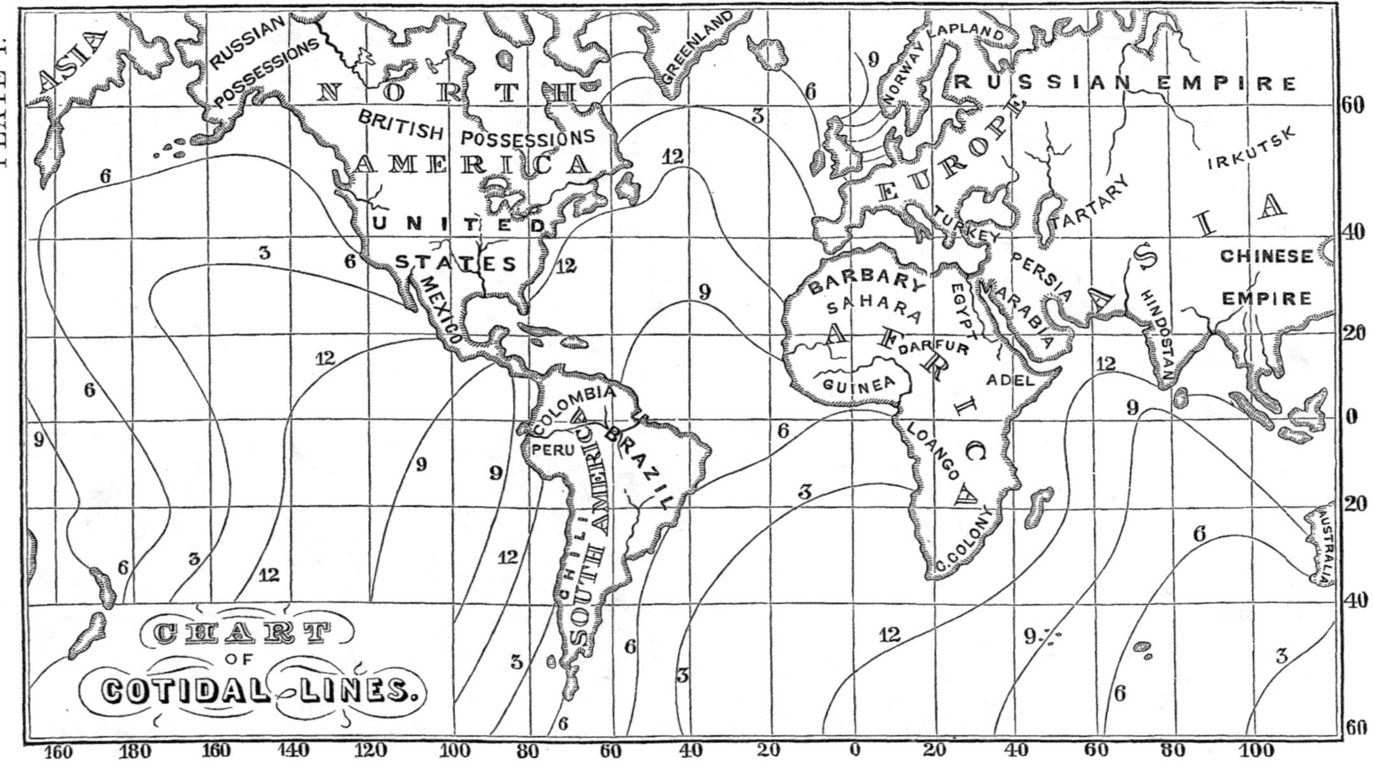

PLATE II.

PLATE III

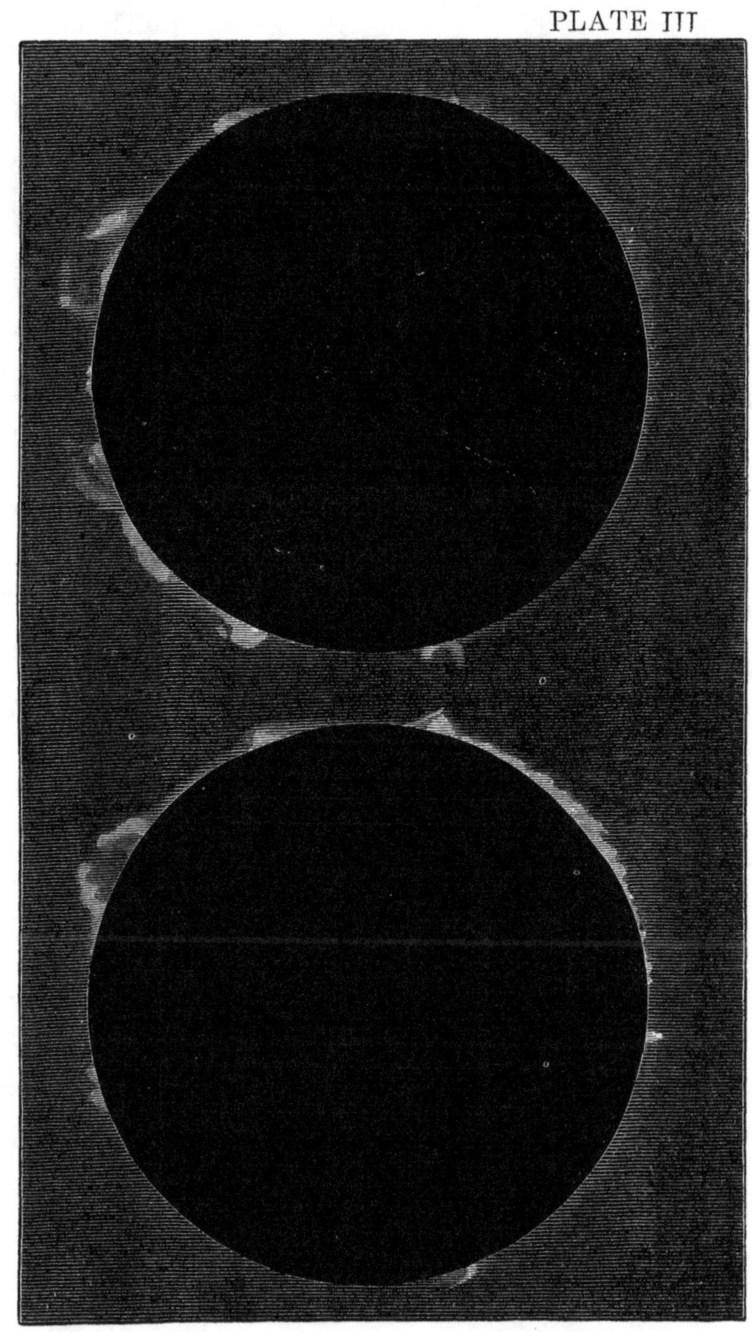

PLATE IV.

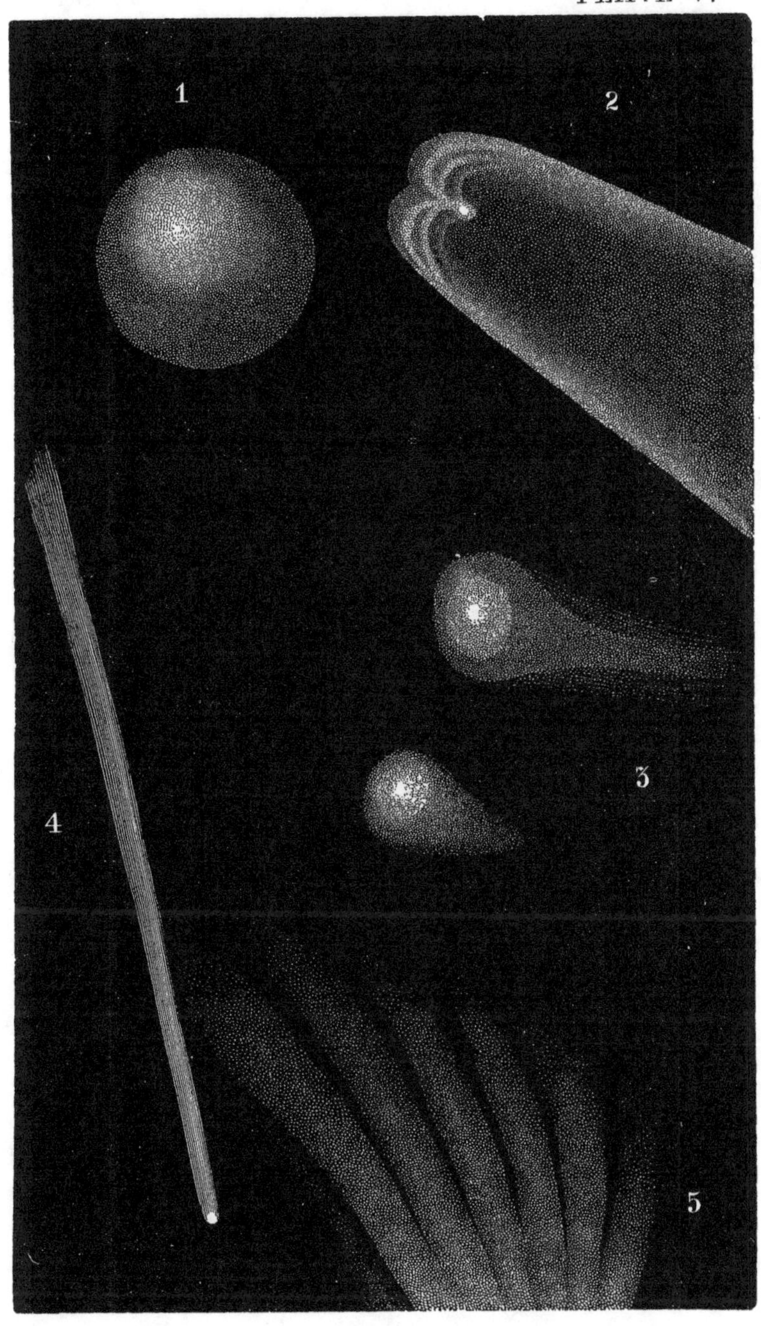
1
2
3
4
5

PLATE VI.

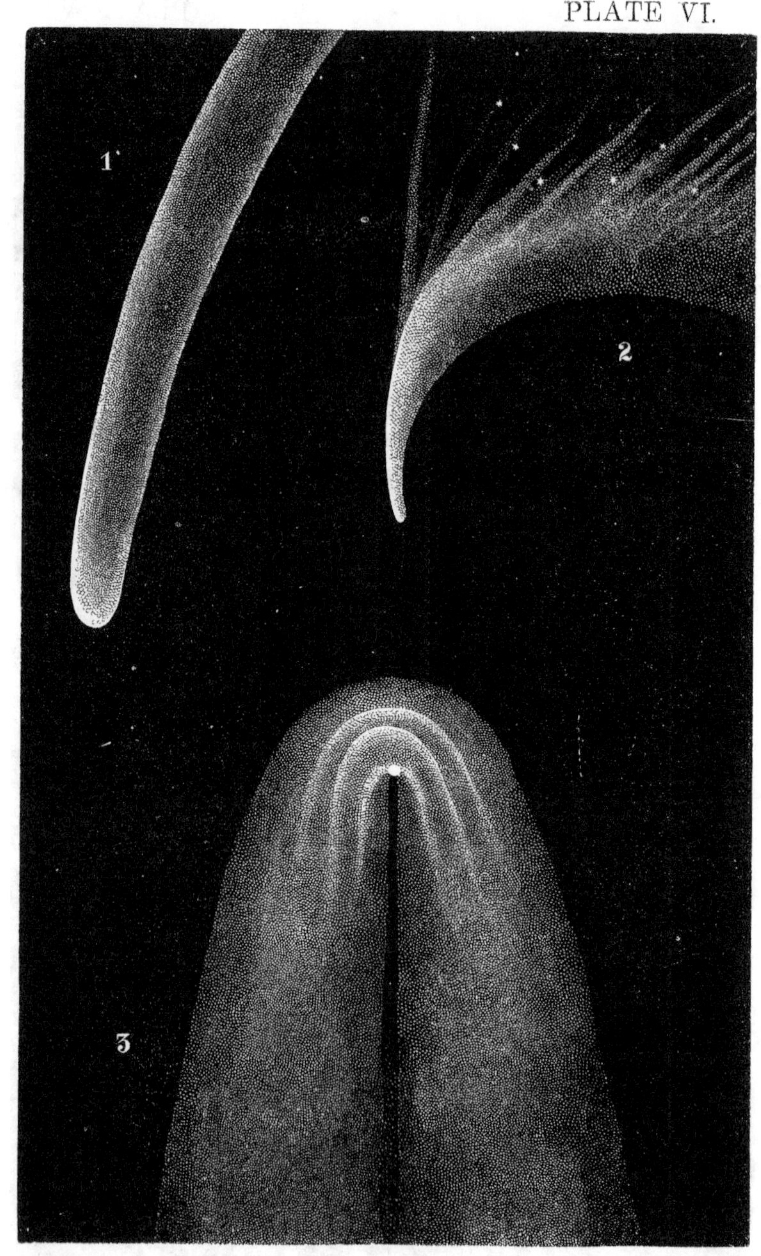

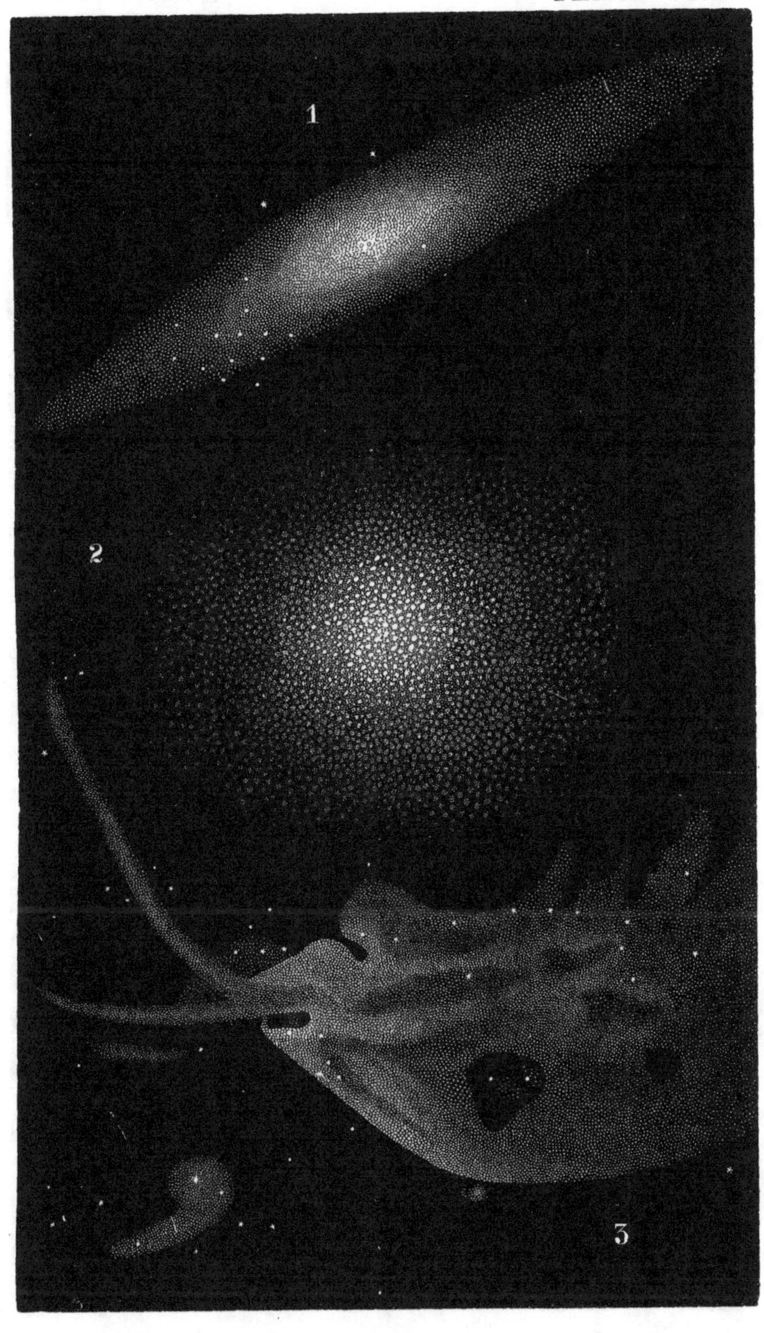
1
2
3

1
2
3
4

www.ingramcontent.com/pod-product-compliance
Lightning Source LLC
LaVergne TN
LVHW010157110826
845151LV00002B/542
9781425538149